Advances in Intelligent and Soft Computing 96

Editor-in-Chief: J. Kacprzyk

Advances in Intelligent and Soft Computing

Editor-in-Chief

Prof. Janusz Kacprzyk
Systems Research Institute
Polish Academy of Sciences
ul. Newelska 6
01-447 Warsaw
Poland
E-mail: kacprzyk@ibspan.waw.pl

Further volumes of this series can be found on our homepage: springer.com

Vol. 86. E. Mugellini, P.S. Szczepaniak, M.C. Pettenati, and M. Sokhn (Eds.)
Advances in Intelligent Web Mastering – 3, 2011
ISBN 978-3-642-18028-6

Vol. 87. E. Corchado, V. Snášel, J. Sedano, A.E. Hassanien, J.L. Calvo, and D. Ślęzak (Eds.)
Soft Computing Models in Industrial and Environmental Applications, 6th International Workshop SOCO 2011
ISBN 978-3-642-19643-0

Vol. 88. Y. Demazeau, M. Pěchoucěk, J.M. Corchado, and J.B. Pérez (Eds.)
Advances on Practical Applications of Agents and Multiagent Systems, 2011
ISBN 978-3-642-19874-8

Vol. 89. J.B. Pérez, J.M. Corchado, M.N. Moreno, V. Julián, P. Mathieu, J. Canada-Bago, A. Ortega, and A.F. Caballero (Eds.)
Highlights in Practical Applications of Agents and Multiagent Systems, 2011
ISBN 978-3-642-19916-5

Vol. 90. J.M. Corchado, J.B. Pérez, K. Hallenborg, P. Golinska, and R. Corchuelo (Eds.)
Trends in Practical Applications of Agents and Multiagent Systems, 2011
ISBN 978-3-642-19930-1

Vol. 91. A. Abraham, J.M. Corchado, S.R. González, J.F. de Paz Santana (Eds.)
International Symposium on Distributed Computing and Artificial Intelligence, 2011
ISBN 978-3-642-19933-2

Vol. 92. P. Novais, D. Preuveneers, and J.M. Corchado (Eds.)
Ambient Intelligence - Software and Applications, 2011
ISBN 978-3-642-19936-3

Vol. 93. M.P. Rocha, J.M. Corchado, F. Fernández-Riverola, and A. Valencia (Eds.)
5th International Conference on Practical Applications of Computational Biology & Bioinformatics 6-8th, 2011
ISBN 978-3-642-19913-4

Vol. 94. J.M. Molina, J.R. Casar Corredera, M.F. Cátedra Pérez, J. Ortega-García, and A.M. Bernardos Barbolla (Eds.)
User-Centric Technologies and Applications, 2011
ISBN 978-3-642-19907-3

Vol. 95. Robert Burduk, Marek Kurzyński, Michał Woźniak, and Andrzej Żołnierek (Eds.)
Computer Recognition Systems 4, 2011
ISBN 978-3-642-20319-0

Vol. 96. A. Gaspar-Cunha, R. Takahashi, G. Schaefer, and L. Costa (Eds.)
Soft Computing in Industrial Applications, 2011
ISBN 978-3-642-20504-0

António Gaspar-Cunha, Ricardo Takahashi,
Gerald Schaefer, and Lino Costa (Eds.)

Soft Computing in Industrial Applications

Springer

Editors

Prof. António Gaspar-Cunha
University of Minho
Department of Polymer Engineering
Campus de Azurém
4800-058 Guimarães
Portugal
E-mail: agc@dep.uminho.pt

Prof. Ricardo Takahashi
Federal University of Minas Gerais (UFMG)
Department of Mathematics
Av. Antonio Carlos 6627
30123970 Belo Horizonte
Brazil
E-mail: taka@mat.ufmg.br

Prof. Gerald Schaefer
Loughborough University
Department of Computer Science
Loughborough, LE11 3TU
U.K.
Phone: +44-1509-635707
Fax: +44-1509-635722
E-mail: Gerald.Schaefer@ieee.org

Prof. Lino Costa
University of Minho
Department of Production and Systems
School of Engineering
4800-058 Guimarães
Portugal
E-mail: lac@dps.uminho.pt

ISBN 978-3-642-20504-0 e-ISBN 978-3-642-20505-7

DOI 10.1007/978-3-642-20505-7

Advances in Intelligent and Soft Computing ISSN 1867-5662

Library of Congress Control Number: 2011925864

Typeset & Cover Design: Scientific Publishing Services Pvt. Ltd., Chennai, India

Printed on acid-free paper
5 4 3 2 1 0
springer.com

WSC 15 Chair's Welcome Message

Dear Colleague,

On behalf of the Organizing Committee, it is our honor to welcome you to WSC15, the 15th Online World Conference on Soft Computing in Industrial Applications.

A tradition started over a decade ago by the World Federation of Soft Computing (http://www.softcomputing.org/), again brings together researchers interested in the ever advancing state of the art in the field. Continuous technological improvements make this online forum a viable gathering format for a world class conference.

The program committee received a total of 86 submissions from 20 countries, which reflects the worldwide nature of this event. Each paper was peer reviewed by typically 3 referees, culminating in the acceptance of 38 papers for publication. Two recognized researchers were invited to give a plenary talk - the first talk is in the field of multi-objective optimization using Particle Swarm Optimization, while the second focuses on practical applications of soft computing in "An Evolutionary Algorithm for the Design and Selection of Direct Load Control Actions".

The organization of the WSC15 conference is entirely voluntary. The review process required an enormous effort from the members of the International Program Committee, and we would therefore like to thank all its members for their contribution to the success of this conference. We would like to express our sincere thanks to the special session organizers, the plenary presenters, and to the publisher - Springer - for their hard work and support in organizing the conference. Finally, we would like to thank all the authors for their high quality contributions. It is all of you who make this event possible!

Finally, we would like to take this opportunity to raise the consciousness of the importance of reducing carbon emissions by participating virtual conferences, such as WSC15.

Sincerely,
António Gaspar Cunha - WSC15 General Chair
Gerald Schaefer, Ricardo Takahashi - WSC15 Program Co-Chairs

Welcome Message from the WFSC Chair

Welcome to the 15th Online World Conference on Soft Computing in Industrial Applications (WSC15). On behalf of the World Federation on Soft Computing (WFSC), it is my privilege to invite you for an active technical discussion on the cyberspace! The online conference provides an opportunity to select good quality papers and have prolonged and detailed discussions on the research.

This year, organizers have selected 38 papers out of 86 submissions. Accepted papers are from 15 different countries around the World. This is an opportunity to publish high quality research in soft computing and its applications without incurring any significant cost. I would like to encourage all of you to get engaged with WFSC and promote the conference for future years.

Soft Computing as a discipline is maturing and we can see wide spread application of the techniques. More and more commercial software tools are becoming available for industrial applications. I would like to encourage the research community to collaborate and develop benchmark data and test function repositories for validation of our research. It is also necessary to develop more soft computing education program and research the pedagogy for application domains, such as engineering and medicine. I also hope the research community will discuss implications of their research and application of soft computing in terms of ethics.

Finally, I would like to thank the organizers of the WSC15 conference for their good work in bring together the researchers in the cyberspace and also for organizing publications afterwards. With their effort and your support, soft computing is taking another step forward.

Chairmen of WFSC
Professor Rajkumar Roy
World Federation on Soft Computing (WFSC)

WSC 15 Organization

General Chair

António Gaspar-Cunha	University of Minho, Portugal

Program Co-chairs

Ricardo Takahashi	Federal University of Minas Gerais, Brazil
Gerald Schaefer	Loughborough University, UK

Special Event Chair

Lino Costa	University of Minho, Portugal

Best Paper Award Chair

Fernanda Costa	University of Minho, Portugal

Web Chair

Fernando Mendes	University of Minho, Portugal

International Advisory Board

Kalyanmoy Deb	Indian Institute of Technology Kanpur, India
Carlos M. Fonseca	University of Algarve, Portugal
Thomas Stützle	University Libre de Bruxelles, Belgium
Jörn Mehnen	Cranfield University, UK
Mario Köppen	Kyushu Institute of Technology, Fukuoka, Japan
Xiao-Zhi Gao	Aalto University, Finland

International Technical Program Committee

Heitor Lopes	UTFPR, Brazil
Ana Maria Rocha	University of Minho, Portugal
Patricia Melin	Tijuana Institute of Technology, Mexico
Sanaz Mostaghim	University of Karlsruhe, Karlsruhe, Germany
Oriane Neto	Federal University of Minas Gerais - UFMG, Brazil
Jae Oh	Syracuse University, United States
Gina Oliveira	UFU, Brazil
Michele Ottomanelli	Politecnico di Bari, Italy
Gisele Pappa	UFMG, Brazil
Marcin Paprzycki	IBS PAN and WSM, Poland
Luís Paquete	Universidade de Coimbra, Portugal
Francisco Pereira	Instituto Superior de Engenharia de Coimbra, Portugal
Rui Pereira	Univ. of Minho, Portugal
S.G. Ponnambalam	Monash University, Malasya
Petrica Pop	North Univcersity of Baia Mare, Romania
Aurora Pozo	UFPR, Brazil
Radu-Emil Precup	Politehnica University of Timisoara, Romania
Fengxiang Qiao	Texas Southern University, USA
Giovanni Semeraro	University of Bari, Italy
Sara Silva	INESC-ID Lisboa, Portugal
Marcone Souza	UFOP, Brazil
Thomas Stuetzle	Université Libre de Bruxelles (ULB), Begium
Yos Sunitiyoso	Institut Teknologi Bandung, Indonesia
Maria Teresa Pereira	Portugal
Eiji Uchino	Yamaguchi University, Japan
Olgierd Unold	Wroclaw University of Technology, Poland
João Vasconcelos	UFMG, Brazil
Ana Viana	Instituto Politécnico do Porto / INESC Porto, Portugal
Fernando Von Zuben	UNICAMP, Brazil
Tobias Wagner	Technische Universität Dortmund, Germany
Elizabeth Wanner	CEFET-MG, Brazil

WSC 15 was supported by

World Federation on Soft Computing (WFSC)

University of Minho

List of Contributors

A. Cubero-Atienza
Area de Proyectos de Ingeniería. Dpto. de Ingeniería Rural,
University of Córdoba
ir1cuata@uco.es

A. Gomes Correia
Department of Civil Engineering/
C-TAC, University of Minho,
4710, Guimarães, Portugal
agc@civil.uminho.pt

A. Prugel Bennett
University of Southampton
apb@ecs.soton.ac.uk

Aboul Ella Hassanien
Faculty of Computers and Information,
Cairo University, 5 Ahmed Zewal St.,
Orman, Giza
aboitcairo@gmail.com

Adrian Sebastian Paul
"Politehnica" University of Timisoara,
Bd. V. Parvan 2, RO-300223
Timisoara, Romania
adi11p@yahoo.com

Alfredo Milani
Department of Mathematics and
Computer Science, University of
Perugia, via Vanvitelli, Perugia,
06123, Italy
milani@unipg.it

Alpha Agape Gopalai
Monash University, Sunway campus, 2,
Jalan Lagoon Selatan, Bandar Sunway,
46150, Malaysia
alpha.agape@ieee.org

Álvaro Gomes
Dept. of Electrical Engineering and
Computers - University of Coimbra,
Coimbra, Portugal
agomes@deec.uc.pt

Ana Rocha
Department of Production and Systems,
School of Engineering, University of
Minho, 4800-058 Guimarães, Portugal,
arocha@dps.uminho.pt

Anabela Tereso
Department of Production and Systems,
University of Minho,
Campus de Gualtar, Braga, 4710-057,
Portugal
anabelat@dps.uminho.pt

Andreas Koenig
Institute of Integrated Sensor System,
University of Kaiserslautern,
Erwin-Schroedinger-Strasse,
Kaiserslautern, D-67663, Germany
koenig@eit.uni-kl.de

Andrew Kho
Institute for Transport Studies,
University of Leeds, 36-40 University
Road, Leeds LS2 9JT United Kingdom
a.koh@its.leeds.ac.uk

Angel Cobo
Department Applied Mathematics and
Computational Sciences,
University of Cantabria, Avda. Los
Castros s/n, Santander, 39005, Spain
acobo@unican.es

António Gomes Martins
Dept. of Electrical Engineering and Computers - University of Coimbra, Coimbra, Portugal
amartins@deec.uc.pt

Antonio Petraglia
PEE / COPPE, Federal University of Rio de Janeiro, Centro de Tecnologia, Cidade Universitária, Ilha do Fundão, Rio de Janeiro, 21945-970, Brazil
antonio@pads.ufrj.br

Arezoo Rajaei
Dept. of Computer Engineering, heikhbahaee University, Baharestan Isfahan, Iran
arezoo_rajaee@yahoo.com

Arosha Senanayake
Monash University, Sunway campus, 2, Jalan Lagoon Selatan, Bandar Sunway, 46150, Malaysia
aroshas@ieee.org

Ashraf Darwish
Faculty of Science, Helwan University, Cairo, Egypt
amodarwish@yahoo.com

Boulmakoul Azedine
LIST Lab., Computer Sciences Department, Faculty of Sciences and Technology (FSTM), Rue Rabat, Mohammedia, 28830, Morocco
boul@uh2m.ac.ma

C. Hervás-Martínez
Department of Computer Science and Numerical Analysis, University of Córdoba, Spain
chervas@uco.es

Carlos A. Coello Coello
CINVESTAV-IPN, Evolutionary Computation Group (EVOCINV), Departamento de Computación, Av. IPN No. 2508, Col. San Pedro Zacatenco, México, D.F. 07360, MEXICO
ccoello@cs.cinvestav.mx

Carlos Henggeler Antunes
Dept. of Electrical Engineering and Computers - University of Coimbra, Coimbra, Portugal
ch@deec.uc.pt

Claudio Meneguzzer
University of Padova - Department of Structural and Transportation Engineering, Via Marzolo 9 35131, Padova, Italy
claudio.meneguzzer@unipd.it

Cosmin Sabo
North University of Baia Mare, Victoriei 76, Baia Mare, Romania
cosmin_sabo@prime-tech.ro

Cristina Santos
Industrial Electronics Department, School of Engineering, University of Minho, 4800-058 Guimarães, Portugal,
cristina@dei.uminho.pt

Domenico Sassanelli
Dept. of Roads and Transportation, Technical University of Bari Via Orabona, 4 - 70125 Bari, Italy
d.sassanelli@poliba.it

Edite Fernandes
Department of Production and Systems, University of Minho, Campus de Gualtar, Braga, 4710-057, Portugal
emgpf@dps.uminho.pt

Emil M. Petriu
University of Ottawa, 800 King Edward, Ottawa, ON, K1N 6N5 Canada
petriu@site.uottawa.ca

Estefhan Wandekoken
Department of Computer Science, Federal University of Espirito Santo, Vitoria, ES, Brazil
estefhan@inf.ufes.br

F. Fernández-Navarro,
Department of Computer Science and Numerical Analysis,
University of Córdoba, Spain
i22fenaf@uco.es;chervas@uco.es

Flavio Varejao
Department of Computer Science, Federal University of Espirito Santo, Vitoria, ES, Brazil
fvarejao@inf.ufes.br

Gerald Schaefer
Department of Computer Science, Loughborough University, U.K.

Giuseppe Iannucci
Dept. of Roads and Transportation, Technical University of Bari - Via Orabona, 4 - 70125 Bari, Italy
g.iannucci@poliba.it

Gregorio Gecchele
University of Padova - Department of Structural and Transportation Engineering, Via Marzolo 9 35131, Padova, Italy
gecchele@dic.unipd.it

H. Mohammadi
University of Southampton

Hao Wu
Department of Computer Science and Engineering, University of Bridgeport
wuhao@bridgeport.edu

Heba F. Eid
Faculty of Science, Al-Azhar University, Cairo, Egypt
heba.fathy@yahoo.com

Hicham Mouncif
Mathematics and Computer Sciences Department, Faculty of Sciences Ain Chock, University Hassan 2, Rue el jadida, Casablanca, 20470, Morocco
hmouncif@yahoo.fr

Hime Oliveira Jr.
National Cinema Agency, Avenida Graça Aranha, 35, Rio de Janeiro, 20030-002, Brazil
hime@engineer.com

Hongwu Qin
Faculty of Computer Systems and Software Engineering,
Universiti Malaysia
Pahang, Lebuh Raya Tun Razak, Malaysia

James F. Peter
Computational Intelligence, Laboratory, Department of Electrical & Computer Engineering,Univ. of Manitoba, E1-526, 75A Chancellor's Circle, Winnipeg, MB R3T 5V6
jfpeters@ee.umanitoba.ca

Joaquim Tinoco
Department of Civil Engineering/ C-TAC, University of Minho, Guimarães, Portugal
jabtinoco@civil.uminho.pt

Keshav Dahal
School of Computing, Informatics and Media, Bradford University, United Kingdom
k.p.dahal@bradford.ac.uk

Kevin Kam Fung Yuen
School of Science and Technology, The Open University of Hong Kong, Ho Man Tin, Kowloon, Hong Kong
kevinkf.yuen@gmail.com

Lamiaa M. El Bakrawy
Faculty of Science, Al-Azhar University, Omar ibn elKHtab, Zhraa Naser City, Cairo-Egypt
lamiaabak@yahoo.com

Lars Nolle
School of Science and Technology, Nottingham Trent University, U.K.

Leandro de Lima
Department of Informatics, PPGI, Federal University of Espírito Santo, Campus de Goiabeiras, Vitória-ES, 29060-970, Brazil
leandromunizlima@yahoo.com.br

Leandro Nunes-de Castro
Natural Computing Laboratory - LCoN, Mackenzie Presbyterian University, Rua da Consolação 930, Sãao Paulo, 01302-907, Brazil
lnunes@mackenzie.br

Leonardo Caggiani
Department of Environmental Engineering and Sustainable Development,
Technical University of Bari, Viale De Gasperi, Taranto, 74100, Italy
l.caggiani@poliba.it

Lino Costa
Department of Production and Systems , School of Engineering, University of Minho, 4800-058 Guimarães, Portugal,
lac@dps.uminho.pt

M. Alamgir Hossain
School of Computing, Informatics and Media, Bradford University, United Kingdom
M.A.Hossain1@bradford.ac.uk

M. H. Tayarani
University of Southampton
mhtn1g09@ece.soton.ac.uk

M.D. Redel-Macías
Area de Proyectos de Ingeniería. Dpto. de Ingeniería Rural.
University of Córdoba, Spain
mdredel@uco.es

Madan Kharat
Principal, Pankaj Laddhad Institute of Technology and Management, Buldana, Sant Gadge Baba Amravati University, Amravati-444605, India.
mukharat@rediffmail.com

Mahboobeh Houshmand
Dept. of Computer Engineering and Information Technology, Amirkabir University of Technology, Hafez street, Tehran, Iran
houshmand@aut.ac.ir

Manuel Ferreira
Industrial Electronics Department, School of Engineering, University of Minho, 4800-058 Guimarães, Portugal,
mjf@dei.uminho.pt

Massimiliano Gastaldi
University of Padova - Department of Structural and Transportation Engineering, Via Marzolo 9 35131, Padova, Italy
massimiliano.gastaldi@unipd.it

Michele Ottomanelli
Dept. of Environmental Engineering and Sustainable Development, Technical University of Bari - Viale De Gasperi - 74100 Taranto, Italy
m.ottomanelli@poliba.it

Miguel Oliveira
Industrial Electronics Department, School of Engineering, University of Minho, 4800-058 Guimarães, Portugal,
mcampos@dei.uminho.pt

Mohamed Rida
Mathematics and Computer Sciences Department, Faculty of Sciences Ain Chock, University Hassan 2, Rue el jadida, Casablanca, 20470, Morocco
mohamed_ridama@yahoo.fr

Mohammad-R Akbarzadeh-T
Ferdowsi University of Mashhad, Department of computer engineering, Azadi street, Mashhad, Iran
akbarzade@ieee.org

Mónica Santos
Department of Production and Systems, University of Minho, Campus de Gualtar, Braga, 4710-057, Portugal
pg13713@alunos.uminho.pt

Monireh Houshmand
Department of Electrical Engineering, Imam Reza University, Asrar street, Mashhad, PostCode:55391735, Iran
monirehhoushmand@gmail.com

Mostafa A. Salama
Department of Computer Science, British University in Egypt, Cairo, Egypt mostafa.salama@gmail.com

Muhammad Akmal Johar
Institute of Integrated Sensor System, University of Kaiserslautern, Erwin-Schroedinger-Strasse, Kaiserslautern, D-67663, Germany
johar@eit.uni-kl.de

Neveen I. Ghali
Faculty of Science, Al-Azhar University, Cairo-Egypt
nev_ghali@yahoo.com

Nitin Choubey
Student, P. G. Department of Computer Science, Sant Gadge Baba Amravati University, Amravati-444605, India.
nschoubey@gmail.com

Norrozila Sulaiman
Faculty of Computer Systems and Software Engineering, Universiti Malaysia
Pahang, Lebuh Raya Tun Razak, Gambang 26300, Kuantan, Malaysia
norrozila@ump.edu.my

Oliviu Matei
North University of Baia Mare, Victoriei 76, Baia Mare, Romania
oliviu.matei@holisun.com

Paulo Cortez
Department of Information systems/R&D Algoritmi Centre, University of Minho, Guimarães, Portugal
pcortez@dsi.uminho.pt

Petrica Pop
North University of Baia Mare,
Victoriei 76, Baia Mare, Romania
petrica.pop@ubm.ro

Pravin Patil
Indian Institute of Technology
Roorkee, Roorkee, Uttarakhand,
247667, India.
pravinppatil2004@gmail.com

Rabie A. Ramadan
Cairo University, Faculty of
Engineering, Computer Engineering
Department, Cairo, Egypt
rabieramadan@gmail.com

Radu-Codrut David
"Politehnica" University of Timisoara,
Bd. V. Parvan 2, RO-300223
Timisoara, Romania
davidradu@gmail.com

Radu-Emil Precup
"Politehnica" University of Timisoara,
Bd. V. Parvan 2, RO-300223
Timisoara, Romania
radu.precup@aut.upt.ro

Rafael Silveira-Xavier
Natural Computing Laboratory - LCoN,
Mackenzie Presbyterian University,
Rua da Consolação 930, São Paulo,
01302-907, Brazil
rafael.xavier@mackenzista.com.br

Razieh Rezayee Saleh
Department of Computer Engineering,
Ferdowsi University of Mashhad,
Azadi Square, Mashhad, PostCode:
9177948974, Iran
r.rezayee@gmail.com

Renato Krohling
Department of Informatics, PPGI,
Federal University of Espírito Santo,
Campus de Goiabeiras, Vitória-ES,
29060-970, Brazil
krohling.renato@gmail.com

Riccardo Rossi
University of Padova - Department of
Structural and Transportation
Engineering
Via Marzolo 9 35131, Padova, Italy
riccardo.rossi@unipd.it

Rocio Rocha
Department Business Administration,
University of Cantabria,
Avda. Los Castros s/n, Santander,
39005, Spain
rochar@unican.es

Rodrigo Batista
Espirito Santo Exploration and
Production Business Unit, Petroleo
Brasileiro S.A. PETROBRAS, Vitoria,
ES, Brazil
rjbatista@inf.ufes.br

Rouhani Modjtaba
Dept. of Electrical Engineering, Islamic
Azad University, Gonabad Branch, Iran
m.rouhani@ieee.org

Satish Jain
Indian Institute of Technology
Roorkee, India.
Roorkee, Uttarakhand, 247667, India.

Satish Sharma
Indian Institute of Technology
Roorkee, Roorkee, Uttarakhand,
247667, India.
sshmefme@iitr.ernet.in

Sophia Bazzi
Sistan and Baluchestan University,
Daneshgah Avenue, Zahedan,
98165-161, Iran
Sophia.bazzi@yahoo.com

Stefan Preitl
"Politehnica" University of Timisoara,
Bd. V. Parvan 2, RO-300223
Timisoara, Romania
stefan.preitl@aut.upt.ro

Sujin Bureerat
Department of Mechanical
Engineering, Faculty of Engineering,
Khon Kaen University, Thailand
sujbur@kku.ac.th

Thomas Rauber
Department of Computer Science,
Federal University of Espirito Santo,
Vitoria, ES, Brazil
thomas@inf.ufes.br

Tutut Herawan
Faculty of Computer Systems and
Software Engineering,
Universiti Malaysia
Pahang, Lebuh Raya Tun Razak,
Gambang 26300, Kuantan, Malaysia
tutut@ump.edu.my

Vahideh Keikha
Sistan and Baluchestan University,
Daneshgah Avenue, Zahedan,
98135-674, Iran
va.keikha@yahoo.com

Valentino Santucci
Department of Mathematics and
Computer Science, University of
Perugia, via Vanvitelli, 1
Perugia, 06123, Italy
valentino.santucci@dmi.unipg.it

Walid Abdo
School of Computing, Informatics and
Media, Bradford University,
United Kingdom
waaabdo@bradford.ac.uk

Xiuqin Ma
Faculty of Computer Systems and
Software Engineering,
Universiti Malaysia
Pahang, Lebuh Raya Tun Razak,
Gambang 26300, Kuantan, Malaysia
xueener@yahoo.com.cn

Zhengping Wu
Department of Computer Science and
Engineering, University of Bridgeport
zhengpiw@bridgeport.edu

Zohreh Davarzani
Ferdowsi university of
Mashhad,Department of computer
engineering,
Azadi street, Mashhad.Iran
z_davarzani@yahoo.com

Contents

Plenary Sessions

Tutorial

Part I: Evolutionary Computation

Part IV: Soft Computing for Modeling, Control, and Optimization

Part V: Soft Computing for Data Mining

Part VIII: Optimization Techniques

Plenary Sessions

An Introduction to Multi-Objective Particle Swarm Optimizers

Carlos A. Coello Coello*

Abstract. This paper provides a discussion on the main changes required in order to extend particle swarm optimization to the solution of multi-objective optimization problems. A short discussion of some potential paths for future research in this area is also included.

Keywords: particle swarm optimization, multi-objective optimization, metaheuristics.

1 Introduction

There is a wide variety of real-world problems which have two or more (normally conflicting) objectives that we aim to optimize at the same time. Such problems are called multi-objective, and their solution involves the search of solutions that represent the best possible compromise among all the objectives.

Particle Swarm Optimization (PSO) is a bio-inspired metaheuristic that simulates the movements of a flock of birds or fish which seek food. Its relative simplicity (with respect to evolutionary algorithms) have made it a popular optimization approach, and a good candidate to be extended for multi-objective optimization.

The first multi-objective particle swarm optimizer (MOPSO) was proposed by Moore and Chapman in an unpublished manuscript from 1999[1] [11]. This

Carlos A. Coello Coello
CINVESTAV-IPN
Evolutionary Computation Group (EVOCINV)
Departamento de Computación, Av. IPN No. 2508
Col. San Pedro Zacatenco, México, D.F. 07360, Mexico
e-mail: ccoello@cs.cinvestav.mx

* The author acknowledges support from CONACyT project no. 103570.

[1] This paper may be found in the EMOO repository located at: http://delta.cs.cinvestav.mx/~ccoello/EMOO/

A. Gaspar-Cunha et al. (Eds.): Soft Computing in Industrial Applications, AISC 96, pp. 3–12.
springerlink.com

paper was published the following year [12], but the actual interest in developing MOPSOs really started in 2002. Due to obvious space limitations, this paper does not intend to provide a survey on MOPSOs (see, for example, [13] for a survey of that sort). Here, we only provide a short review of PSO and the way in which it has to be modified so that it can solve multi-objective optimization problems.

The remainder of this paper is organized as follows. In Section 2, we provide some basic concepts from multi-objective optimization required to make the paper self-contained. Section 3 presents an introduction to the PSO strategy and Section 4 presents a brief discussion about extending the PSO strategy for solving multi-objective problems. In Section 5, possible paths of future research are discussed and, finally, we present our conclusions in Section 6.

2 Basic Concepts

We are interested in solving the so-called *multi-objective optimization problem* (MOP) which has the following form[2]:

$$\text{minimize } \mathbf{f}(\mathbf{x}) := [f_1(\mathbf{x}), f_2(\mathbf{x}), \ldots, f_k(\mathbf{x})] \tag{1}$$

subject to:

$$g_i(\mathbf{x}) \leq 0 \quad i = 1, 2, \ldots, m \tag{2}$$

$$h_i(\mathbf{x}) = 0 \quad i = 1, 2, \ldots, p \tag{3}$$

where $\mathbf{x} = [x_1, x_2, \ldots, x_n]^T$ is the vector of decision variables, $f_i : \mathbb{R}^n \to \mathbb{R}$, $i = 1, ..., k$ are the objective functions and $g_i, h_j : \mathbb{R}^n \to \mathbb{R}$, $i = 1, ..., m$, $j = 1, ..., p$ are the constraint functions of the problem.

Definition 1. Given two vectors $\mathbf{x}, \mathbf{y} \in \mathbb{R}^k$, we say that $\mathbf{x} \leq \mathbf{y}$ if $x_i \leq y_i$ for $i = 1, ..., k$, and that $\mathbf{x}$ **dominates** $\mathbf{y}$ (denoted by $\mathbf{x} \prec \mathbf{y}$) if $\mathbf{x} \leq \mathbf{y}$ and $\mathbf{x} \neq \mathbf{y}$.

Definition 2. We say that a vector of decision variables $\mathbf{x} \in \mathcal{X} \subset \mathbb{R}^n$ is **nondominated** with respect to $\mathcal{X}$, if there does not exist another $\mathbf{x}' \in \mathcal{X}$ such that $\mathbf{f}(\mathbf{x}') \prec \mathbf{f}(\mathbf{x})$.

Definition 3. We say that a vector of decision variables $\mathbf{x}^* \in \mathcal{F} \subset \mathbb{R}^n$ ($\mathcal{F}$ is the feasible region) is **Pareto-optimal** if it is nondominated with respect to $\mathcal{F}$.

Definition 4. The **Pareto Optimal Set** $\mathcal{P}^*$ is defined by:

$$\mathcal{P}^* = \{\mathbf{x} \in \mathcal{F} | \mathbf{x} \text{ is Pareto-optimal}\}$$

[2] Without loss of generality, we will assume only minimization problems.

Definition 5. The **Pareto Front** $\mathcal{PF}^*$ is defined by:

$$\mathcal{PF}^* = \{\mathbf{f}(\mathbf{x}) \in \mathbb{R}^k | \mathbf{x} \in \mathcal{P}^*\}$$

We thus wish to determine the Pareto optimal set from the set $\mathcal{F}$ of all the decision variable vectors that satisfy (2) and (3). In practice, it is normally the case that only some elements of the Pareto optimal set is required, although multi-objective metaheuristics normally aim to find as many elements of the Pareto optimal set as possible [2].

3 An Introduction to Particle Swarm Optimization

PSO is a population-based metaheuristic which was originally introduced by James Kennedy and Russell C. Eberhart in the mid-1990s [8]. PSO was originally adopted for balancing weights in neural networks, but it soon became a very popular global optimizer, maybe in problems in which the decision variables are real numbers [6, 9].

Although some authors consider PSO as another evolutionary algorithm (EA), other authors such as Angeline [1], make important distinctions between them:

1. EAs rely on three main mechanisms: parents encoding, selection of individuals and fine tuning of their parameters. In contrast, PSO only relies on two mechanisms, since it does not adopt an explicit selection function (this is compensated by the use of leaders to guide the search, but there is no notion of offspring generation in PSO as in EAs).
2. PSO adopts a velocity value for each particle, and this is used to guide the search. The velocity can be seen as a directional mutation operator in which the direction is defined by both the particle's personal best and the global best (of the swarm). In contrast, EAs use a randomized mutation operator that can set an individual in any direction. Clearly, PSO has a more limited operator, and such limitations have led to several researchers to incorporate a randomized mutation operator.

In order to make this paper self-contained, we provide next a small glossary of terms used by the PSO community:

- **Swarm:** Number of particles adopted (i.e., population size).
- **Particle:** One member (or individual) of the swarm. Each particle represents a potential solution to the problem being solved. The position of a particle is determined by the solution that it currently represents.
- ***pbest*** (*personal best*)**:** The best position that a given particle has achieved so far. That is, the position of the particle that has provided the greatest success (measured in terms of a scalar value defined by the user, which is analogous to the fitness value adopted in EAs).
- ***lbest*** (*local best*)**:** Position of the best particle member belonging to the neighborhood of a given particle.

- ***gbest*** (*global best*): Position of the best particle of the entire swarm.
- **Leader:** Particle that is used to guide another particle towards better regions of the search space.
- **Velocity (vector):** This vector drives the optimization process, that is, it determines the direction in which a particle needs to "fly" (move), in order to improve its current position.
- **Inertia weight:** The inertial weight (denoted by W) is adopted to control the impact of the previous history of velocities on the current velocity of a given particle.
- **Learning factor:** It represents the attraction that a particle has towards either its own best previous value or that of its neighbors. Two learning factors are adopted in PSO: C_1, which is the *cognitive* learning factor and represents the attraction that a particle has toward its own success, and C_2, which is the *social* learning factor and represents the attraction that a particle has toward the success of its neighbors. Both of them are normally defined as constants.
- **Neighborhood topology:** It determines the way in which particles are interconnected and thus defines the way in which they contribute to the computation of the *lbest* value of a given particle.

In PSO, the position of a particle (within the search space being explored) changes based on its own experience and the success of its neighbors.

Let $\mathbf{x}_i(t)$ denote the position of particle p_i, at time step t. The position of p_i is then changed by adding a velocity $\mathbf{v}_i(t)$ value to the current position of the particle, i.e.:

$$\mathbf{x}_i(t) = \mathbf{x}_i(t-1) + \mathbf{v}_i(t) \tag{4}$$

The velocity vector reflects the exchanged information and, in general, is defined in the following way:

$$\begin{aligned}\mathbf{v}_i(t) = W\mathbf{v}_i(t-1) + C_1 r_1(\mathbf{x}_{pbest_i} - \mathbf{x}_i(t))\\ +C_2 r_2(\mathbf{x}_{leader} - \mathbf{x}_i(t))\end{aligned} \tag{5}$$

where and $r_1, r_2 \in [0,1]$ are randomly generated values.

Particles are influenced by the success of any particle connected to them. It is worth noting, however, that the way this influence information is propagated depends on the neighborhood topology adopted. Any of the possible neighborhood topologies that can be adopted in PSO can be represented as a graph. The following are the most commonly adopted neighborhood topologies:

- **Empty graph:** In this topology, each particle is connected only with itself, and it compares its current position only to its own best position found so far (*pbest*) [5]. In this case, $C_2 = 0$ in equation (5).
- **Local best:** In this topology, each particle is affected by the best performance of its k immediate neighbors (*lbest*), as well as by their own past experience (*pbest*) [5]. When $k = 2$, this structure is equivalent to a ring topology. In this case, *leader=lbest* in equation (5).

- **Fully connected graph:** This topology connects all members of the swarm to one another. This structure is also called *star* topology in the PSO community [5]. In this case, *leader=gbest* in equation (5).
- **Star network:** In this topology, one particle is connected to all others and they are connected to only that one (called *focal* particle) [5]. This structure is also called *wheel* topology in the PSO community. In this case, *leader=focal* in equation (5).
- **Tree network:** In this topology, all particles are arranged in a tree and each node of the tree contains exactly one particle [7]. This structure is also called *hierarchical* topology in the PSO community. In this case, $leader = pbest_{parent}$ in equation (5).

The neighborhood topology is likely to affect the rate fo convergence as it determines how much time it takes to the particles to find out about the location of good (better) regions of the search space. For example, since in the *fully connected* topology all particles are connected to each other, all particles receive the information of the best solution from the entire swarm at the same time. Thus, when using such topology, the swarm tends to converge more rapidly than when using *local best* topologies, since in this case, the information of the best position of the swarm takes a longer time to be transferred. However, for the same reason, the *fully connected* topology is also more susceptible to suffer premature convergence (i.e., to converge to local optima) [6].

Figure 1 shows the way in which the general (single-optimization) PSO algorithm works. First, the swarm (both positions and velocities) is randomly initialized. The corresponding *pbest* of each particle is initialized and the leader is located (usually the *gbest* solution is selected as the leader). Then, for a maximum number of iterations, each particle flies through the search space updating its position (using equations (4) and (5)) and its *pbest* and, finally, the leader is updated too.

```
Begin
    Initialize swarm
    Locate leader
    g = 0
    While g < gmax
            For each particle
                    Update position (flight)
                    Evaluation
                    Update pbest
            EndFor
            Update leader
            g++
    EndWhile
End
```

Fig. 1 Pseudocode of the general PSO algorithm for single-objective optimization.

4 Particle Swarm Optimization for Multi-Objective Problems

In order to apply the PSO strategy for solving MOPs, it is obvious that the original scheme has to be modified. As we saw in Section 2, in multi-objective optimization, we aim to find not one, but a set of different solutions (the so-called Pareto optimal set). In general, when solving a MOP, the main goals are to converge to the true Pareto front of the problem (i.e., to the solutions that are globally nondominated) and to have such solutions as well-distributed as possible along the Pareto front. Given the population-based nature of PSO, it is desirable to produce several (different) nondominated solutions with a single run. So, as with any other evolutionary algorithm, the main issues to be considered when extending PSO to multi-objective optimization are [2]:

1. How to select particles (to be used as leaders) in order to give preference to nondominated solutions over those that are dominated?
2. How to retain the nondominated solutions found during the search process in order to report solutions that are nondominated with respect to all the past populations and not only with respect to the current one? Also, it is desirable that these solutions are well spread along the Pareto front.
3. How to maintain diversity in the swarm in order to avoid convergence to a single solution?

As we just saw, when solving single-objective optimization problems, the leader that each particle uses to update its position is completely determined once a neighborhood topology is stablished. However, when dealing with MOPs, each particle might have a set of different leaders from which just one can be selected in order to update its position. Such set of leaders is usually stored in a different place from the swarm, that we will call *external archive* [3]: This is a repository in which the nondominated solutions found so far are stored. Only solutions that are nondominated with respect to the contents of the entire archive are retained. The solutions contained in the external archive are used as leaders when the positions of the particles of the swarm have to be updated. Furthermore, the contents of the external archive is also usually reported as the final output of the algorithm.

Figure 2 shows the way in which a general MOPSO works. We have marked with *italics* the processes that make this algorithm different from the general PSO algorithm for single objective optimization shown before. First, the swarm is initialized. Then, a set of leaders is also initialized with the nondominated particles from the swarm. As we mentioned before, the set of leaders is usually stored in an external archive. Later on, some sort of quality measure is calculated for all the leaders in order to select (usually) one leader for each particle of the swarm. At each generation, for each particle, a leader is selected and the flight is performed. Most of the existing MOPSOs apply some

[3] This *external archive* is also used by many Multi-Objective Evolutionary Algorihtms (MOEAs).

```
Begin
    Initialize swarm
    Initialize leaders in an external archive
    Quality(leaders)
    g = 0
    While g < gmax
            For each particle
                    Select leader
                    Update Position (Flight)
                    Mutation
                    Evaluation
                    Update pbest
            EndFor
            Update leaders in the external archive
            Quality(leaders)
            g++
    EndWhile
    Report results in the external archive
End
```

Fig. 2 Pseudocode of a general MOPSO algorithm.

sort of mutation operator[4] after performing the flight. Then, the particle is evaluated and its corresponding *pbest* is updated. A new particle replaces its *pbest* particle usually when this particle is dominated or if both are incomparable (i.e., they are both nondominated with respect to each other). After all the particles have been updated, the set of leaders is updated, too. Finally, the quality measure of the set of leaders is re-calculated. This process is repeated for a certain (usually fixed) number of iterations.

As we can see, and given the characteristics of the PSO algorithm, the issues that arise when dealing with multi-objective problems are related with two main algorithmic design aspects [14]:

1. **Selection and updating of leaders**

    - How to select a single leader out of set of nondominated solutions which are all equally good? Should we select this leader in a random way or should we use an additional criterion (to promote diversity, for example)?
    - How to select the particles that should remain in the external archive from one iteration to another?

[4] The mutation operators adopted in the PSO literature have also been called *turbulence* operators.

2. **Creation of new solutions**

   - How to promote diversity through the two main mechanisms to create new solutions: updating of positions (equations (4) and (5)) and a mutation (turbulence) operator.

Regarding the selection of leaders, the most simple approach is to adopt aggregating functions (i.e., weighted sums of the objectives) or approaches that optimize each objective separately. However, most researchers redefine the concept of leader, incorporating the definition of Pareto optimality. However, since all the nondominated solutions currently available can be considered as potential leaders, a quality measure is required in order to choose one of them at a given time. Several authors have proposed the use of density measures for this sake. The two most commonly adopted are:

- **Nearest neighbor density estimator** [4]. The nearest neighbor density estimator gives us an idea of how crowded are the closest neighbors of a given particle, in objective function space. This measure estimates the perimeter of the cuboid formed by using as vertices the nearest neighbors.
- **Kernel density estimator** [3]: When a particle is sharing resources with others, its fitness is degraded in proportion to the number and closeness to particles that surround it within a certain perimeter. A neighborhood of a particle is defined in terms of a parameter that defines the radius of the neighborhood. Such neighborhoods are called *niches*.

As indicated before, most MOPSOs adopt an external archive that retains solutions that are nondominated with respect to all the previous populations (or swarms). Such an archive will allow the entrance of a solution only if: (a) it is nondominated with respect to the contents of the archive or (b) it dominates any of the solutions within the archive (in this case, the dominated solutions have to be deleted from the archive).

Mainly due to practical reasons, archives tend to be bounded [2], which makes necessary the use of an additional criterion to decide which nondominated solutions to retain, once the archive is full. In evolutionary multi-objective optimization, researchers have adopted different techniques to prune the archive (e.g., clustering [15] and geographical-based schemes that place the nondominated solutions in cells in order to favor less crowded cells when deleting in-excess nondominated solutions [10]).

It is worth noting that, strictly speaking, three archives should be used when implementing a MOPSO: one for storing the global best solutions, one for the personal best values and a third one for storing the local best (if applicable). However, in practice, few authors report the use of more than one archive in their MOPSOs.

In a MOPSO, diversity can be promoted through the selection of leaders. However, this can be also done through the two main mechanisms used for creating new solutions: (a) updating of positions (topologies that define neighborhoods smaller than the entire swarm for each particle can also preserve diversity within the swarm a longer time), and (b) through the use of a mutation (or turbulence) operator (this will help a MOPSO to escape from local optima).

5 Future Research Paths

Most of the work in this are has focused on algorithm development, but we believe that there are several other topics that constitute very promising paths for future research:

- **Self-Adaptation:** The design of MOPSOs with no parameters that have to be fine-tuned by the user is a topic that is worth studying. The design of a parameterless MOPSO requires a in-depth knowledge of the relationship between its parameters and its performance in problems with different features.
- **Theoretical Developments:** There is an evident lack of research on even the most basic aspects of a MOPSO (e.g., convergence properties, run-time analysis, population dynamics, etc.), but it is expected that some work in this direction will be conducted in the next few years.
- **Applications:** The applications of MOPSOs have steadily grown in the last few years, but more are expected to arise, as MOPSOs become better developed and widespread multi-objective optimization tools.

6 Conclusions

This paper has provided a review of the basic concepts of the PSO algorithm, including its basic equation, neighborhood topologies and leader selection mechanisms. Then, the main changes required to extend PSO to the solution of MOPs were briefly described, including the use of external archives, the mechanisms to select leaders and the promotion of diversity in a swarm. In the final part of the paper, some of the possible paths for future research on MOPSOs were briefly addressed.

References

1. Angeline, P.J.: Evolutionary optimization versus particle swarm optimization: Philosophy and performance differences. In: Porto, V., Saravanan, N., Waagen, D., Eiben, A. (eds.) EP 1998. LNCS, vol. 1447, pp. 601–610. Springer, Heidelberg (1998)
2. Coello Coello, C.A., Lamont, G.B., Van Veldhuizen, D.A.: Evolutionary Algorithms for Solving Multi-Objective Problems, 2nd edn. Springer, New York (2007); ISBN 978-0-387-33254-3

3. Deb, K., Goldberg, D.E.: An investigation of niche and species formation in genetic function optimization. In: Schaffer, J.D. (ed.) Proceedings of the Third International Conference on Genetic Algorithms, pp. 42–50. George Mason University, Morgan Kaufmann Publishers, San Mateo, California (1989)
4. Deb, K., Pratap, A., Agarwal, S., Meyarivan, T.: A fast and elitist multiobjective genetic algorithm: NSGA–II. IEEE Transactions on Evolutionary Computation 6(2), 182–197 (2002)
5. Engelbrecht, A.P. (ed.): Computational Intelligence: An Introduction. John Wiley & Sons, England (2002)
6. Engelbrecht, A.P.: Fundamentals of Computational Swarm Intelligence. John Wiley & Sons, Chichester (2005)
7. Janson, S., Middendorf, M.: A hierarchical particle swarm optimizer. In: Congress on Evolutionary Computation (CEC 2003), pp. 770–776. IEEE Press, Camberra (2003)
8. Kennedy, J., Eberhart, R.C.: Particle swarm optimization. In: Proceedings of the 1995 IEEE International Conference on Neural Networks, pp. 1942–1948. IEEE Service Center, Piscataway (1995)
9. Kennedy, J., Eberhart, R.C.: Swarm Intelligence. Morgan Kaufmann Publishers, San Francisco (2001)
10. Knowles, J.D., Corne, D.W.: Approximating the nondominated front using the pareto archived evolution strategy. Evolutionary Computation 8(2), 149–172 (2000)
11. Moore, J., Chapman, R.: Application of particle swarm to multiobjective optimization, department of Computer Science and Software Engineering. Auburn University (1999)
12. Moore, J., Chapman, R., Dozier, G.: Multiobjective Particle Swarm Optimization. In: Turner, A.J. (ed.) Proceedings of the 38th Annual Southeast Regional Conference, pp. 56–57. ACM Press, Clemson (2000)
13. Reyes-Sierra, M., Coello Coello, C.A.: Multi-Objective Particle Swarm Optimizers: A Survey of the State-of-the-Art. International Journal of Computational Intelligence Research 2(3), 287–308 (2006)
14. Toscano Pulido, G.: On the use of self-adaptation and elitism for multiobjective particle swarm optimization. PhD thesis, Computer Science Section, Department of Electrical Engineering, CINVESTAV-IPN, Mexico (2005)
15. Zitzler, E., Thiele, L.: Multiobjective evolutionary algorithms: A comparative case study and the strength pareto approach. IEEE Transactions on Evolutionary Computation 3(4), 257–271 (1999)

Direct Load Control in the Perspective of an Electricity Retailer – A Multi-Objective Evolutionary Approach

Álvaro Gomes, Carlos Henggeler Antunes, and Eunice Oliveira

Abstract. The judicious use of end-use electric loads in the framework of demand-side management programs as valuable resources to increase the global efficiency of power systems as far as economical, technical and quality of service aspects are concerned, is a relevant issue in face of the changes underway in the power systems industry. This paper presents the results of a multi-objective optimization model, in the perspective of an electricity retailer, which is aimed at designing load control actions to be applied to groups of electric loads. An evolutionary algorithm is used to compute solutions to this problem.

Keywords: Load control, load management, multi-objective optimization, evolutionary algorithms.

1 Introduction

The recent changes in the structure of power systems, in which traditional vertically integrated utilities have been replaced by several entities in the different functions, with competition at the generation and the retail levels, associated with the dissemination of advanced metering technologies, dynamic pricing tariff systems, micro-generation, electric storage devices and smart grids, are pushing forward demand- side management (DSM) activities, namely price-based demand response programs. During the last decades, DSM programs encompassing a set of activities designed to change the shape of the load diagram and the amount of electricity used have significantly evolved. DSM has moved from actions mainly

Álvaro Gomes · Carlos Henggeler Antunes
Dept. of Electrical Engineering and Computers, University of Coimbra,
Polo II, 3030 Coimbra, Portugal and R&D Unit INESC Coimbra
e-mail: agomes@deec.uc.pt, ch@deec.uc.pt

Eunice Oliveira
School of Technology and Management, Polytechnic Institute of Leiria,
Morro do Lena, Ap. 4163, 3411-901 Leiria, Portugal and R&D Unit INESC Coimbra
e-mail: eunice@ipleiria.pt

A. Gaspar-Cunha et al. (Eds.): Soft Computing in Industrial Applications, AISC 96, pp. 13–26.
springerlink.com

focused on energy conservation and load management measures to activities responsive to prices. Different entities have different objectives, and besides operational benefits such as increasing the load factor, reducing peak power demand, reducing cost and losses, as well as reliability concerns, the possibility of increasing profits with the selling of electricity becomes an attractive objective to be pursued by several players [1][2][3].

Some of the DSM programs being offered are dynamic pricing programs, which give consumers incentives to reduce peak demand thus reducing their electricity bills, and load curtailment programs, which give customers some financial stimuli for reducing peak demand during critical periods. Load curtailment programs encompass direct load control (DLC) actions and interruptible rates, which are programs typically based on a up-front incentive payment, and demand bidding or buyback programs, in which customers are paid a certain amount of money for each MWh reduced (curtailed) during critical time periods.

The implementation of load curtailment programs triggered by emergency signals or price signals usually reduce peak load in those critical periods by turning off some end-use loads. Suitable loads for these activities are loads to which some storage capability is associated, that is loads providing an energy service that can be temporarily interrupted or deferred in time without decreasing the quality of the service provided. Examples of loads with these characteristics are thermostatic loads such as air conditioners, heat pumps, electric water heaters and electric space heaters, and loads associated with other types of storage capacity such as water pumping or conveyors filling storage silos. If there is a certain period of time (designated as time constant) during which the load can be turned off without decreasing too much the quality of energy service provided then it is possible to shift the load operation to a lower demand period, therefore contributing to reduce the peak demand while increasing energy consumption in low demand periods. The working cycle of thermostatic loads is determined by a thermostat according to the end-use needs (desired temperature, heat load, space characteristics, etc.) and according to the physical characteristics of the loads (for instance, the nominal power). Therefore, when many thermostatic loads are running simultaneously the aggregate demand of all loads is lower than the sum of the maximum demand of every load due to diversified demand patterns. It should be noticed that when the regular working cycle of these loads is changed, by deferring the demand in time, the diversified demand changes in a way that may cause a higher peak demand during power restoration than it would be if no load curtailment had occurred (the so-called payback effect) (Figure 1 and Figure 2). This is one of the potential undesirable effects of such load management actions if they are not properly designed and scheduled. A second unwanted effect is a possible strong decrease in profits or discomfort caused to consumers associated with a decrease of the quality of the energy service provided.

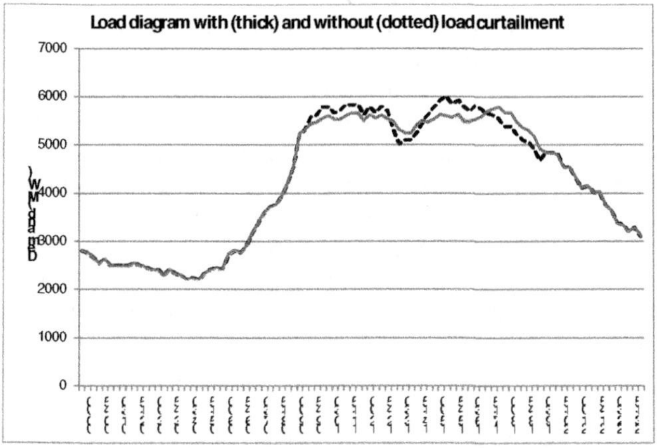

Fig. 1 Impact of a load control action on the load diagram.

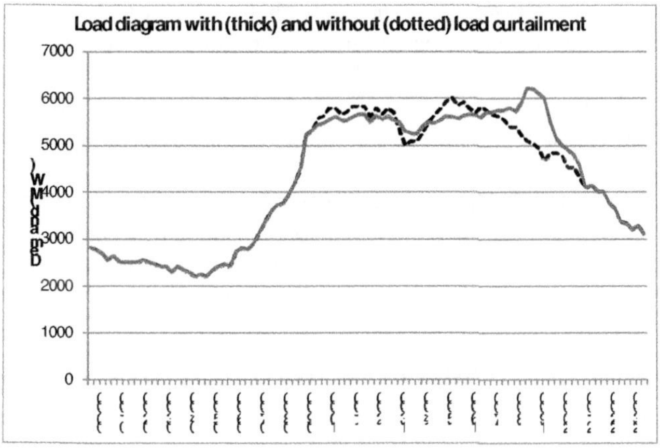

Fig. 2 The payback effect.

Besides the prevention of potential undesirable effects, load curtailment programs should be implemented taking into account a careful balance of the objectives pursuit. These objectives include economical, technical and quality of service aspects of evaluation of the merits of different measures.

Distinct load management programs are being offered by electric utilities in many countries [4][5][6]. Electric utilities, namely distribution utilities, retailers and aggregators are some of the entities currently operating in the power systems sector that are directly interested in implementing DSM activities. Electric utilities responsible for the network management might be interested in DSM activities aimed at increasing the network reliability [7][8], reducing maximum demand and losses, reducing costs and emissions, and increasing revenues [9][10][11]. Retailers and aggregators might be interested in exploiting the differences between the retail market and the wholesale market electricity prices in order to increase profits. Also they might be interested in reducing the peak demand of their customers as the electricity purchase prices or the costs of using networks may be a function of the maximum demand [10][12].

Furthermore, other issues are increasing the interest in DSM activities. One is the integration of renewable sources into the grid. It is well known that a high penetration of renewable sources, namely wind power, affects the planning and operation of power systems and the impacts on costs and reliability may not be neglected. In face of the existing forces towards increasing the contribution of renewable energy, properly designed load curtailment programs can be attractive tools to mitigate the impacts of such energy sources. Another issue pushing forward DSM activities is the overall efficiency of the power system by making an integrated management of all available resources: traditional generation, distributed and local generation, storage, electric vehicle, and demand-side resources. Load management should also play an essential role in the stand-alone operation of parts of the power system. From a consumer point of view, the possibility of reducing the electricity bill by an adequate management of the loads is a motivating objective. With the penetration of micro-generation and storage technologies, consumers become *prosumers*, in the sense that not just displaying a more proactive and intelligent consumer behavior but also being active producers, meaning that even at individual customer level there is the need for an adequate management of all resources available. The ongoing development and dissemination of information and communication technologies, advanced metering and control systems lays the technical and operational foundations for an integrated management of all energy resources. The interest and motivation of this study has been provided in this section. In section 2 an overview is made of a multi-objective model for the design of load control actions. The case study is presented in section 3 as well as illustrative results. Finally some conclusions are drawn in section 4.

2 A Multi-Objective Model for the Design of Load Control Actions

The design of load control actions involves determining the time intervals throughout a given control period in which the loads are turned off, namely when they occur and how long they last. This characterization of load control actions has been often done based on information from past implementations, (costly) field experiments or pilot programs. Moreover, a cycling strategy has been frequently used with pre-determined on/off patterns applied to the loads under control. However, if the different (dynamic) usages of energy services are not taken into consideration the on/off constant patterns of cycling strategies increase the probability of causing discomfort to the end-users. Usually the demand imposed by loads presents a daily behavior and the effects of power curtailment actions last for significant periods of time. The use of physically-based load models [10][13] for the identification of control actions enable to mitigate these risks. These models mimic the physical phenomena happening in end-use loads, making possible the characterization of changes in the demand derived from the control actions. It is important to notice that the use of physically-based load models enables to tackle real-world situations in a very detailed way but it imposes an additional computational burden. However, this does not hinder its use in a normal situation of one day ahead planning.

The problem consists in designing load control actions to be applied to load groups under control to optimize economical, technical and quality of service objectives. The objectives to be achieved with the implementation of load curtailment actions include:

- Minimization of peak power demand. In this case, the electricity retailer wants to assess the load management impacts at three different demand aggregation levels: the aggregated level, representing the demand of all the customers; the residential customer's level, representing the demand of all residential customers; and at a power distribution transformer level, feeding mainly service sector customers. Besides enabling the retailer to continuously exploiting the differences between purchasing and selling prices, the evaluation carried out at these different levels may also be used for a better design of the load management actions.
- Profit maximization. Profits in the selling of electricity are influenced by the amount of electricity sold, the time of day/season/year, and the rate of using energy (power). In the presence of demand and wholesale price forecasts, the retailer wants to design adequate load curtailment actions in order to maximize profits.
- Loss factor minimization. This objective is related with operational and economic goals.
- Minimize discomfort caused to customers. The changes on the regular load working cycle may eventually cause discomfort to customers that must be minimized so that those actions become also attractive from the customers' point of view (with possible reduction in their electricity bill) and/or at least not decrease their willingness to accept them. Discomfort is evaluated through objective functions related with the time a state variable (controlled by loads) is over or under a pre-specified threshold level: the maximum continuous time interval in which this condition has occurred and the total time it has occurred.

A multi-objective mathematical model has been developed [10][13] in which the decision variables refer to the on/off patterns to be applied to each group of loads. Since total flexibility exists in the definition of the on/off periods determining the normal operation/curtailment of loads, this is a large-scale combinatorial problem due to the vast number of feasible combinations of on/off patterns for the groups of loads. A multi-objective evolutionary algorithm (EA) has been developed and tailored for this problem [10][13]. Besides not requiring any specific form of the mathematical model, EAs are particularly suited for combinatorial multi-objective optimization since they work with a set (population) of potential solutions and the aim is generally the identification of the non-dominated front [14][15][16]. In complex real-world problems it is not usually possible to assert the true non-dominance of a set of solutions because no theoretical tools exist to guarantee that a solution found in the optimization process is indeed non-dominated as it is generally the case in mathematical programming approaches.

The structure of the solution (on/off patterns) asks for a binary representation. The solutions (individuals) are the codification of a load control strategy, which represents the on/off patterns to be applied to every group of loads during one day. An individual is represented by a binary string: "1" = a load curtailment action is applied to a group of loads; "0" = no load curtailment occurs (Figure 3).

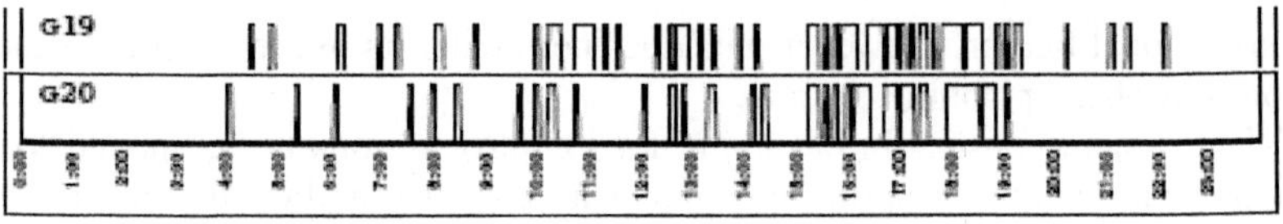

Fig. 3 Examples of control actions applied to two groups of loads.

Figure 3 shows the load control actions applied to two groups of loads. The ability of "tuning" the control actions to the characteristics of the loads in every group makes this methodological approach very attractive as it allows exploiting the amount of controllable load available and maximizing the reduction of peak demand available for control without increasing the discomfort caused to consumers. This ability is beyond the capabilities of traditional cycling strategies with equal and pre-defined on/off periods that do not take into consideration the utilization of the energy service.

In the approach followed in this work an EA is used for the design and the selection of load control strategies. The demand of controllable loads - air conditioners (ACs) and electric water heaters (EWHs) - is estimated by using Monte Carlo simulations of physically-based load models that have been experimentally validated. One of the main advantages of using physically-based load models is that changes in individual power demand due to external load curtailment actions are automatically taken into account in individual and global demand, due to their ability reproduce the changed demand of controlled loads.

3 A Case Study and Illustrative Results

A common objective for the different entities competing in the retail of electricity is the maximization of profits. These entities may buy electricity in the wholesale market, from a generator or any other entity selling electricity. Depending on the tariff systems being used for electricity transactions and the purchase contracts established between retailers and sellers of electricity, the energy acquisition costs as seen by retailers can change over time, and probably costs change in a way that is very close to what is known as real-time price. Acquisition costs may also take into consideration costs with the use and management of the networks, and can be a function of the amount of energy and the peak power flow over the wires. On the other hand, for most consumers, at least residential consumers, the electricity prices are fixed over a significant time length, one trimester or one year. For instance under static time-of-use tariffs, which are very common nowadays, the prices of kWh vary over the day but are fixed during one trimester or one year. Figure 4 shows the Iberian Electricity Market (MIBEL) spot indices for 4 days, in which prices vary over the day and are also different for different days.

In this work, a case study taking into consideration the perspectives of a retailer is presented. The retailer sells electricity to about 5000 customers whose average daily load diagram is shown in Figure 5 ("Aggregate level" curve). Most customers are residential, services and small commerce clients. Besides assessing the changes at the aggregate demand level (all customers), the retailer also wants also to analyze the impacts of load management activities on the demand of their

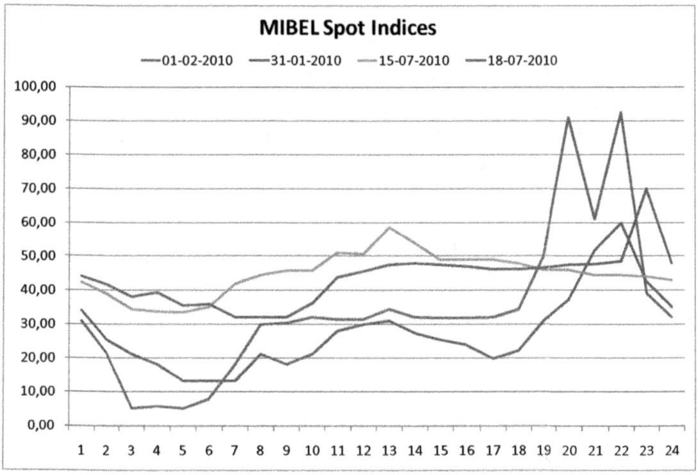

Fig. 4 MIBEL spot indices.

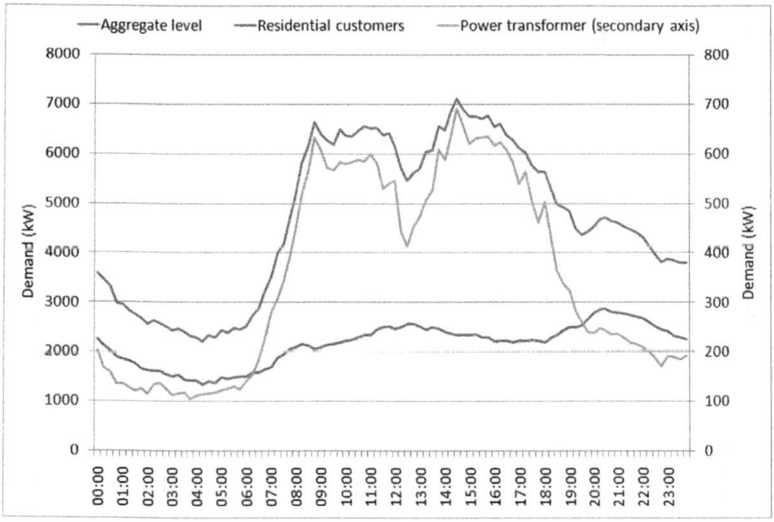

Fig. 5 Demand at three aggregation levels.

residential consumers ("Residential customers" curve) and a specific group of service sector customers fed by a distribution power transformer owned by the retailer that faces some capacity constraints ("Power transformer" curve).

Figure 6 shows the demand of all the retailer's customers and the demand of all controllable end-use loads. Figure 7 shows the demand of a specific group of service sector customers and the corresponding controllable load. The demand of all residential customers is shown in Figure 8 together with the controllable load associated with these consumers.

The peak power demand of all the retailer's customers is 7105 kW, occurring at 14:30h, while the peak demand of all residential consumers is 2872 kW at 20:30h, and the peak demand at the power transformer is 691kW at 14:30h (Table 1). The total controllable load when maximum demand at aggregate level occurs is 445 kW, which is 6.25% of peak demand. The controllable load for residential consumers is 141 kW (4.9%) of the maximum value of the residential load diagram, while for service sector customers the controllable peak load is 139kW (20.1%).

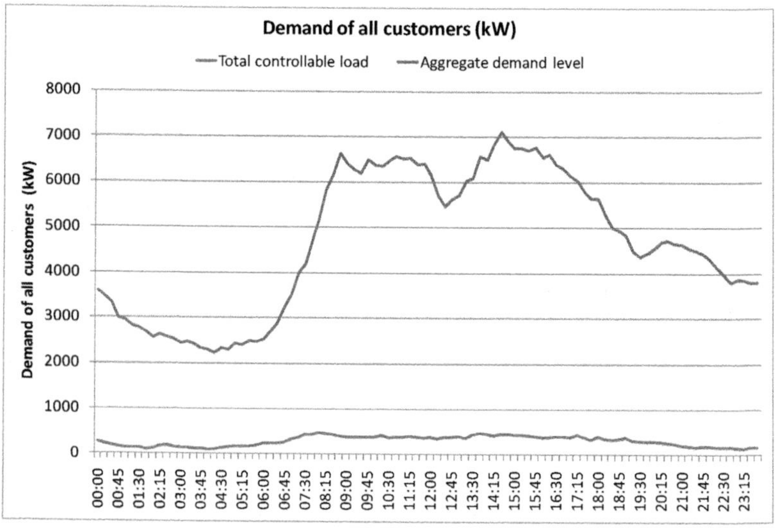

Fig. 6 Total power demand and total controllable load.

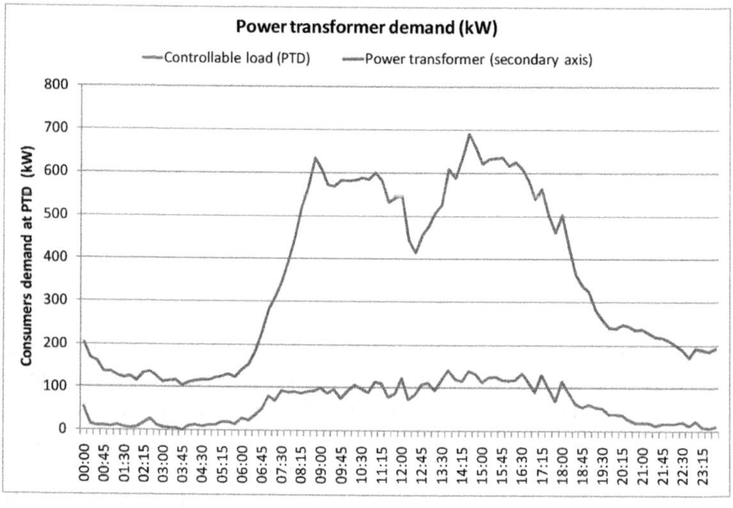

Fig. 7 Demand at power transformer level.

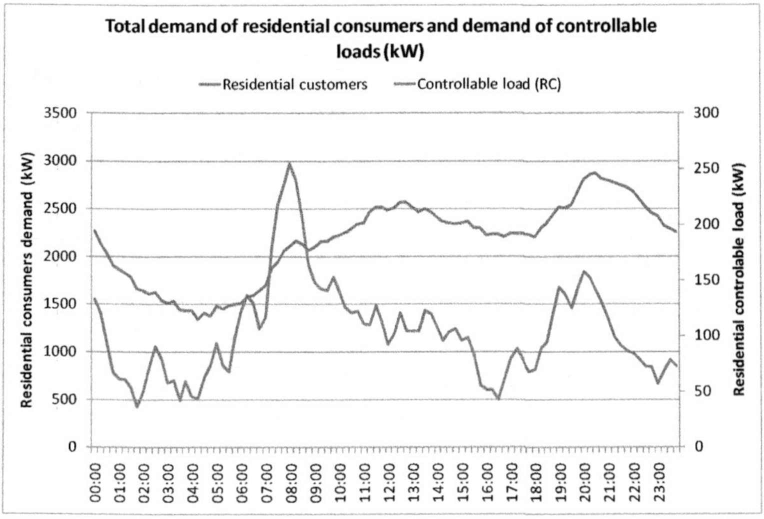

Fig. 8 Demand of all residential customers and controllable residential demand.

Table 1 Characteristics of total demand and controllable demand

	Maximum demand		Controllable load when maximum demand occurs		Profits	Loss factor
	kW	Hour	kW	%	€	–
Total	7105	14:30	445	6.28	2224.9	0.48434
Residential	2872	20:30	141	4.9	–	–
PTD	691	14:30	139	20.1	–	–

Controllable loads, in this case electric water heaters and air conditioners, have been grouped in 20 groups. Table 2 shows some details of the groups. Each group contains only one type of load grouped according to some physical and geographical characteristics.

Average unit profits per kWh sold are shown in Figure 9.

Besides the combinatorial nature and the size of the search space, also the need to assess the impacts at different demand aggregation levels with different shapes contributes to increase the difficulty of this type of problems.

Table 2 Number of loads per group and types of loads.

Group	Type	LD	# Loads	Group	Type	LD	# Loads
1	EWH	RC	20	11	EWH	PTD	20
2	EWH	RC	35	12	EWH	PTD	15
3	EWH	RC	30	13	EWH	A	30
4	EWH	RC	20	14	EWH	RC	15
5	EWH	RC	30	15	AC	PTD	20
6	EWH	RC	20	16	AC	A	30
7	EWH	A	25	17	AC	A	50
8	EWH	RC	30	18	AC	PTD	20
9	EWH	A	25	19	AC	A	40
10	EWH	RC	20	20	AC	PTD	30

A = Aggregate demand level.
RC = Residential demand level.
PTD = Power transformer demand level.
EWH = Electric water heaters.
AC = Air conditioners.

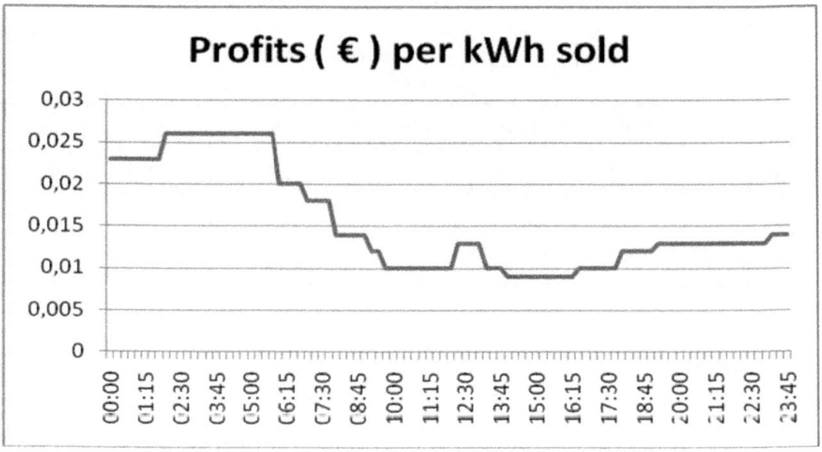

Fig. 9 Profits per unit of energy sold.

The results for 7 non-dominated solutions are shown in Table 3. The algorithm was able to identify solutions that could reduce maximum demand at the three demand levels (aggregate, residential, power transformer), and at the same time increase profits and reduce the loss factor, without decreasing too much the quality of energy services provided.

Table 3 Results in the seven evaluation axis for 7 non-dominated solutions.

Solutions	1	2	3	4	5	6	7
Maximum demand at aggregate level	6793,11	6807,08	6800,72	6796,15	6819,28	6811,94	6821,45
Residential consumers demand	2759,63	2760,59	2766,18	2760,76	2767,65	2761,66	2767,25
Power transformer demand	641,79	644,47	650,91	639,19	642,42	644,47	642,42
No. minutes	14	17	28	13	14	14	18
Maximum interval	1	3	6	2	2	1	2
Loss factor	0,482318	0,482312	0,482315	0,482576	0,482279	0,482327	0,482279
Profits	2293,285	2289,895	2291,477	2293,203	2286,782	2288,784	2286,273

Table 4 shows the extreme values encountered for these seven solutions shown in Table 3 and the maximum improvement in each objective. According to the results in Table 4 it was possible to reduce up to 312/112/52 kW the maximum demand at aggregate/residential/power transformer levels, which are 4.39% /3.91% /7.49%, respectively, of the original maximum demand values. Profits increased up to 3.07%.

Table 4 Details about the results obtained.

	Minimum	Maximum	Difference	Original Values	Maximum improvement
Maximum demand at aggregate level	6793,11	6821,44	28,3345	7105	311,89
Residential consumers demand	2759,63	2767,65	8,026273	2872	112,37
Power transformer demand	639,19	650,91	11,72003	691	51,81
No. minutes	13	28	15	0	–
Maximum interval	1	6	5	0	–
Loss factor	0,48228	0,48258	0,000297	0,48434	0,00206
Profits	2286,27	2293,29	7,012812	2224,9	68,39

A more detailed analysis of the results corresponding to solution 4 (Figure 10) can be done using the values in Table 5. Figure 10 shows the on/off patterns corresponding to this solution.

Table 5 Results for solution 4.

	Solution 4	Original Values	Difference	
Maximum demand at aggregate level	6796,154	7105	308,8464	4,35%
Residential consumers demand	2760,762	2872	111,238	3,87%
Power transformer demand	639,1904	691	51,80963	7,50%
No. minutes	13	0	13	–
Maximum interval	2	0	2	–
Loss factor	0,482576	0,48434	0,001764	0,36%
Profits	2293,203	2224,9	68,30289	3,07%

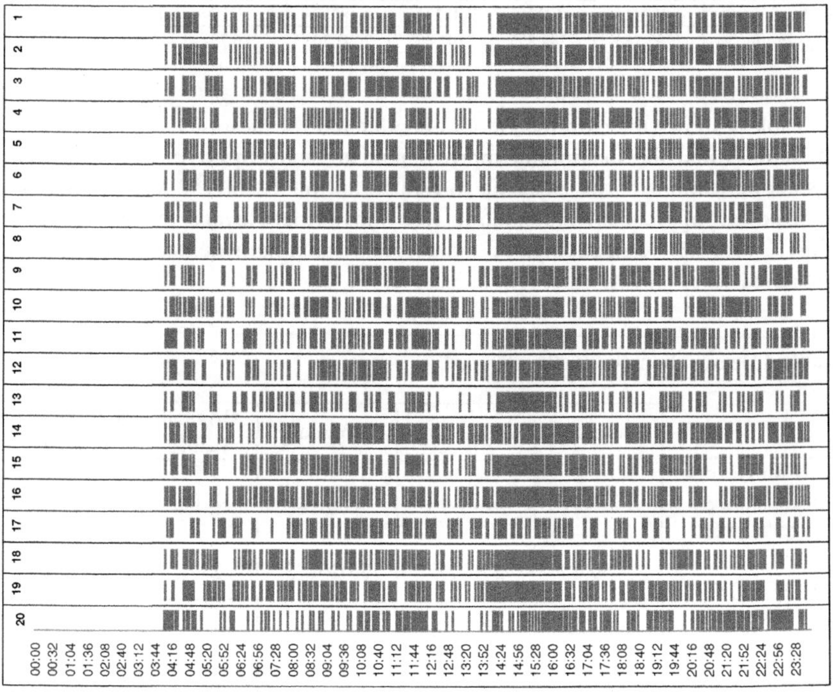

Fig. 10 On/off patterns to be applied to the 20 groups of loads (solution 4).

4 Conclusions

The concurrent development, dissemination and implementation of new technologies are changing the paradigm of electric power systems sector: smart metering allowing a two-way communication between suppliers and consumers; distributed and micro-generation systems; storage including the electric vehicle allowing to make electricity a service/good with characteristics similar to other negotiable services/goods; control systems, such as smart thermostats, allowing to explore different operating modes of some end-use loads. Load resources should play an essential role in this new power systems paradigm, to be exploited in the framework of demand-side management activities. These resources can be used as a way to increase the system global efficiency, to reduce the costs with the acquisition of electricity, to increase profits with the selling of electricity, and as tool to maximize the integration of renewable and micro-generation systems. In this work, the perspective of a retailer has been used as a starting point to explore the potential of LM activities in a distribution network. Load control actions computed using an EA designed to tackle a multi-objective model have proved their effectiveness in the attainment of the retailer's objectives.

Acknowledgments. This research has been partially supported by FCT and FEDER under Project Grant MIT/SET/0018/2009.

References

[1] International Energy Agency, The Power to Choose - Demand Response in Liberalised Electricity Markets (2003)

[2] Heffner, G.C., Goldman, C.A.: Demand Responsive Programs - An Emerging Resource for Competitive Electricity Markets? In: Proceedings of the International Energy Program Evaluation Conference, IEPEC 2001 (2001)

[3] Hirst, E., Kirby, B.: Retail-Load Participation in Competitive Wholesale Electricity Markets, Edison Electric Institute (2001)

[4] Greening, L.A.: Demand response resources: Who is responsible for implementation in a deregulated market? Energy 35, 1518–1525 (2010)

[5] Cappers, P., Goldman, C., Kathan, D.: Demand response in U.S. electricity markets: Empirical evidence. Energy 35, 1526–1535 (2010)

[6] Torriti, J., Hassan, M.G., Leach, M.: Demand response experience in Europe: Policies, programmes and implementation. Energy 35, 1575–1583 (2010)

[7] 2007 Assessment of Demand Response and Advanced Metering, Federal Energy Regulatory Commission (2007)

[8] Salehfar, H., Patton, A.D.: A Production Costing Methodology for Evaluation of Direct Load Control. IEEE Trans. on Power Systems 6(1), 278–284 (1991)

[9] Lachs, W.R., Sutano, D., Logothetis, D.N.: Power System Control in the Next Century. IEEE Trans. on Power Systems 11(1), 11–18 (1996)

[10] Gomes, A., Antunes, C.H., Martins, A.G.: A multiple objective evolutionary approach for the design and selection of load control strategies. IEEE Transactions on Power Systems 19(2), 1173–1180 (2004)

[11] Horn, J., Nafpliotis, N.: Multiobjective Optimization Using The Niched Pareto Genetic Algorithm, IlliGAl Report 93005, Department of Computer Science, University of Illinois at Urbana-Champaign, Urbana, Illinois, USA (1993)

[12] Chen, J., Lee, F.N., Breipohl, A.M., Adapa, R.: Scheduling Direct Load Control to Minimize System Operation Cost. IEEE Trans. on Power Systems 10(4), 1994–2001 (1995)

[13] Gomes, A., Antunes, C.H., Martins, A.G.: A multiple objective approach to electric load management using an interactive evolutionary algorithm. IEEE Transactions on Power Systems 22, 1004–1011 (2007)
[14] Fonseca, C.M., Fleming, P.J.: An Overview of Evolutionary Algorithms in Multiobjective Optimization. Evolutionary Computation 3(1), 1–16 (1995)
[15] Jones, D.F., Mirrazavi, S.K., Tamiz, M.: Multi-objective meta-heuristics: An overview of the current state-of-the-art. European Journal of Operational Research 137, 1–9 (2002)
[16] Deb, K.: Multi-Objective Optimization using Evolutionary Algorithms. John Wiley & Sons Ltd., Chichester (2001)

Tutorial

Evolutionary Approaches for Optimisation Problems

Lars Nolle and Gerald Schaefer

Abstract. Many problems can be formulated as optimisation problems. Among the many classes of algorithms for solving such problems, one interesting, biologically inspired group is that of evolutionary optimisation techniques. In this tutorial paper we provide an overview of such techniques, in particular of Genetic Algorithms and Genetic Programming and its related subtasks of selection, crossover, mutation, and coding. We then also explore Ant Colony Optimisation and Particle Swarm Optimisation techniques.

1 Evolutionary Computing

Many scientific problems can be viewed as search or optimisation problems, where an optimum input parameter vector for a given system has to be found in order to maximise or to minimise the system response to that input vector. Often, auxiliary information about the system, like its transfer function and derivatives, is not known and the measures might be incomplete and distorted by noise. This makes such problems difficult to be solved by traditional mathematical methods. Here, evolutionary optimisation algorithms, which are based on biological principles borrowed from nature, can offer a solution. These algorithms work on a population of candidate solutions, which are iteratively improved so that an optimal solution evolves over time.

This tutorial paper discusses the general problem of search and optimisation before it introduces the system's view, followed by a definition of search space and fitness landscape. It then explains the process of optimisation and the concept of optimisation loops. It continuous with introducing biological-inspired evolutionary optimisation algorithms, namely Genetic Algorithms and Genetic Programming. Other evolutionary inspired optimisation techniques, namely Ant Colony Optimisation and Particle Swarm Optimisation are also discussed.

Lars Nolle
School of Science and Technology, Nottingham Trent University, U.K.

Gerald Schaefer
Department of Computer Science, Loughborough University, U.K.

A. Gaspar-Cunha et al. (Eds.): Soft Computing in Industrial Applications, AISC 96, pp. 29–52.
springerlink.com

1.1 Systems

Every process or object can be seen as a system. Fenton and Hill (1993) define a system as "... an assembly of components, connected together in an organised way, and separated from its environment by a boundary. This organised assembly has an observable purpose which is characterised in terms of how it transforms input from the environment into output to the environment." By definition, a system has exactly one input channel x and exactly one output channel y (Figure 1). All interactions with the environment have to be made through these interfaces.

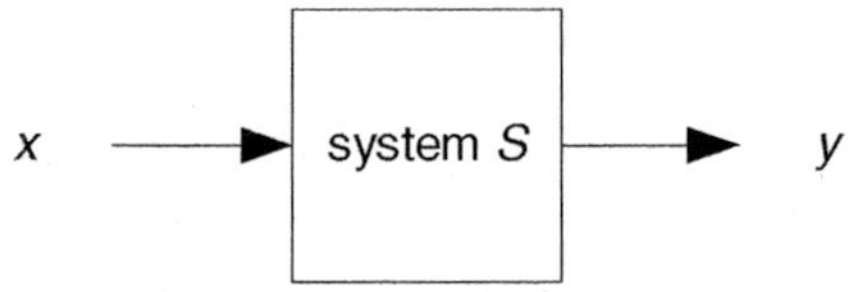

Fig. 1 Generic system.

Both input and output can be vectors or scalars. The input is called the *independent variable* or *parameter*, because its value(s) can be chosen freely, and results in the output y, the so-called *dependent variable*. If the present state of the system does not depend on previous states but only on the current input, the system is said to be a *steady state system*, the output of the system can be described as a function of the input $y = f(x)$.

1.2 Objective Function

In order to rate the quality of a candidate solution x, it is necessary to transform the system response to x into an appropriate measure, called the *objective* or *fitness*. If the system has only one output variable, the system output y equals the fitness. If y has more than one component the output variables of the system have to be combined into a single value, computed by the so called *objective function* or *fitness function*. In general, there are four approaches to judge the system output: *aggregation*, the *Changing Objectives Method*, the *Use of Niche Techniques* and *Pareto Based Methods* (Fonseca, 1995). The most often used method is aggregation. In its simplest case, the fitness function $F(x)$ equals the weighted sum of the components $y_i = c_i \cdot F_i(x)$ of y, where c_i is the weight for component i:

$$F(x) = c_0 + c_1 \cdot F_1(x) + \ldots + c_n \cdot F_n(x) \tag{1}$$

1.3 Search Space and Fitness Landscape

If all the possible candidate solutions are collected in an ordered way, this collection is called the search space. Sometimes, this space is also referred to as input space. For an optimisation problem of dimension n, i.e. a system with n independent parameters, the search space also has dimension n. By adding the dimension Fitness or Costs to the search space, one will get the (n+1) dimensional fitness landscape (Wright, 1931).

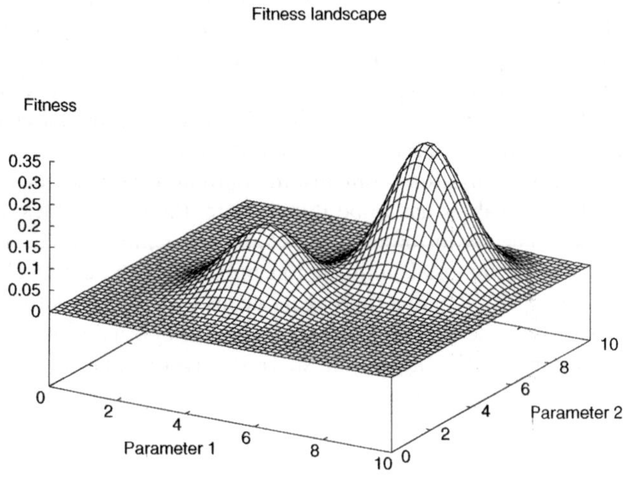

Fig. 2 Example of a fitness landscape for a system with two input parameters.

1.4 Optimisation

Optimisation (Schwefel, 1995) is the process of selecting the best candidate solution from a range of possibilities, i.e. the search space. In other words, a system *S*, that has to be optimised in terms of a quality output value *y*, is brought into a new state that has a better quality output value *y* than the previous state. This is done by changing the independent input parameters *x*. The *error function* describes the difference between the predefined objective $y_{desired}$ and systems response *f*(*x*) to the input *x*.

$$Error(x) = y_{desired} - f(x) \tag{2}$$

Usually, the aim is to find the vector x' that leads to a minimal error for the system *S*, i.e. the minimal departure from the optimal output value:

$$Error(x') = 0 \tag{3}$$

Often, a predefined target value is not known. In this case one tries to gain a fitness value that is as high as possible in case of maximisation, or as low a possible in the case of minimisation.

Ideally, one would evaluate all possible candidates and choose the best one. This is known as exhaustive search. However, often it is not feasible to consider all possible solutions, for example if the search space is too large and the evaluation of a single candidate is too expensive. In this case, only a subset of the solutions can be evaluated.

Optimisation problems can be either function optimisation problems or combinatorial problems. The first class of problems can be divided in continuous optimisation and discrete optimisation problems. In continuous function optimisation, the independent variables are real numbers whereas for discrete function optimisation, the independent variables can only be chosen from a predefined set of allowed and somehow ordered numbers, for example $\{10, 20, 30, 40\}$.

In combinatorial optimisation problems, the optimum sequence or combination of a fixed set of input values has to be found. Here, the input values are symbols and might not be connected or ordered, for example {*apple*, *orange*, *strawberry*}. An example of a combinatorial optimisation problem is the classical Travelling Salesman Problem (TSP), where a sales agent needs to visit a predefined set of cities and return to base. The problem here is to find an optimal route, i.e. the route that connects all cities whilst having the shortest travel distance, by choosing the order in which the cities are visited.

1.5 Optimisation Loop

Mathematical or *calculus-based* methods use known functional relationships between variables and objectives to calculate the optimum of the given system. Therefore, an exact mathematical model of the process must exist. Edelbaum (1962) introduced the differentiation of calculus-based methods in *direct methods* and *indirect methods*.

Direct methods solve the optimisation problem by iterative calculation and derivation of the error function and moving in a direction to the maximum slope gradient. Indirect methods solve the optimisation problem in one step - without testing - by solving a set of equations (usually non-linear). These equations result from setting the derivative of the error function equal to zero.

Both classes of methods are local in scope, i.e. they tend to find only local optima. Therefore, they are not robust. They depend on the existence of derivatives. Real problem functions tend to be perturbed by noise and are not smooth, i.e. derivations may not exist for all points of functions. This class of problem cannot be solved by mathematical methods.

If the functional relations between input variables and objectives are not known, one can experiment on the real system (or a model of this system) in order to find the optimum. Access to the independent variables must exist for the whole multi-dimensional search space, i.e. the collection of all possible candidate solutions. Also a possibility of measuring the independent variable and the objective must be given. The optimisation process is iterative, i.e. it has to be done in a closed *optimisation loop* (Figure 3).

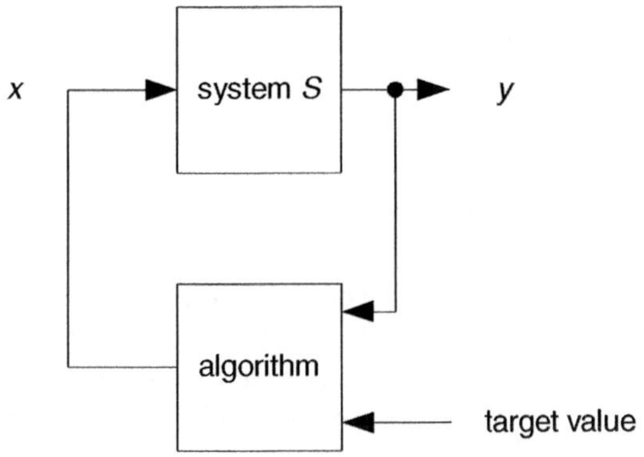

Fig. 3 Closed optimisation loop consisting of a system and an optimisation algorithm.

Experimental optimisation methods can therefore be seen as a search for the optimum by traversing over the fitness landscape.

2 Genetic Algorithms

As Darwin's theory of natural selection articulates, nature is very effective at optimisation, e.g. to enable life-forms to survive in a unfriendly and changing environment, only by means of the simple method of trial and error. Genetic Algorithms (GAs) simulate this evolutionary mechanism by using *heredity* and *mutation*. They were first introduced in 1975 by Holland (1975) who also provided a theoretical framework for Genetic Algorithms, the *Schemata Theorem* (Goldberg, 1989).

For genetic algorithms, the independent input parameters of a system S (Figure 4) are coded into a binary string, the *genotype* of an individual (Figure 5).

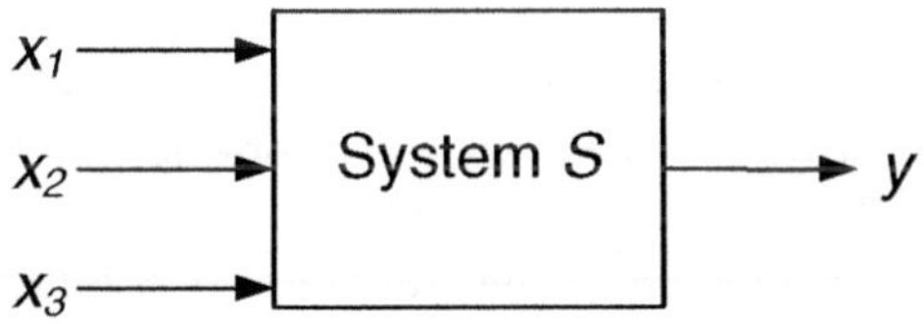

Fig. 4 System to be optimised.

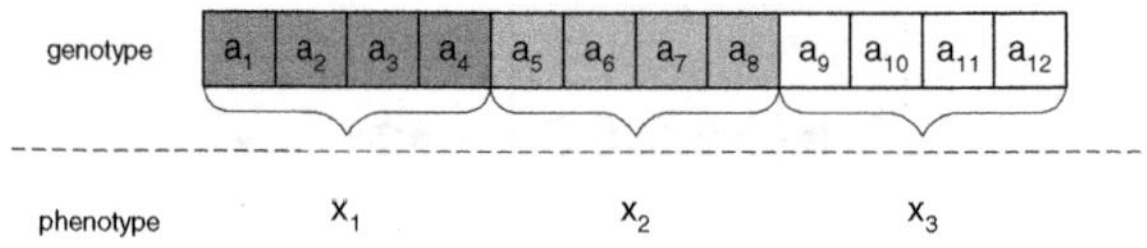

Fig. 5 Binary string representing one input pattern of the system.

The individual represented by genotype is called a *phenotype*. This phenotype has a certain quality or *fitness* to survive which can be determined by presenting the phenotype to the system S and measuring the system response.

The search is not only undertaken by one individual but by a population of n genotypes, the *genepool*. Therefore, the search space is tested at n points in parallel. All the individuals of the genepool at a time t_n are called a *generation*.

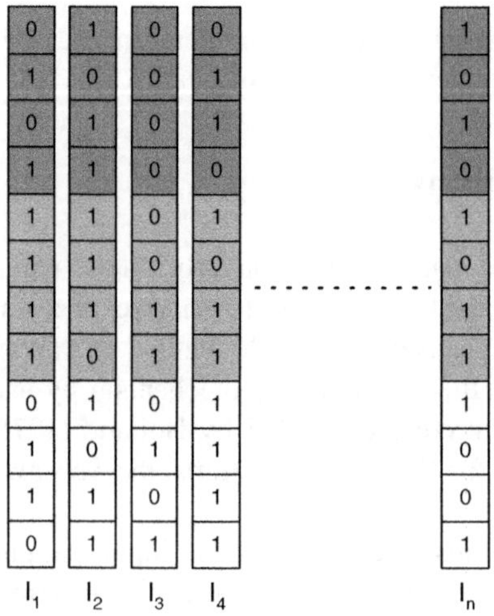

Fig. 6 Genepool consisting of individuals $I_1 \ldots I_n$.

A new generation for time t_{n+1} is generated by selecting N individuals from the current population for breeding. They are copying into the genepool of the next generation and their genetic information is then recombined, using the *cross-over* operator (see 0), with a predefined cross-over probability p_c. The resulting offspring is then copied into the new genepool and mutation is applied to the offspring. Figure 7 shows the flowchart of a simple Genetic Algorithm.

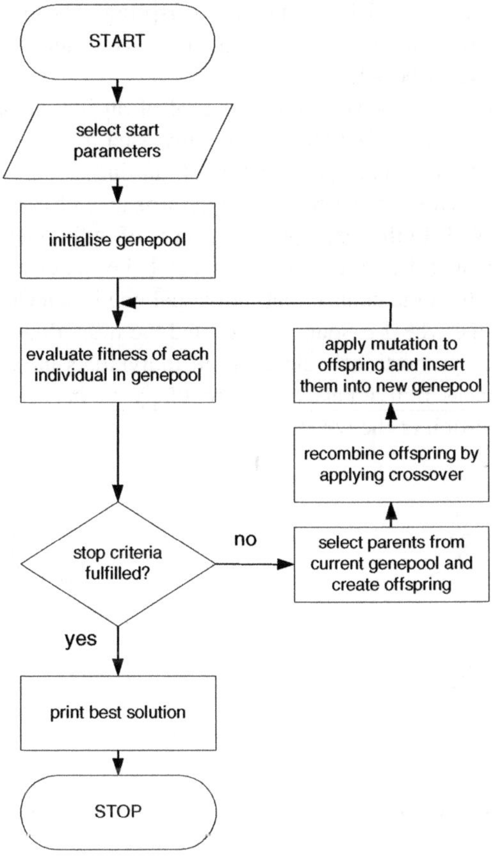

Fig. 7 Flowchart of basic GA algorithm.

The search will be carried out until at least one individual has a better fitness than the defined minimum fitness, or a maximum number of generations have been reached

2.1 Selection

In general, there are three different approaches to choose individuals from the current generation for re-production, namely *Tournament Selection*, *Fitness Proportional Selection* and *Rank Based Selection.* In Tournament Selection, two or more individuals are randomly selected from the current generation of N genotypes to compete with each other. The individual with the highest fitness of this set is the winner and will be selected for generating offspring. The process is repeated *N* times in order to create the new population. Using Tournament Selection, the least fit individual can never be selected.

In fitness proportional selection, the chance of an individual to be selected is related to its fitness value. The most commonly used method of this type is Roulette Wheel Selection. Here, proportions of an imaginary roulette wheel are distributed in proportion to the relative fitness of an individual. Figure 8 shows and example for N=3. In this example, the fitness of individual 3 is approximately four times higher than the fitness of individual 1, i.e. its chance to be selected is four times greater then the chance that individual one is selected. For a population of *N* individuals, the wheel is spun *N* times and the individual under the pointer is selected. In fitness proportional selection, all individuals have a chance of selection but high fitness individuals are more likely to be selected, because they occupy a larger portion of the wheel.

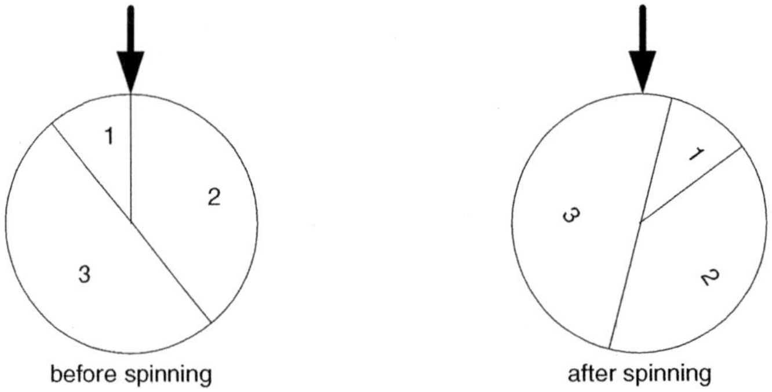

Fig. 8 Roulette Wheel selection.

However, there is the statistical chance that the actual selected distribution might not reflect the expected distribution based on the fitness values. If the selection is too strong, it can lead to premature convergence, i.e. the population would converge before it has found the region of the search space that contains the global optimum. In other words, the exploitation would start before the search space is fully explored. On the other hand, if the selection is too weak, it can lead to stalled evolution, which means the search is reduced to randomly walking through search space.

These effects are overcome using Stochastic Universal Selection (SUS). Here, the same roulette wheel is used, but instead of using a single pointer, *N* equally-spaced pointers are used for a population of *N* individuals and the wheel is spun only once (Figure 9).

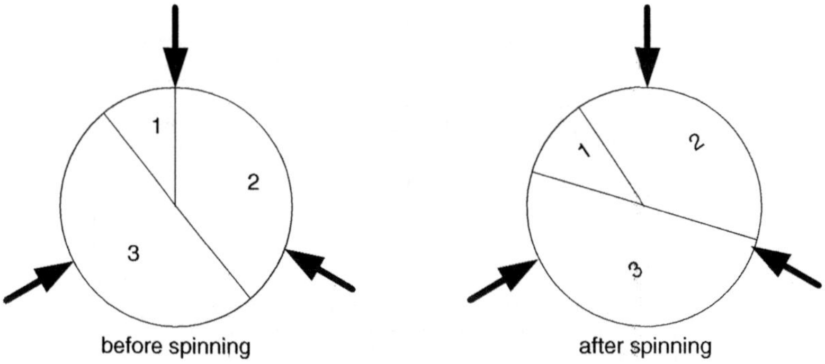

Fig. 9 SUS selection

Instead of using the fitness of an individual for selection, a selective s value can be used, which is based on the rank position of an individual in the population (Equation 4).

$$s_i = Min + (Max - Min)\frac{rank_i - 1}{N - 1} \quad (4)$$

where

Min: minimum fitness within a generation
Max: maximum fitness within a generation
$rank_i$: rank of individual i within the population in a generation
N: number of individuals within population

So, instead of using the raw fitness to determine the proportion for an individual, the rank of the individual within the generation is used.

Sometimes the m fittest individuals in a generation are cloned into the next generation in order to make sure to preserve their genetic material. This is known as *elitism*.

2.2 *Cross-Over*

The most important operator in terms of robustness of the algorithm is the cross-over operator. Figure 10 shows the so-called *one-point cross-over* operator, which combines the information of two parents. They are aligning and then both cut at a randomly chosen cross-over point and the tails are swapped successively.

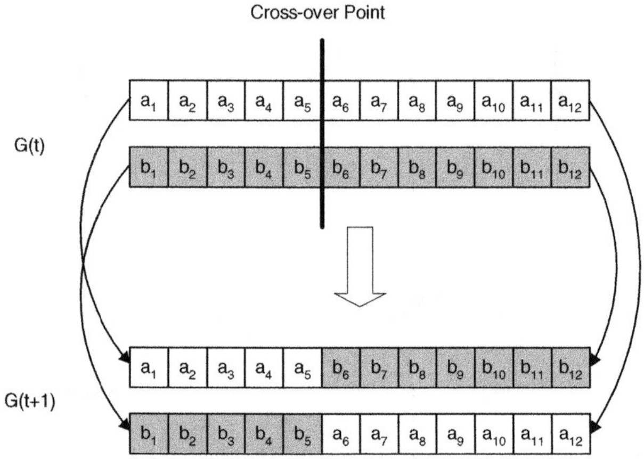

Fig. 10 Cross-over operator.

Instead of a single cross-over point, two or more random cross-over points can be used for recombining the genetic information of the parents.

Another form of cross-over is called *Uniform cross ov*er (Syswerda, 1989). Here, every component of a parent individual X is randomly passed on either to offspring A or offspring B. If X passes on its component to A, the position in B is filled using the component from parent Y and vice versa.

2.3 Mutation

After the genetic information of the parents is recombined using cross-over, mutation is applied to every individual of the new generation. Here, every bit of the offspring is inverted (*mutated*) with probability p_m. The mutation operator is important for restoring lost information and therewith to result in a better effectiveness of the GA.

2.4 Discussion

The advantages of GAs are that they use payoff (objective function) information, not derivatives or other auxiliary knowledge, i.e. they are black box optimisation methods. GAs tend to converge towards the global optimum rather than getting stuck in a local optimum and therefore they are very robust. On the other hand, it is not always straightforward to find the right GA parameters for a particular optimisation problem, e.g. a suitable genepool size or mutation probability. Also, the efficiency of GAs relies heavily on the right coding of the input parameters, i.e. the chosen mapping function from phenotype to genotype, and they tend to fail if the inputs of the system are heavily correlated.

2.5 Schemata Theorem

Holland provided a theoretical foundation of GAs, i.e. a theoretical proof of convergence, which he called the *Schemata Theorem*. A schema is a template for binary strings, but built from a three letter alphabet containing the symbols *, *0* and *1*. The * symbol is the 'don't care symbol' which either stands for 0 or 1. Figure 11 shows an example of a schema for chromosomes consisting of 12 bits, of which three are set to the don't care symbol and the remaining nine bits are set to fixed values.

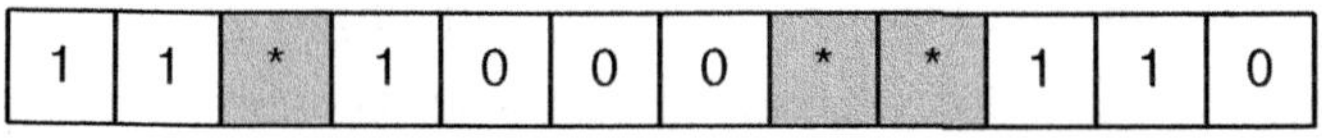

Fig. 11 Example of a schema in GA.

The distance between the first and the last fixed bit is called the *defined length* of the schema and the number of fixed bits is called the *order* of the schema. Figure 12 shows an example of a schema *H* and the different instances it represents.

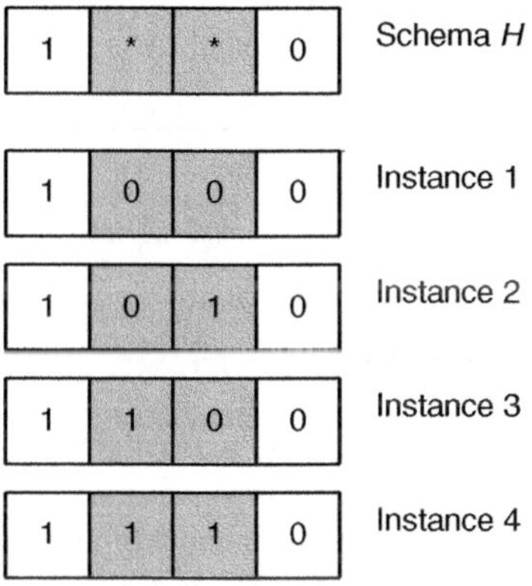

Fig. 12 Example of a schema H and the instances it represents.

A binary string *s* is an instance of a schema *H* if it fits into the template. Therefore, any binary string of length *l* does not just represent one candidate solution, it is also an instance of 2^l schemata a the same time. As a consequence, a GA with the genepool of size *n* does not only test *n* different solutions at the same time, but also a high number of different schemata. This is known as *implicit parallelism* in GA and provides an explanation for their effectiveness and efficiency:

According to Holland, the number of instances *m* of a schema *H* that are contained in the population at generation *t+1* can be determined as follows:

$$m(H,t+1) = m(H,t) \cdot \frac{\bar{f}(H)}{\bar{f}} \tag{5}$$

where:

H: Schema or "Building Block" with at least one instance in the last generation,

$m(H,t)$: number of instances of H at time t,

$m(H,t+1)$: number of instances of H at time t+1,

$\bar{f}(H)$: average fitness of the instances of schema H,

$\bar{f}$: average fitness of the whole population.

This is a simplified version of the schemata theorem, because it does not take into account the effects of the cross-over and the mutation operator. However, it is sufficient to demonstrate the basic idea. A more detailed description can be found, for example, in (Goldberg, 1989).

Suppose that a particular schemata H remains above-average an amount $c \cdot \bar{f}$ with c being a constant factor, equation 4 can be rewritten as follows:

$$m(H,t+1) = m(H,t) \cdot \frac{\bar{f} + c \cdot \bar{f}}{\bar{f}} = (1+c) \cdot m(H,t) \tag{6}$$

Assuming c is stationary and starts at $t = 0$, equation 5 can be rewritten as follows:

$$m(H,t) = m(H,0) \cdot (1+c)^t \tag{7}$$

It can be seen that this equation is similar to the formula of interest: the number of instances of a schema H with a fitness above-average grows exponentially to generation t. Hence, schemata with good fitness will survive and ones with a fitness below average will eventually die out. Therefore, the fitter building blocks, i.e. the better partial solution, will take over the genepool within finite time. However, the schemata theorem is controversial, because it assumes that the factor c is constant over time.

2.6 Coding Problem

Traditionally, GAs use binary stings. However, if an input variable is coded using standard binary coding, this can lead to the problem that a small change in the phenotype would require a large number of bits of the genotype to be inverted. An example of the coding problem is given in Figure 13.

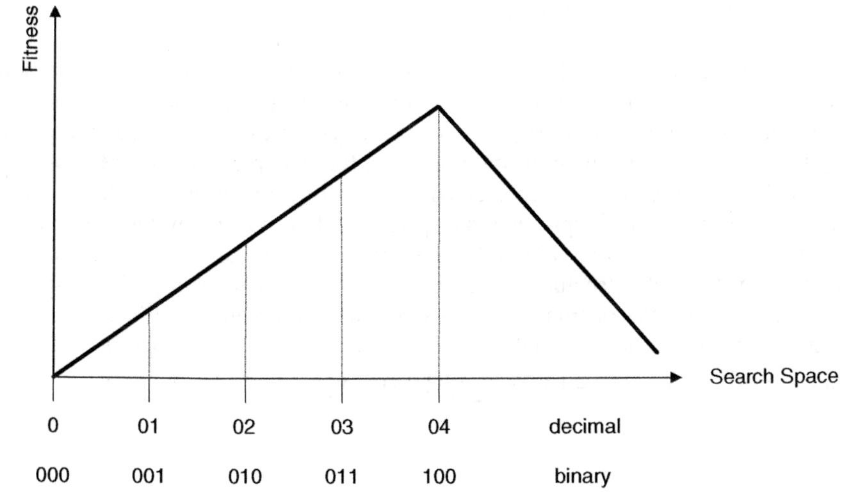

Fig. 13 Differences between decimal and standard binary coding.

As it can be seen from the figure, a step from 3_{10} to 4_{10} requires flipping 3 bits in binary representation whereas it only changes the least significant digit in decimal representation by one. One solution is to use Gray Code, which has the advantage that only one bit changes between any two positions, i.e. it has a constant Hamming Distance of one.

Decimal	Binary	Gray Code
0	0000	0000
1	0001	0001
2	0010	0011
3	0011	0010
4	0100	0110
5	0101	0111
6	0110	0101
7	0111	0100
8	1000	1100
9	1001	1101
10	1010	1111

Fig. 14 Gray Code.

3 Genetic Programming

Genetic Programming (GP) was introduced by Koza (1992) and is a machine learning technique that uses a Genetic Algorithm for the automated generation of computer programs. These programs model a system using sample data provided by that system. A typical application of GP is, for example, symbolic regression.

In GP, the programs are represented as tree structures where a node represents an operator and a leaf represents an operand. Often, GP programs are coded in computer languages like LISP, because they can be used straightforward to implement tree-like structures. The operators are chosen for a problem specific function set and the operands are chosen from a problem specific terminal set. Figure 15 shows an example of a tree representing a program to calculate $y = f(x_1, x_2, x_3) = (x_1 + x_2)/2 * \sin(x_3)$:

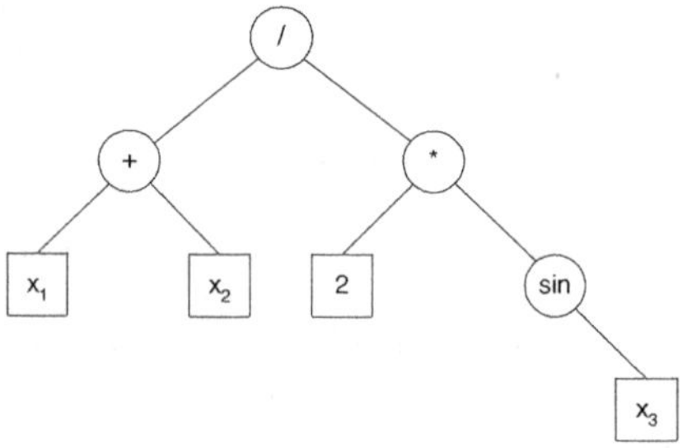

Fig. 15 Example of a program represented by a tree.

The function set depends on the problem at hand. It may contain unitary or binary operators and has to be chosen so that the function to be learned can be approximated as accurately as possible. The terminal set contains variables and constants and is again problem dependent. In the example above, the function set contains at least the operators +, /, *, *sin*, whereas the terminal set contains at least the variables x_1, x_2, x_3 and the constant *2*. The quality of a program can be evaluated by applying the training data to the program and measuring either the average (often quadratic) error or by counting how many instances of the training set are reproduced correctly by the program.

GP works similar to GA. It works on a population of programs, the gene pool. The individuals, i.e. the programs, are randomly created and then evaluated. The algorithm applies genetic operators to the population in each generation. These operators are selection, cross-over and mutation.

3.1 Selection

All the different selection methods from the GA domain can be employed, e.g. fitness proportional selection, tournament selection, etc.

3.2 Cross-Over

For each of the selected parents, a random node is chosen to be the cross-over point. Figure 16 shows two individuals before cross-over. The black-circled nodes are the randomly selected cross-over points.

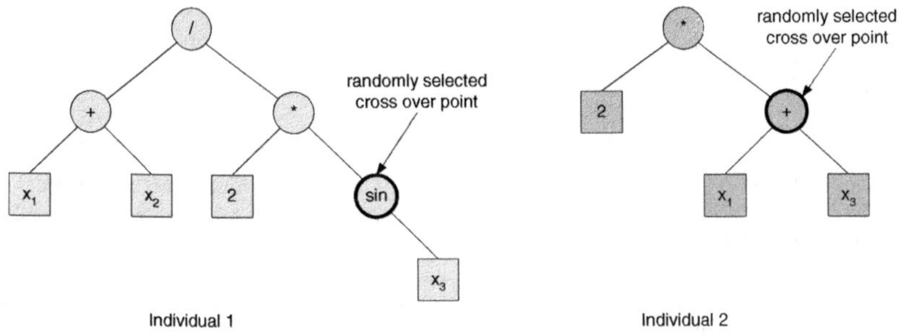

Fig. 16 Individuals before cross-over.

The nodes and their subtrees are subsequently swapped to create the off-spring individuals (Figure 17)

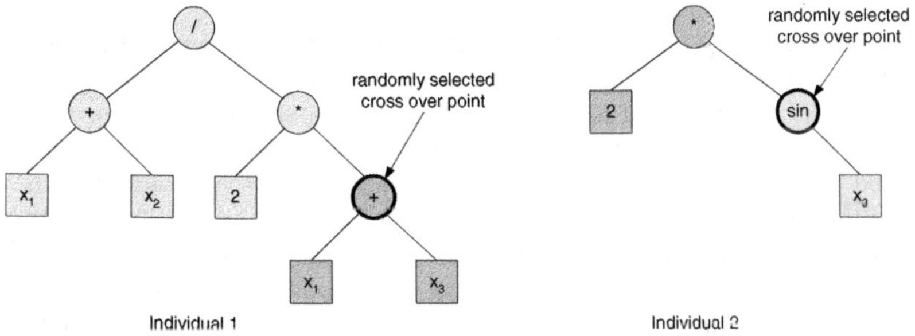

Fig. 17 Individuals after cross-over.

As it can be seen from Figure 16 and Figure 17, the resulting individuals are quite different from their parent individuals and hence the cross-over operator is very effective. It can also be seen from the figures that the trees have all different sizes, which requires variable tree sizes. This is a major difference to GA, where individuals usually have a fixed size.

3.3 Mutation

The other genetic operator commonly used in GP is mutation. Here, either a random node is deleted from the tree or its content is replaced. In the latter case it is important to maintain integrity. For example, if a binary operator would be replaced by an unary operator, one of the operands would become obsolete. Figure 18 shows an example of mutation.

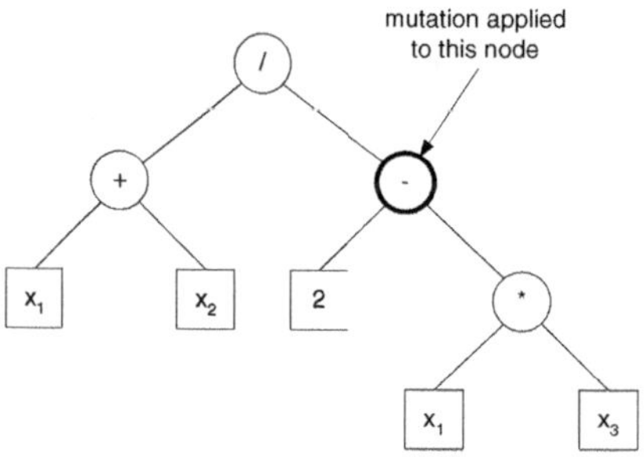

a) Individual before mutation.

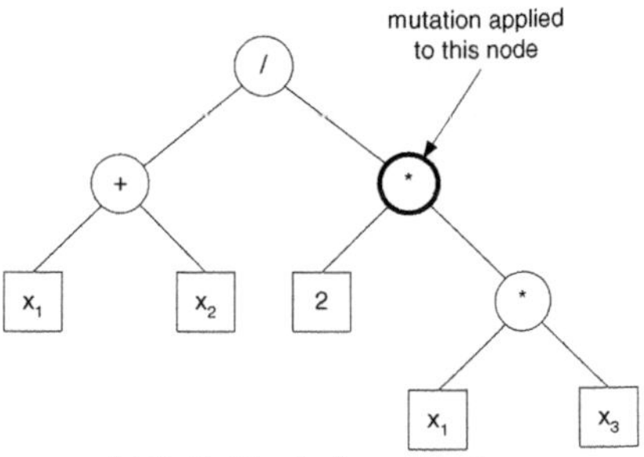

b) Individual after mutation.

Fig. 18 Example of mutation in GP.

It can be seen that the operator - is replaced by the operator * in this example. Because both operators are binary, the integrity is maintained.

4 Ant Colony Optimisation

Ant Colony Optimisation (ACO) (Dorigo et al., 1999) refers to a class of discrete optimisation algorithms, i.e. a meta-heuristic, which is modelled on the collective behaviour of ant colonies.

Real ants are very limited in their individual cognitive and visual capabilities, but an ant colony as a social entity is capable of solving complex problems and tasks in order to survive in an ever-changing hostile environment. For example, ants are capable of finding the shortest path to a food source (Goss et al., 1989). If the food source is depleted, the ant colony adapts itself in a way that it will explore the environment and discover new food sources.

Ants communicate indirectly with other ants by depositing a substance called pheromone on the ground while they are walking around. This pheromone trail can then be used by the ant to find its way back to its nest after the ant has found a food source and other ants can also sense it. Ants have the tendency to follow existing paths with high pheromone levels. If there is no existing pheromone trail, they walk around in a random fashion. If an ant has to make a decision, for example to choose a way around an obstacle in its way, it follows existing paths with a high probability. However, there is always a chance that the ant explores a new path or a path with a lower pheromone level. If an ant has chosen an existing path, the pheromone level of this path will be increased because the ants deposit new pheromone on top of the existing one. This makes it more likely that other ants will also follow this path, increasing the pheromone level again. This positive feedback process is known as autocatalysis (Dorigo et al., 1991). Although the pheromone evaporates over time, the entire colony builds up a complex solution based on this indirect form of communication, called stigmergy (Dorigo et al., 1999).

Figure 19 demonstrates the basic principle of the ACO meta-heuristic, which is modelled after the behaviour described above. In this example, the system S that has to be optimised has three independent variables $x_1 \ldots x_3$ and the quality of the solution can be measured by the achieved fitness value y. Each input can have one of five different discrete alternative values s_{ij}, where i represents the input and j the chosen alternative for that input. Each alternative has an associated probability value, which is randomly initialised. The collection of probability distributions can be seen as a global probability matrix. Each artificial ant in the colony has to choose randomly a 'path' through state space, i.e. the input value for each independent variable. In the example below, the ant chooses alternative s_{12} for input x_1, s_{24} for input x_2 and s_{33} for input x_3. The chosen path depends on the probabilities associated with the states, i.e. a state with a high probability is more likely to be selected for a trial solution than states with a low probability value. This probability values are refereed to as the pheromone level τ.

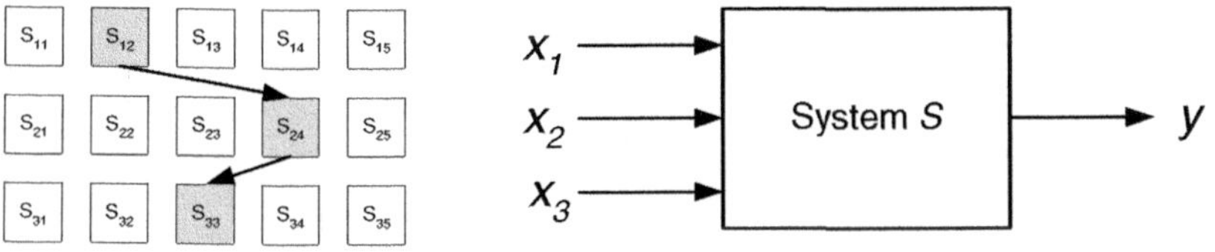

Fig. 19 Example of an artificial ant constructing a trial vector by traversing through state space.

A chosen path represents one candidate solution, which is evaluated and the probabilities of the states that the ant has visited on that trail is updated based on the achieved fitness. In the next generation, the updated probability matrix is used, which means that states that have proven fit in the past are more likely to be selected for the subsequent trail. However, it should be pointed out that a 'path' is not actually traversing through the state space; it simply refers to the collection of chosen alternatives for a particular candidate solution. The order in which the states are selected does not have any effect on the candidate solution itself, i.e. one could start with determining the input for x_1 first or, alternatively, with x_2 or x_3. The resulting candidate solutions would still be the same.

A major advantage of ACO is that adjacent states in the neighbourhood do not need to show similarities, i.e. the state space does not need to be ordered. This is different to most optimisation heuristics, which rely on ordered collections of states, i.e. fitness landscapes.

Figure 20 shows a flowchart of the basic ACO meta-heuristic for a colony consisting of n artificial ants. During one iteration, called a time-step, every ant generates a trial solution, which is evaluated and based on the fitness of the solution the pheromone level of the states involved in the trail is updated in a local probability matrix for the ant. After one iteration, i.e. time-step, all the local probability matrices are combined and added to the global one, which is usually scaled down in order to simulate the evaporation process of real pheromone trails. This helps to avoid search stagnation and ensures that ants maintain their ability to explore new regions of the state space.

The main principle of ACO is that a colony of artificial ants builds up discrete probability distributions for each input parameter of a system to be optimised. Figure 21 shows an example of a probability distribution for an input i with ten alternative states.

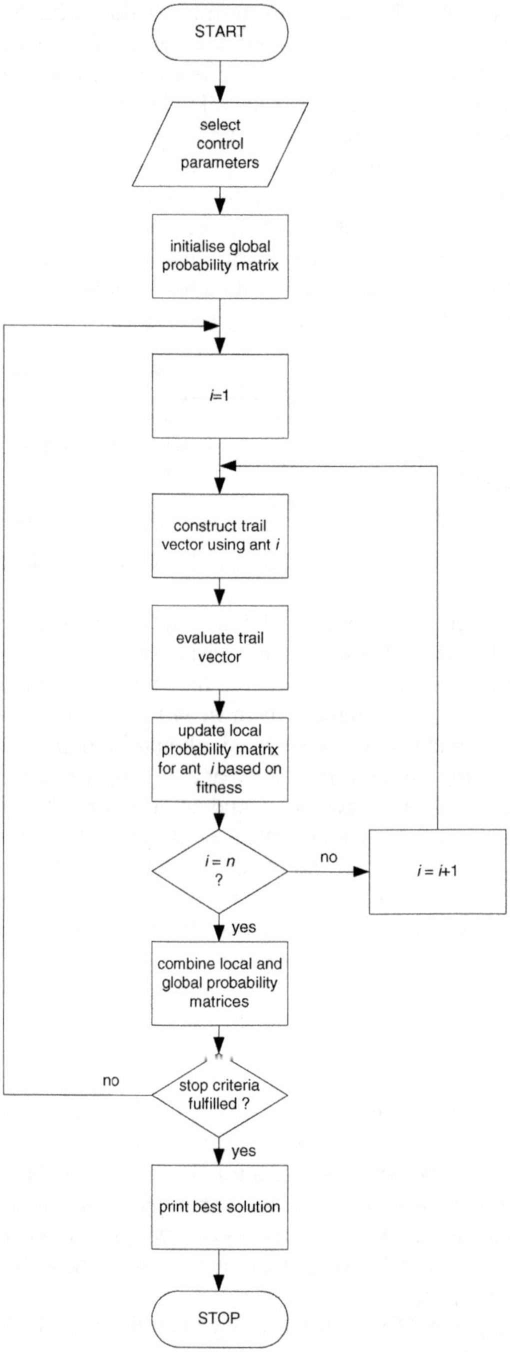

Fig. 20 Flowchart of ACO meta-heuristic.

It can be seen that state s_{i8} has the highest pheromone level, i.e. probability, and hence has a high chance to be selected for a trial. States s_{i7} and s_{i8}, on the other hand, have a pheromone level of zero and can never be selected. However, even states with a low pheromone level, e.g. s_{i3} in Figure 21, have a certain chance to be selected.

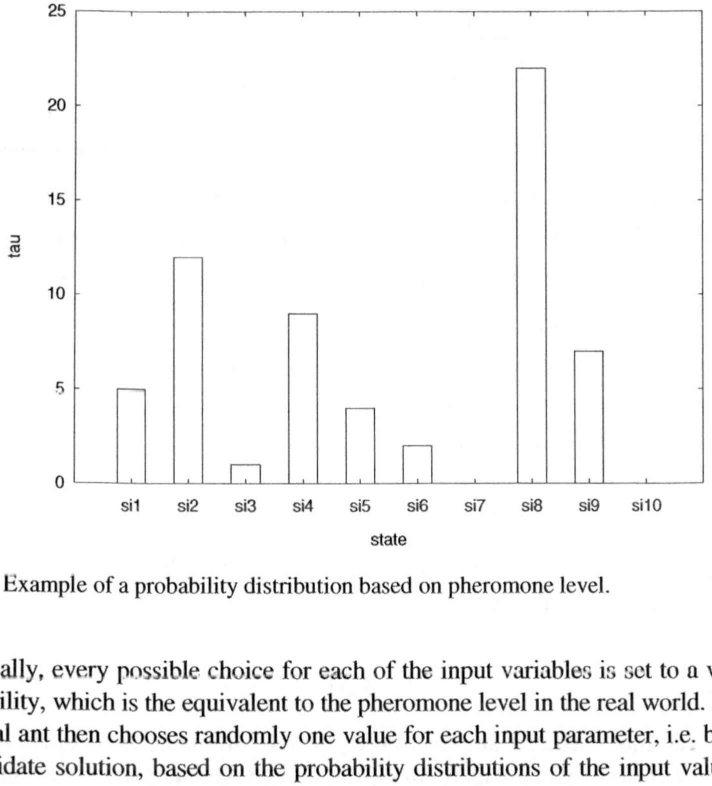

Fig. 21 Example of a probability distribution based on pheromone level.

Initially, every possible choice for each of the input variables is set to a very low probability, which is the equivalent to the pheromone level in the real world. Each individual ant then chooses randomly one value for each input parameter, i.e. builds up a candidate solution, based on the probability distributions of the input values. Depending on the quality of the resulting candidate solution, the probability values of the chosen input values are updated. The whole process is repeated in iterations called time-steps until a suitable solution is found or the algorithm has converged, i.e. has reached a stable set of probability distributions. It has been proved, for example by Stützle and Dorigo (2002) and Gutjahr (2000), that ACO algorithms are capable of converging towards the global optimum within finite time.

The first computational optimisation algorithm based on ant colonies was the Ant System (AS) algorithm (Dorigo, 1992). It was successfully applied to the Travelling Salesman Problem and the Quadratic Assignment Problem. This was later followed by the Ant Colony System (ACS) (Dorigo and Gambardella, 1997), the Max-Min Ant System (MMAS) (Stützle and Hoos, 2000) and the Rank-Based Ant System (RBAS) (Bullheimer et al., 1999,).

For ACO the probability of a state s_{ij} to be chosen as input parameter i can be calculated using the following transition rule (Equation 8):

$$p(s_{ij}) = \begin{cases} \dfrac{\tau_{ij}^{\alpha} \cdot \eta_{ij}^{\beta}}{\sum_{j=1}^{m} \tau_{ij}^{\alpha} \cdot \eta_{ij}^{\beta}} & if\ s_{ij} \in N_i \\ 0 & otherwise \end{cases} \tag{8}$$

Where τ_{ij} is the pheromone level for state s_{ij}, η_{ij} is a heuristic value related to the fitness of the solution, α and β are control parameters that determine the relative importance of pheromone versus fitness, m is the number of alternatives for input parameter i, and N_i is the set of possible alternatives for input i. If the heuristic value η_{ij} is set to a constant value of one, the algorithm becomes the Simple Ant Colony Optimization algorithm (SACO) (Dorigo and Stützle, 2004).

The evaporation after time-step t can be computed using Equation 9, where $\rho \in (0,1]$ is the evaporation rate.

$$\tau_{ij}(t+1) = (1-\rho) \cdot \tau_{ij}(t) \tag{9}$$

The pheromone updating rule is given in Equation 10, where $\Delta\tau_{ij}(t) = f(y_1, y_2, \hbar\ , y_n)$:

$$\tau_{ij}(t+1) = \tau_{ij}(t) + \Delta\tau_{ij}(t) \tag{10}$$

Unlike real ants, artificial ants can be equipped with additional capabilities, for example with look ahead capabilities (Michel and Middendorf, 1998) and back-tracking (Di Caro and Dorigo, 1998) in order to improve efficiency. They can also be combined with local search methods, for example see Dorigo and Di Caro (1999) or Shmygelska and Hoos (2005).

However, one problem related to ACO is that it is not a straightforward task to find optimum control parameter settings for an ACO application (Gaertner and Clark, 2005).

5 Particle Swarm Optimisation

Particle Swarm Optimisation (PSO) is a simple but effective algorithm that was original developed by Kennedy and Eberhart (1995) for continuous function optimisation. It is based on the social behaviour of a collection of animals that can be observed, for example, in fish schooling and bird flocking. PSO uses a population of agents where the population is referred to as swarm and the agents are called particles. Each particle represents an input vector for the system and is randomly initialised.

Each particle i has a position $x_{ij}(t)$ and a velocity $v_{ij}(t)$ for each dimension j of the search space. In every iteration of the algorithm, i is 'flying' through search space by adjusting the position vector $x_i(t)$ using the velocity vector $v_i(t)$ as follows:

$$x_{ij}(t+1) = x_{ij}(t) + v_{ij}(t+1) \tag{11}$$

It should be stressed that, in the physical world, a velocity and a position cannot be added. The velocity would need to be multiplied with a time interval in order to get a distance that could then be added to the original position. However, if one thinks of an iteration as a time step, the velocity vector could be multiplied with one time unit, which would not change the actual value but it would change the unit. The velocity vector itself is determined using the following equation:

$$v_{ij}(t+1) = v_{ij}(t) + c_1 r_2 (x_{i\_best} - x_{ij}(t)) + c_2 r_2 (x_{global\_best} - x_{ij}(t)) \tag{12}$$

Here, r_1 and r_2 are random numbers, c_1 and c_2 are tuning constants, x_{i_best} is the best position that particle i found during the search so far and x_{global_best} is the best position the swarm found so far. The second term in Equation 12 is called the component cognitive component whereas the third one is called the social component. Figure 22 shows a flow-chart of the basic PSO algorithm.

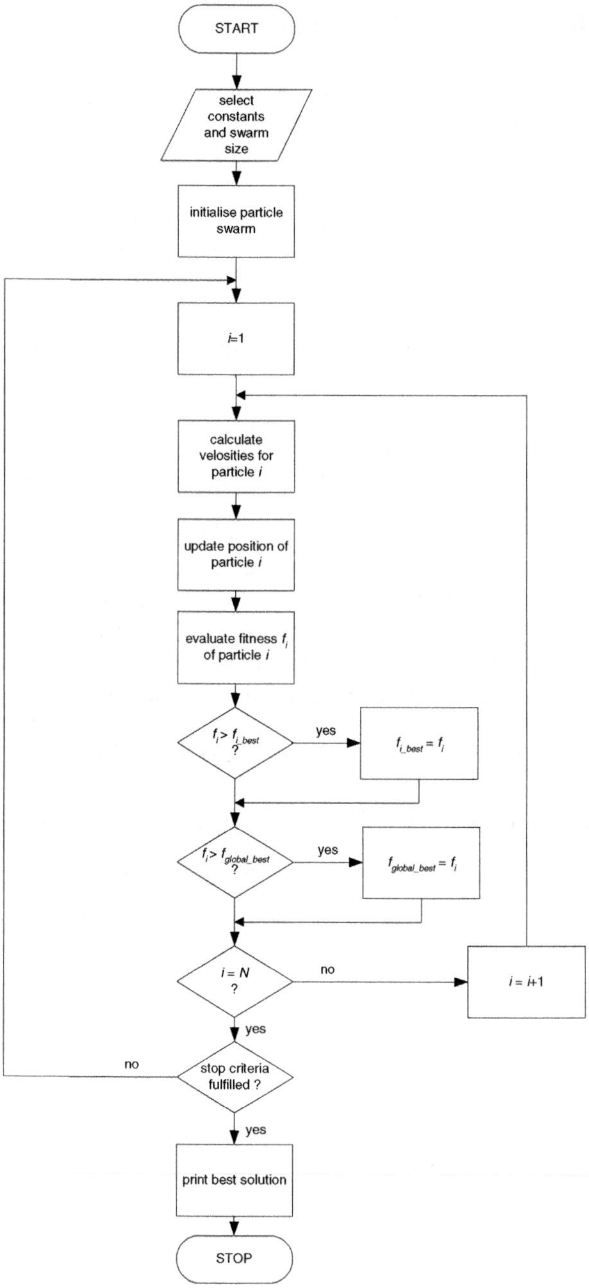

Fig. 22 Flowchart of PSO algorithm.

One variation of the basic PSO algorithm is that, instead of using the global best position, the best position of the neighbourhood of particle i is used in the social component.

6 Conclusions

In this tutorial paper we provided a brief overview of evolutionary approaches to optimisation. In particular, we discussed Genetic Algorithms and Genetic Programming and their subtasks of selection, cross-over, mutation and coding. In addition, we gave an overview of other evolutionary optimisation approaches, namely Ant Colony Optimisation and Particle Swarm Optimisation.

References

Bullheimer, B., Hartl, R.F., Strauss, C.: A new rank-based version of the Ant System: A computational study. Central European Journal for Operations Research and Economics 7(1), 25–38 (1999)

Dorigo, M.: Optimization, Learning and Natural Algorithms. PhD Thesis, Politecnico di Milano, Italy (1992)

Di Caro, G., Dorigo, M.: AntNet: Distributed Stigmergetic Control for Communications Networks. Journal of Artificial Intelligence Research 9, 317–365 (1998)

Dorigo, M., De Caro, G.D., Gambardella, L.M.: Ant Algorithms for Discrete Optimization. Artificial Life 5, 137–172 (1999)

Dorigo, M., Maniezzo, V., Colorni, A.: Positive Feedback as a Search Strategy. Technical Report No 91-016, Politecnico di Milano (1991)

Dorigo, M., Gambardella, L.: Ant Colony System: A Cooperative Learning Approach to the Travelling Salesman Problem. IEEE Transactions on Evolutionary Computation 1(1), 53–66 (1997)

Dorogo, M., Stützle, T.: Ant Colony Optimization. MIT Press, Cambridge (2004)

Edelbaum, T.N.: Theory of Maxima and Minima. In: Leitmann (ed.) Optimization Techniques with Applications to Aerospace Systems, pp. 1–32. Academic Press, New York (1962)

Fenton, N., Hill, G.: Systems construction and analysis: a mathematical and logical framework. McGraw-Hill, New York (1993)

Fonseca, C.M., Fleming, P.J.: An Overview of Evolutional Algorithms in Multiobjective Optimization. Evolutionary Computation (31), 1–16 (1995)

Holland, J.H.: Adaptation in Natural and Artificial Systems. University of Michigan Press (1975)

Gaertner, D., Clark, K.: On Optimal Parameters for Ant Colony Optimization algorithms. In: Proceedings of the International Conference on Artificial Intelligence 2005, Las Vegas, USA, pp. 83–89 (2005)

Goldberg, D.E.: Genetic Algorithms in Search, Optimization, and Machine Learning. Addison-Wesley, Reading (1989)

Goss, S., Aron, S., Deneubourg, J.L., Pasteels, J.M.: Self-organized shortcuts in the Argentine ant. Naturwissenschaften 76, 579–581 (1989)

Gutjahr, W.: A graph-based ant system and its convergence. Future Generation Computer Systems 16(8), 873–888 (2000)

Kennedy, J., Eberhart, E.: Particle Swarm Optimization. In: IEEE International Conference on Neural Networks, vol. 4, pp. 1942–1948. IEEE Press, Los Alamitos (1995)

Koza, J.R.: Genetic Programming: On the Programming of Computers by Means of Natural Selection. MIT Press, Cambridge (1992)

Michel, R., Middendorf, M.: An Island Model Based Ant System with Lookahead for the Shortest Supersequence Problem. In: Proceedings of the Fifth International Conference on Parallel Problem Solving from Nature, Amsterdam, The Netherlands, pp. 692–701 (1998)

Shmygelska, A., Hoos, H.H.: An Ant Colony Optimisation Algorithm for the 2D and 3D Hydrophobic Polar Protein Folding Problem. BMC Bioinformatics 6, 6–30 (2005)

Stützle, T., Dorigo, M.: A short convergence proof for a class of ant colony optimisation algorithms. IEEE Transactions on Evolutionary Computation 6(4), 358–365 (2002)

Stützle, T., Hoos, H.H.: MAX-MIN Ant System. Future Generation Computer Systems 16(8), 889–914 (2000)

Schwefel, H.-P.: Evolution and Optimum Seeking. John Wiley & Son, Inc., New York (1995)

Syswerda, G.: Uniform Crossover in Genetic Algorithms. In: Proceedings of International Conference on Genetic Algorithm 1989, ICGA 1989, pp. 2–9 (1989)

Wright, S.: Evolution in Mendelian populations. Genetics 16, 97–159 (1931)

Part I: Evolutionary Computation

Approaches for Handling Premature Convergence in CFG Induction Using GA

Nitin S. Choubey and Madan U. Kharat

Abstract. Grammar Induction (or Grammar Inference or Language Learning) is the process of learning of a grammar from training data of the positive and negative strings of the language. Genetic algorithms are amongst the techniques which provide successful result for the grammar induction. The paper is an extended approach to the earlier work by the authors regarding using stochastic mutation scheme based on Adaptive Genetic Algorithm for the induction of the grammar. Optimization by Genetic Algorithm often comes with premature convergence. The paper suggests two approaches, Elite Mating Pool and generating the population with the Dynamic Application of Reproduction Operator, for handling local convergence by considering a set of eleven different languages and their comparison. The algorithm produces successive generations of individuals, computing their 'fitness value'at each step and selecting the best of them when the termination condition is reached. The paper deals with the issues in implementation of the algorithm,chromosome representation and evaluation, selection and replacement strategy, and the genetic operators for crossover and mutation. The model has been implemented, and the results obtained for the set of eleven languages are shown in the paper.

1 Introduction

The field of evolutionary computing has been applying problem-solving techniques similar in intent to the Machine Learning recombination methods. Most evolutionary computing approaches hold in common that they try and find a solution to a

Nitin S. Choubey
Student, P. G. Department of Computer Science, S. G. B. A. University, Amravati, India
e-mail: nschoubey@gmail.com

Madan U. Kharat
Principal, Pankaj Laddhad Institute of Technology and Management, Buldana,
S. G. B. A. University, Amravati, India
e-mail: mukharat@rediffmail.com

A. Gaspar-Cunha et al. (Eds.): Soft Computing in Industrial Applications, AISC 96, pp. 55–66.
springerlink.com

particular problem, by recombining and mutating individuals in a society of possible solutions[12]. John Holland invented Genetic Algorithms(GAs) in 1960s. In contrast with Evolution Strategies and Evolutionary Programming, Holland's original goal was not to design algorithms to solve specific problems, but rather to formally study the phenomenon of adaptation as it occurs in nature and to develop ways in which the mechanisms of natural adaptation might be utilized into computer systems. Holland's 1975 book 'Adaptation in Natural and Artificial Systems'presented the GA as an abstraction of biological evolution and gave a theoretical framework for adaptation under the GA. Many problems in engineering and related areas require the simultaneous genetic optimization of a number of, possibly competing, objectives have been solve by combining the multiple objectives in to single scalar by some linear combination. The combining coefficients, however, usually based on heuristic or guesswork and can exert an unknown influence on the outcome of the optimization. A more satisfactory approach is to use the notion of Pareto optimality [5] in which an optimal set of solutions prescribe some surface 'The Pareto front'in the vector space of the objectives. For a solution on the Pareto front no objective can be improved without simultaneously degrading at least one other.

The perennial problem with GA is that of premature convergence, a non-optimal genotype taking over a population resulting in every individual being either identical or, the consequences of which is a population that does not contain sufficient genetic diversity to evolve further. To avoid the premature convergence, in a GA is imperative to preserve the population diversity during the evolution. An approach to increase the population size may not be enough to avoid the problem, as any increase in population size will incur the two fold cost of both extra computation time and more generations to converge on an optimal solution. Several approaches are adapted to avoid the premature convergence such as The Pygmy Algorithm, use of Incest Prevention, Crowding, Sharing, restricted mating, introducing a random chromosome in every generation[13], Adaptive mutation rate, Random Offspring Generation[10], Immoderate crossover greediness and low influence of random factors[8], Social Disaster Technique, The Population Partial Re-initialization, Dynamic Application of Crossover and Mutation Operators[11].

Inductive Inference is the process of making generalizations from sample. Wyard [14] explored the impact of different grammar representations and experimental results show that an evolutionary algorithm using standard context-free grammars (BNF) outperformed other representations.A formal language is context-free if some context-free grammar generates it[6].

These languages are exactly recognized by a non-deterministic pushdown automata. In the conventional grammatical induction, a language generator is constructed to accept all the positive examples. Learning from positive examples is called text learning.A more powerful technique uses negative examples as well is called as informant learning, the language acceptor is constructed so as to accept all the positive examples and reject all the negative examples. A positive sentence is defined as a sentence represented by the grammar of a language and hence included in the language. A negative sentence is defined accordingly. The problem discussed here is finding generalizations for Context Free Languages from finite sets of positive and negative examples.

2 Methodologies Adapted

The GA produces successive generations of individuals, computing their "fitness" at each step and selecting the best of them when the termination condition arises. In GA application the choice of the chromosome structure is an important decision. When dealing with grammars, however, the number of parameters required by the model is unknown, and hence (ideally) the chromosomes can be of variable lengths, making the operation of the crossover operator less straightforward than before. Furthermore, the interaction of the individual genes is profound: the flipping of a single bit (and the corresponding removal or addition of a production) can render a previously perfect grammar utterly useless. Perhaps as a result of these problems, only relatively simple (and deterministic) Context Free Grammars have been inferred (e.g.[1, 2, 3, 14]) using GAs. A sequential structured chromosome[3] is used in the implementation consists of random sequence of 0s and 1s. The decoding procedure of the grammar maps the random chromosome according to bit sequence based on the number of terminals available in given sample data. The symbolic grammar equivalent to the bit sequence/chromosome is traced from the Starting symbol S through terminal symbol to eliminate all useless productions from the symbolic chromosome. The production length of 5 (one L.H.S. and four R.H.S. symbol.) is taken for the experiment. The generated grammar is then processed with the left recursion removal and left factoring method. The string to be checked from the sample set is then passed to the PDA simulator (Fig. 1) to verify its acceptability to the generated grammar[2].

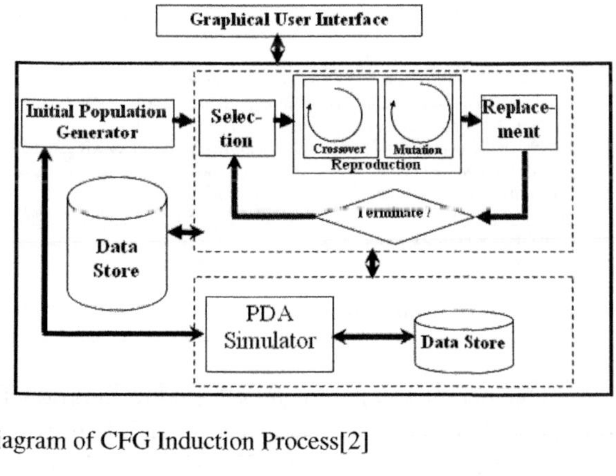

Fig. 1 Block diagram of CFG Induction Process[2]

The positive and negative string set required for the experiment is generated by using the minimum length principle[7]. The strings with the terminals for the given language are generated for the length, L, starting with L=0 and gradually increasing L to get the required number of strings. The validity of the generated string is checked with the best known grammar for the languages. The invalid string generated during this process is considered as negative strings. The objective of the method is to find a generalized best grammar fit for accepting the positive string set and reject the negative string set. The fitness function is formulated to achieve the multiple objectives to accept all the positive strings and to reject all the negative string form the corpus with the grammar having minimum number of production rules. The Grammar which accepts all the positive samples and rejects the entire negative sample set from the corpus is considered to be the best grammar. Due to the stochastic nature of GA, the result is obtained as the average of 10 GAs runs for Languages considered for experimentation[12].

The approaches for handling the premature convergence problem includes the Elite Mating Pool Approach (EMPA) and Dynamic Application of Reproduction Operators (DARO) for generating the proportionate amount of population based on the performance of the different operators.

2.1 *Elite Mating Pool (EMP) Approach*

The premature convergence problem of Genetic Algorithm can be handled by the effective use of the operators, crossover and reproduction, which cause convergence even by encouraging population to converge on a solution while still maintaining diversity. This approach is equivalent to simply taking the best solution after multiple executions of the Simple GA on different initial populations. The approach suggested in the paper is shown in the Fig. 2.

1. Create mating pool of 'n' individuals (n, very small than population size N)
2. Calculate the fitness of all individuals in the population.
3. Sort the mating pool on fitness of the individuals.
4. Perform crossover of the first parent with the elite member of the mating pool and create two offspring.
5. Replace old weaker individuals in mating pool with newly created offspring in sorted order.
6. Perform mutation on the newly created offspring.
7. Replace old weaker individuals in mating pool with newly created offspring in sorted order.
8. Once repeat steps 4 through 7.

Fig. 2 Elite Mating Pool (EMP) Approach

Two individuals in the parent population are selected for reproduction and passed to the elite mating pool. Two elite members from the pool are placed in the child population as newly generated offspring. The diversity is maintained by generating the elite members more capable of satisfying the multiple objectives. The mating pool approach is used for resolving the issue of premature convergence in case of the population converges before reaching to the threshold value. The entire generation of the chromosome is generated by the application of elite mating pool until either the population with a new elite chromosome is evolved or a minimum number of generations (the number of generation used in the experiment is 20.) are completed with the converged population.

2.2 Dynamic Application of Reproduction Operator (DARO)

The dynamic method for the reproduction operator selection is proposed to get the effective utilization of the various operators. The working of the DARO approach is shown in the Fig. 3.

1. Create initial population of random individuals and calculate the fitness of all individuals.
2. Assign the initial probability values to the operator combination ($m \times n$, m Crossover operators × n Mutation combination) by the formula

$$p(u) = 1/(m * n), \quad \& \quad \sum_{u=1}^{m*n} p(u) = 1$$

3. Sort the current population in increasing order of fitness.
4. Select the first crossover operator, $X_{i=1}$. Select the first mutation operator, $M_{j=1}$ and set $u = 1$.
5. Set Resultant Effectivity (RE) to 0.
6. Select Parents, say P_1 & P_2 from current population and produce two offspring, say C_1 & C_2 with the crossover operator X_i
7. Perform mutation on the newly generated offspring with the mutation operator M_i
8. Calculate the fitness of the mutated offspring, say fitness(C_1) & fitness(C_2).
9. Update the Resultant Effectivity (RE) as

$$RE \leftarrow RE + \frac{\left([fitness\,(C_1) + fitness\,(C_2)] / 2 \right)}{\left([fitness\,(P_1) + fitness\,(P_2)] / 2 \right)}$$

10. Insert the newly generated offspring in to child population
11. Repeat step 6 through step 10 until $p(u)$ proportion of population is generated through the current operator combination.
12. update $p(u)$ as

$$p(u) = p(u) + RE$$

13. Select next mutation operator & increment u.
14. If all the mutation operators are applied with current crossover operator then go to step 15 else go to step 6.
15. Select next crossover operator. If all the crossover operators are applied then go to step 12 else go to step 4.
16. Update $p(u)$ for next generation as

$$p(u)_{next} = \frac{p(u)_{current}}{\sum_{u=1}^{m*n} p(u)_{current}},$$

$$\& \quad \sum_{u=1}^{m*n} p(u)_{next} = 1$$

17. Create new current population by overlapping the current population with child population.
18. If termination criterion is not reached, go to step 3.
19. Stop.

Fig. 3 The DARO approach

It is an extension of the Dynamic Application of crossover and mutation Operator Approach given in Elena Simona Nicoară[11]. The GA run begins with the allocation of the equal probability to every reproduction operator combination (Crossover-Mutation operator combination, CMOC). The probability value determines the proportion of the child population to be generated with the application the CMOC. The probability value for the CMOC is updated after every generation based on the ratio of the average fitness of the generated children to the average fitness value of the selected parent. The each CMOC will have chance to generate individuals in child population in proportionate to their probability value.

Three crossover operators and four mutation operators are used in the process. The Crossover Operators used are Two Point crossover (TPC), Two Point crossover with internal swapping (TPCIS) and Uniform Crossover (UC) whereas the Mutation Operators are Stochastic Mutation operator, Inverse Mutation Operator (IMO), Block Copier with Fixed Length Mutation Operator (BCFLMO), and Block Copier with Random Length Mutation Operator (BCRLMO). In the Two Point crossover (TPC), the individual is cut from two random points and offspring are created by copying the middle part of one parent to another. The Point crossover with internal

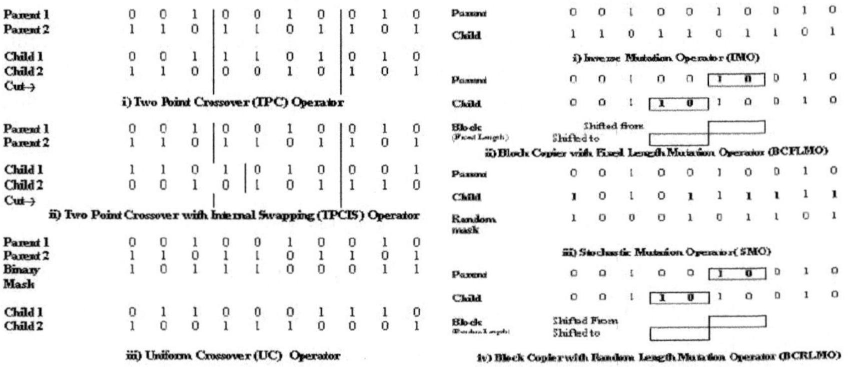

Fig. 4 Reproduction Operators Used

swapping (TPCIS) is the variation of two point crossover where internal left shifting is done after the offspring are created whereas the Uniform Crossover (UC) utilizes a binary mask to create offspring[13]. Inverse Mutation Operator (IMO) complements the parent to get offspring[13] , The Stochastic Mutation operator uses a binary mask for offspring creation, where as the Block Copier with Fixed Length Mutation Operator (BCFLMO), and Block Copier with Random Length Mutation Operator (BCRLMO) copies the block with fixed and variable length, respectively, to get new offspring [4]. The working of the Crossover and Mutation operators is shown in Fig. 4.

3 The Language Set Used

The Languages used for the purpose of experiment corresponding to various GAs runs are listed in the Table 1. The languages chosen for the experiment are the collection of Context Free Language as well as Regular Language.

4 Experimental Setup and Outcome

Experiment is done with JDK 1.4 on an Intel Core^{TM}2 CPU with 2.66 GHZ and 2 GB RAM. The Corpus size = 50, Chromosome size = 240, The Population size = 120, the maximum number of generation considered are 400 for the experiment. The Resultant Best Equivalent Grammar received and its fitness value is shown in the Table 2.

The EMP approach maintains a small pool of elite members. This approach is equivalent to taking the best solution after multiple executions of the SGA on different initial populations. Execution time required for such approach is assumed to be greater than that of a normal SGA approach. To check the effective utilization of the EMP method the generations are processed with the different probability

Table 1 The Languages Used

Languages	Language Description
L1	All string not containing 000 over $(0+1)*$.
L2	$0*1$ over $0+1*$.
L3	$(00)*(111)*$ over $0+1*$.
L4	Any String with even 0 and odd 1 over $0+1*$.
L5	$0(00)*1$ over $0+1*$.
L6	All strings with even number of 0 over $0+1*$.
L7	$(00)*10*$ over $0+1*$.
L8	Balanced Parentheses Problem.
L9	$0^n, 1^n, n \geq 0$ over $0+1*$.
L10	$0^n 1^{2n}, n \geq 0$ over $0+1*$
L11	Even Length Palindrome over $a, b*$

Table 2 The Resultant Best Equivalent Grammar and Its Fitness Value

Languages	Fitness Value	The Resultant Equivalent Grammar
L1	1009	$\langle\{S,M,K\},\{0,1\},\{S \rightarrow M, S \rightarrow 0K, M \rightarrow \varepsilon, M \rightarrow 1SM, K \rightarrow M, K \rightarrow 0M\},S\rangle$
L2	1013	$\langle\{S\},\{0,1\},\{S \rightarrow 1, S \rightarrow 0S\},S\rangle$.
L3	1011	$\langle\{S,M\},\{0,1\},\{S \rightarrow M, S \rightarrow 00SM, M \rightarrow \varepsilon, M \rightarrow 111M\},S\rangle$.
L4	1008	$\langle\{S,M,K\},\{0,1\},\{S \rightarrow 1K, S \rightarrow 0SM0, M \rightarrow \varepsilon, M \rightarrow 0M0, K \rightarrow 1S1M, K \rightarrow M, K \rightarrow 0M0\},S\rangle$.
L5	1011	$\langle\{S,L,C\},\{0,1\},\{S \rightarrow 0L, L \rightarrow C, L \rightarrow 0S, C \rightarrow 1\},S\rangle$.
L6	1010	$\langle\{S,M\},\{0,1\},\{S \rightarrow M, S \rightarrow 1SSM, S \rightarrow 0S0M, M \rightarrow \varepsilon, M \rightarrow 1M\},S\rangle$.
L7	1011	$\langle\{S,M\},\{0,1\},\{S \rightarrow 1M, S \rightarrow 00S, M \rightarrow \varepsilon, M \rightarrow 0M\},S\rangle$.
L8	1011	$\langle\{S,M\},\{(,)\},\{S \rightarrow (M, M \rightarrow S)M, M \rightarrow \varepsilon, M \rightarrow)M\},S\rangle$.
L9	1012	$\langle\{S,M\},\{0,1\},\{S \rightarrow M, S \rightarrow 0S1M, M \rightarrow \varepsilon\},S\rangle$.
L10	1013	$\langle\{S,M\},\{0,1\},\{S \rightarrow ?, S \rightarrow 0S11\},S\rangle$.
L11	1010	$\langle\{S,J\},\{a,b\},\{S \rightarrow bJ, S \rightarrow aSa, S \rightarrow \varepsilon, J \rightarrow b, J \rightarrow Sb\},S\rangle$.

of EMP. The probabilities with the entire generation is processed with EMP approach are 0%, 25% and 50%.The Statistical Analysis of the EMP approach over probabilities such as 0%, 25% and 50% are given in Table 3, Table 4 and Table 5 respectively. The analysis of the Local Optimum Convergence (LOC) and the success rate (SUCC %) attended by the method it is found that the DARO approach is more suitable for generating the context free grammar from the sample corpus. In all the cases EMP is also used to break the local optimum convergence situation of the GA run whereas in case of DARO approach no any such method is applied for handling the local optimum convergence situation. The Statistical Analysis of the DARO approach is given in Table 6. The proposed approaches found to result in the superset of the grammar required. The threshold value is considered as the fitness value of the grammar sufficient to accept all the strings from positive sample set and reject the strings from negative sample set. The language sets used are the example of the lightweight grammars.

Table 3 The Data Analysis for EMP-00 Approach

L	Total Runs	Generations				Success Rate	
		Range	Threshold	Mean μ	Std.dev. σ	LOC	SUCC%
L1	11	30±18	36	26.90	12.08	01	90.90%
L2	10	08±06	09	08.70	03.83	00	100%
L3	26	42±25	38	35.80	15.88	16	38.46%
L4	10	17±10	22	17.10	06.04	00	100%
L5	18	12±10	15	11.90	05.54	08	55.55%
L6	13	16±07	22	14.40	04.69	03	76.92%
L7	14	20±14	32	22.50	08.91	04	71.42%
L8	10	12±09	11	09.20	05.49	00	100%
L9	10	11±08	18	08.50	06.51	00	100%
L10	10	31±24	32	21.90	14.67	00	100%
L11	28	23±13	36	19.90	09.13	18	35.71%

Table 4 The Data Analysis for EMP-25 Approach

L	Total Runs	Generations				Success Rate	
		Range	Threshold	Mean μ	Std.dev. σ	LOC	SUCC%
L1	10	30±13	34	27.55	09.24	00	100%
L2	10	07±05	07	08	03.05	00	100%
L3	40	29±15	31	25.40	09.45	30	25%
L4	10	15±06	21	14.60	04.35	00	100%
L5	10	12±07	14	11	03.57	00	100%
L6	11	16±07	20	13.80	05.26	01	90.90%
L7	13	18±14	28	15.90	09.46	03	76.92%
L8	10	06±03	09	05.80	02.34	00	100%
L9	10	08±05	13	06.60	03.30	00	100%
L10	10	23±18	24	20.41	12.55	00	100%
L11	31	15±08	23	17.80	05.05	21	32.25%

Results obtained for first ten successful run on each language set, which does not results in premature convergence and the comparative chart for the execution time (in seconds) per generation for the languages considered over the approaches, are summarized in the Fig. 5. The generation chart displays the average of best fitness values received over GAs run for the approached used in the experiment. It is found that, the population is converged to the best value earlier in case of the simple grammar, whereas it has converged late for the relatively complex Context Free Grammar. The EMP approach found to converge earlier than the DARO approach whereas the execution time taken by the EMP approach is more as compared to DARO approach.

Table 5 The Data Analysis for EMP-50 Approach

L	Total Runs	Generations				Success Rate	
		Range	Threshold	Mean μ	Std.dev. σ	LOC	SUCC%
L1	10	21±09	27	22.30	06.21	00	100%
L2	10	09±05	05	07.40	04.69	00	100%
L3	50	21±15	23	20.30	10.02	40	20%
L4	32	12±04	16	13	03.39	22	31.25%
L5	10	09±06	09	06.90	04.72	00	100%
L6	11	11±07	16	10.60	04.62	01	90.90%
L7	11	16±09	16	14.10	05.54	01	90.90%
L8	10	05±03	07	03	01.63	00	100%
L9	10	10±05	14	09.50	04.06	00	100%
L10	10	24±21	14	11.50	12.92	00	100%
L11	30	15±08	20	15.30	05.59	20	33.33%

Table 6 The Data Analysis for DARO Approach

L	Total Runs	Generations				Success Rate	
		Range	Threshold	Mean μ	Std.dev. σ	LOC	SUCC%
L1	10	23±16	27	20.90	08.41	00	100%
L2	10	05±03	06	05.70	02.05	00	100%
L3	10	28±18	39	28.20	13.35	00	100%
L4	10	35±26	51	29.60	16.42	00	100%
L5	10	18±15	18	14	09	00	100%
L6	10	15±06	18	14.20	03.96	00	100%
L7	10	13±03	16	12	01.90	00	100%
L8	10	12±10	22	15.70	6.96	00	100%
L9	10	07±05	9	05.70	03.20	00	100%
L10	10	17±10	18	12	07	00	100%
L11	10	19±16	27	18	09.20	00	100%

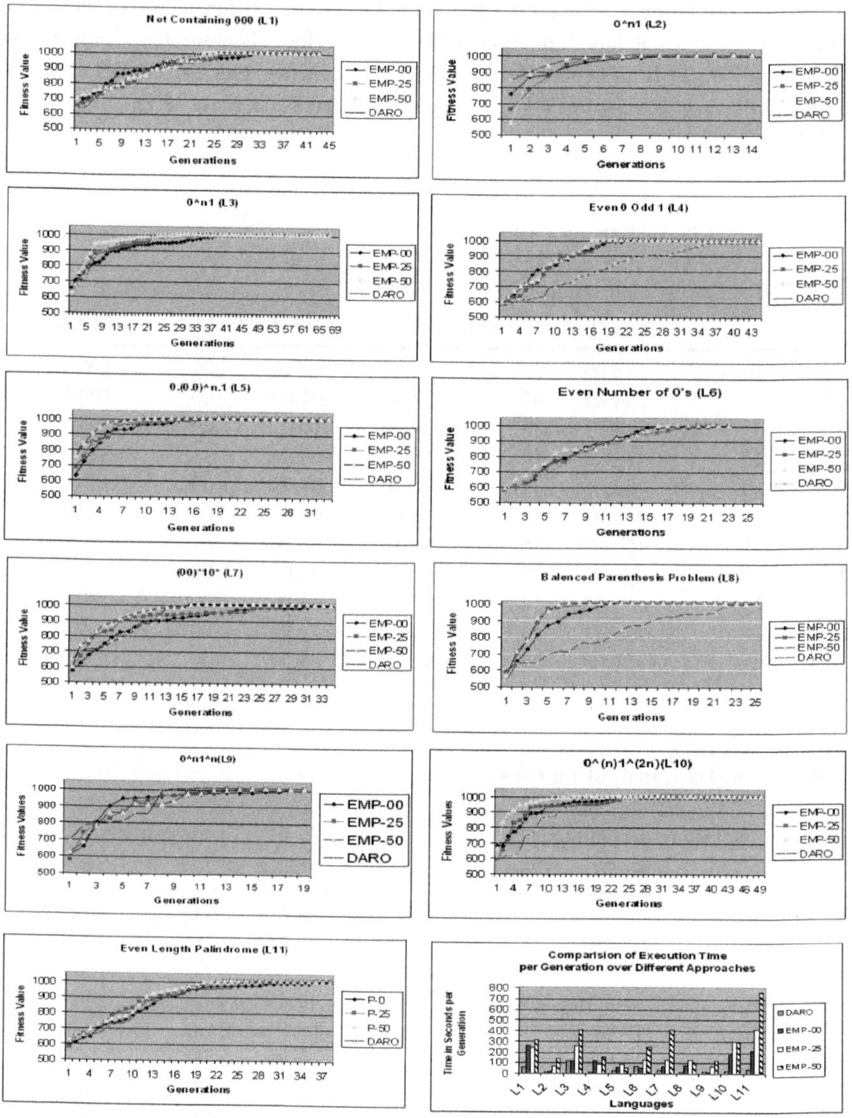

Fig. 5 Generation charts for the Languages used and the comparison chart for the execution time required in seconds/generation

5 Conclusion

The Experiment is successfully conducted and the adapted method is found to be working successfully on the language set considered. There is further scope for adoption of the same method for more complex grammar sets. It is found that the EMP method have shown higher tendency to get in to local optimum convergence as compared to the DARO method used in the experiment. It is also found that the execution time taken by the EMP method much higher as compared to DARO approach. The methods used have shown fast convergence for the induction of grammar for languages. Minimum length description principle[7] is found to be more effective in the selection of the corpus. The selection of the good quality corpus (positive and negative string inputs) has resulted into induction of good quality grammar for the languages considered. Results have shown tendency towards the local optimum convergence which requires special attention in future work.

Acknowledgements. Authors thank to Dr. V. M. Thakare, Sant Gadge Baba Amravati University, Amravati, Maharastra, for his kind support in providing Laboratory infrastructural facility required for the conduction of the experiment.

References

1. Choubey, N.S., Kharat, M.U.: Grammar Induction and Genetic Algorithms-An Overview. Pacific Journal of Science and Technology 10(2), 884–888 (2009)
2. Choubey, N.S., Kharat, M.U.: PDA Simulator for CFG Induction Using Genetic Algorithm. In: Proceedings of UKSim-AMSS 12th International Conference on Computer Modelling and Simulation, UKSim 2010 (2010)
3. Choubey, N.S., Kharat, M.U.: Sequential Structuring Element for CFG Induction Using Genetic Algorithm. International Journal of Futuristic Computer Application 1, Article 2 (2010)
4. Choubey, N.S., Kharat, M.U.: Stochastic Mutation Approach for Grammar Induction Using Genetic Algorithm. In: The 2nd International Conference on Electronic Computer Technology, ICECT 2010, Kuala Lumpur, Malaysia, May 7-10 (2010)
5. Goldberg, D.E.: Genetic Algorithms in Search Optimizations and Machine Learning. Addison-Wesley, Reading (1989)
6. Hopcroft, J.E., Motwani, R., Ullman, J.D.: Introduction to Automata Theory, Languages and Computation, 3rd edn. Addison-Wesley, Reading (2007)
7. Keller, B., Lutz, R.: Evolving stochastic context-free grammars from examples using a minimum description length principle. Paper presented at the Workshop on Automata Induction Grammatical Inference and Language Acquisition, ICML 1997 (1997)
8. Kureichick, V.M., Melikhov, A.N., Miagkikh, V.V., et al.: Some New Features in Genetic Solution of the travelling salesman problem. In: Proceedings of ACEDC 1996 PEDC, University of Plymouth, UK (1996)
9. Lankhorst, M.: A Genetic Algorithm for the Induction of Nondeterministic Pushdown Automata. Technical Report CS-R 9502, University of Groningen, The Netherlands (1995)
10. Neves, J., Rocha, M.: Preventing Premature Convergence to Local Optima in Genetic Algorithms via Random Offspring Generation. In: Proceedings of the 12th International Conference on Industrial and Engineering Applications of Artificial Intelligence and Expert Systems: Multiple Approaches to Intelligent Systems, IEA/AIE, Cairo, Egypt (1999)

11. Nicoar, E.S.: Mechanisms to Avoid the Premature Convergence of Genetic Algorithms. Universitatea Petrol-Gaze din Ploieti. Matematica-Informatic-Fizica LXI(1), 87–96 (2009)
12. Pauw, G.D.: Evolutionary Computing as a Tool for Grammar Development. In: Cantú-Paz, E., Foster, J.A., Deb, K., Davis, L., Roy, R., O'Reilly, U.-M., Beyer, H.-G., Kendall, G., Wilson, S.W., Harman, M., Wegener, J., Dasgupta, D., Potter, M.A., Schultz, A., Dowsland, K.A., Jonoska, N., Miller, J., Standish, R.K. (eds.) GECCO 2003. LNCS, vol. 2723, pp. 549–560. Springer, Heidelberg (2003)
13. Sivanandam, D.: Introduction to Genetic Algorithm. Springer, Heidelberg (2008)
14. Wyard, P.: Representational Issues for Context-Free Grammar Induction Using Genetic Algorithm. In: Carrasco, R.C., Oncina, J. (eds.) ICGI 1994. LNCS, vol. 862, pp. 222–235. Springer, Heidelberg (1994)

A Novel Magnetic Update Operator for Quantum Evolutionary Algorithms

Mohammad H. Tayarani, N., Adam Prugel Bennett, and Hosein Mohammadi

Abstract. Quantum Evolutionary Algorithms (QEA) are novel algorithms proposed for class of combinatorial optimization problems. The probabilistic representation of possible solutions in QEA helps the q-individuals to represent all the search space simultaneously. In QEA, Q-Gate plays the role of update operator and moves q-individuals toward better parts of search space to represent better possible solutions with higher probability. This paper proposes an alternative magnetic update operator for QEA. In the proposed update operator the q-individuals are some magnetic particles attracting each other. The force two particles apply to each other depends on their fitness and their distance. The population has a cellular structure and each q-individual has four neighbors. Each q-individual is attracted by its four binary solution neighbors. The proposed algorithm is tested on Knapsack Problems, Trap problem and fourteen numerical function optimization problems. Experimental results show better performance for the proposed update operator than Q-Gate.

1 Introduction

Quantum Evolutionary Algorithms are new optimization algorithms proposed for class of combinatorial optimization problems [1]. QEA uses probabilistic representation for possible solutions and this characteristic helps the q-individuals to represent all the search space simultaneously. Several works try to improve the performance of QEA. Combining the concepts of Immune systems and QEA, [2] proposes an immune quantum evolutionary algorithm. In another work [3] proposes a novel particle swarm quantum evolutionary algorithm. A new adaptive rotation gate

Mohammad H. Tayarani, N.
University of Southampton
e-mail: mhtn1g09@ece.soton.ac.uk

Adam Prugel Bennett
University of Southampton
e-mail: apb@ecs.soton.ac.uk

A. Gaspar-Cunha et al. (Eds.): Soft Computing in Industrial Applications, AISC 96, pp. 67–76.
springerlink.com

is proposed in [4] which uses the probability amplitude ratio of the corresponding states of quantum bits. Inspired by the idea of hybrid optimization algorithms, [5] proposes two hybrid-QEA based on combining QEA with PSO. In [6] a novel Multi-universe Parallel Immune QEA is proposed. In the algorithm all the q-individuals are divided into some independent sub-colonies, called universes. Since QEA is proposed for the class of combinatorial optimization problems, [7] proposes a new version of QEA for numerical function optimization problems. A novel quantum coding mechanism for QEA is proposed in [8] to solve the travelling salesman problem. In another work [9] points out some weaknesses of QEA and explains how hitching phenomena can slow down the discovery of optimal solutions. In this algorithm, the attractors moving the population through the search space are replaced at every generation. A new approach based on Evolution Strategies is proposed in [10] to evolve quantum unitary operators which represents the computational algorithm a quantum computer would perform to solve an arbitrary problem. In order to preserve the diversity in population and empower the search ability of QEA, [11] proposes a novel diversity preservation operator for QEA. Reference [12] proposes a sinusoid sized population QEA that makes a tradeoff between exploration and exploitation. While QEA is suitable for combinatorial problems and is relatively weak for real coded problems like numerical function optimization problems, several works have focused on this foible. Reference [13] proposes a probabilistic optimization algorithm, which similar to QEA uses a probabilistic representation for possible solutions.

In QEA, Q-Gate plays the role of update operator and moves the q-individuals toward better parts of the search space. Each q-individual moves toward its best observed possible solution and the only interaction among the q-individuals is the simple copying of best observed binary solutions (see local and global migrations in [1]). This paper proposes a more complicated update operator for QEA, inspiring magnetic field theory which offers more interaction among q-individuals and binary solutions to help the q-individuals extract more information from each other. In the proposed algorithm, the binary solutions attract q-individuals and the binary solutions with higher fitness apply more force to the q-individuals. The proposed algorithm has a parameter and this paper tries to investigate the effect of the parameter on the performance of the proposed algorithm. The proposed algorithm is tested on several benchmark functions including Knapsack problem, Trap problem and numerical function optimization problems. Experimental results show better performance for the proposed update operator than Q-Gate.

This paper is organized as follows. Section 2 introduces Quantum Evolutionary Algorithm and its representation. In section 3 the proposed algorithm is proposed and its parameter is investigated. Experimental results are performed in section 4 and finally section 5 concludes the paper.

2 QEA

QEA is inspired from the principles of quantum computation, and its superposition of states is based on qubits, the smallest unit of information stored in a two-state quantum computer. A qubit could be either in state "0" or "1", or in any superposition of the two as described below:

$$|\psi\rangle = \alpha|0\rangle + \beta|1\rangle \tag{1}$$

Where α and β are complex number, which denote the corresponding state appearance probability, following below constraint:

$$|\alpha|^2 + |\beta|^2 = 1 \tag{2}$$

This probabilistic representation implies that if there is a system of m qubits, the system can represent 2^m states simultaneously. At each observation, a qubits quantum state collapses to a single state as determined by its corresponding probabilities.

Consider $i-th$ individual in $t-th$ generation defined as an m-qubit as below:

$$\begin{bmatrix} \alpha_{i1}^t & \alpha_{i2}^t & \cdots & \alpha_{ij}^t & \cdots & \alpha_{im}^t \\ \beta_{i1}^t & \beta_{i2}^t & \cdots & \beta_{ij}^t & \cdots & \beta_{im}^t \end{bmatrix} \tag{3}$$

Where $\left|\alpha_{ij}^t\right|^2 + \left|\beta_{ij}^t\right|^2 = 1$, $j = 1,2,,m,m$ is the number of qubits, i.e., the string length of the qubit individual, $i = 1,2,,n,n$ is the number of possible solution in population and t is generation number of the evolution.

2.1 QEA Structure

In the initialization step of QEA, $[\alpha_{ij}^t \;\; \beta_{ij}^t]^T$ of all q_i^0 are initialized with $\frac{1}{\sqrt{2}}$. This implies that each qubit individual q_i^0 represents the linear superposition of all possible states with equal probability. The next step makes a set of binary instants; x_i^t by observing $Q(t) = \{q_1^t, q_2^t, ..., q_n^t\}$ states, where $X(t) = /x_1^t, x_2^t, ..., x_n^t/$ at generation t is a random instant of qubit population. Each binary instant, x_i^t of length m, is formed by selecting each bit using the probability of qubit, either $|\alpha_{ij}^t|$ or $|\beta_{ij}^t|$ of q_i^t. Each instant x_i^t is evaluated to give some measure of its fitness. The initial best solution $b = max_{i=1}^n \{f(x_i^t)\}$ is then selected and stored from among the binary instants of $X(t)$. Then, in 'update' $Q(t)$, quantum gates U update this set of qubit individuals $Q(t)$ as discussed below. This process is repeated in a while loop until convergence is achieved. The appropriate quantum gate is usually designed in accordance with problems under consideration.

2.2 Quantum Gates Assignment

The common mutation is a random disturbance of each individual, promoting exploration while also slowing convergence. Here, the quantum bit representation can be simply interpreted as a biased mutation operator. Therefore, the current best individual can be used to steer the direction of this mutation operator, which will speed up the convergence. The evolutionary process of quantum individual is completed through the step of "update $Q(t)$". A crossover operator, quantum rotation gate, is described below. Specifically, a qubit individual q_i^t is updated by using the rotation gate $U(\theta)$ in this algorithm. The $j-th$ qubit value of $i-th$ quantum individual in generation t, $[\alpha_{ij}^t \ \beta_{ij}^t]^T$ is updated as:

$$\begin{bmatrix} \alpha_{ij}^t \\ \beta_{ij}^t \end{bmatrix} = \begin{bmatrix} cos(\Delta\theta) & -sin(\Delta\theta) \\ sin(\Delta\theta) & cos(\Delta\theta) \end{bmatrix} \begin{bmatrix} \alpha_{ij}^{t-1} \\ \beta_{ij}^{t-1} \end{bmatrix} \quad (4)$$

Where $\Delta\theta$ is rotation angle and controls the speed of convergence and determined from Table 1. Reference [14] shows that these values for $\Delta\theta$ have better performance.

Table 1 Lookup Table of $\Delta\theta$, the rotation gate. x_i is the $i-th$ bit of the observed binary solution and b_i is the $i-th$ bit of the best found binary solution.

x_i	b_i	$f(x) \geq f(b)$	$\Delta\theta$
0	0	false	0
0	0	true	0
0	1	false	0.01π
0	1	true	0
1	0	false	-0.01π
1	0	true	0
1	1	false	0
1	1	true	0

3 Magnetic Update Operator

As it is seen in previous section, Q-Gate has the role of update operator in QEA and moves the q-individuals toward better parts of the search space to represent better possible solutions with higher probability. In each iteration, Q-Gate moves the q-individuals to their best observed possible solutions with the certain value of $\Delta\theta_i$. This kind of update operator has two weaknesses. First regardless of the fitness of the best observed possible solution, the q-individuals are moved with a constant value of $\Delta\theta_i$, which is the same for various best observed possible solutions with various values of fitness. Second, each q-individual is affected with only one possible solution and other possible solutions have not any effect on the q-individual. In Q-Gate update operator the only interaction among q-individuals is the local and global migrations which are a simple copying of binary solutions. This paper proposes a novel update operator for QEA inspiring from the attraction among magnetic particles. Recently we proposed a novel optimization algorithm called Magnetic Optimization Algorithm [14]. In MOA the possible solutions are some magnetic particles attracting each other. Each magnetic particle applies a force to its neighbors, and the amplitude of force is determined by the fitness of the particles and the distance the particles have. Here we propose a similar update operator for QEA. In the proposed update operator, the q-individuals are attracted toward all of their binary solution neighbors. In the proposed update operator even the inferior binary solutions attract the q-individuals and have effect on searching process. The pseudo code of the proposed update operator is as follows:

Procedure Basic MQEA
begin
  $t = 0$
1. initialize Q^0
2. while not termination condition do
  begin
    $t = t + 1$
3. make X^t by observing the states of Q^{t-1}
4. evaluate the particles in X^t and store their performance in magnetic fields B^t
5. normalize B^t according to 6
6. evaluate the mass M^t for all particles according to 7
7. for all q-individuals q^t_{ij} in Q^t do
    begin
8. F_{ij}=0
9. find N_{ij}
10. for all x^t_{uv} in N_{ij} do
11. $F_{ij,k} = F_{ij,k} + \frac{(x^t_{uv,k} - (\beta^t_{ij,k})^2)\times B^t_{uv}}{D(x^t_{ij}, x^t_{uv})}$
    end
12. for all q-individuals q^t_{ij} in Q^t do
    begin
13. $v^{t+1}_{ij,k} = \frac{F_{ij,k}}{\eta \times M_{ij}}$
14. $q^{t+1}_{ij,k} = q^t_{ij,k} + v^{t+1}_{ij,k}$
    end
  end
end

The description of the proposed algorithm is as follows: Step 1. This paper uses a cellular structure for population. In the initialization step, the quantum-individuals q^0_{ij} are located in a lattice-like population. Then $[\alpha^0_{ij,k}\ \beta^0_{ij,k}]^T$ of all q^0_{ij} are initialized with $1/\sqrt{2}$, where $i, j = 1,2,,S$ is the location of the q-individuals in the lattice, $k = 1,2,...,m$, and m is the number of qubits in the individuals. This implies that each qubit individual q^0_{ij} represents the linear superposition of all possible states with equal probability.

Step 3. This step makes a set of binary instants $X^t = \{x^t_{ij} | i, j = 1,2,...,S\}$ at generation t by observing $Q^{t-1} = \{x^{t-1}_{ij} | i, j = 1,2,...,S\}$ states, where X^t at generation t is a random instant of qubit population and S is the size of lattice. Each binary instant, x^t_{ij} of length m, is formed by selecting each bit using the probability of qubit, either $|\alpha^{t-1}_{ij,k}|^2$ or $|\beta^{t-1}_{ij,k}|^2$ of q^{t-1}_{ij}. Observing the binary bit $x^t_{ij,k}$ from qubit $[\alpha^t_{ij,k}\ \beta^t_{ij,k}]^T$ performs as:

$$x^t_{ij,k} = \begin{cases} 0 & if\ R(0,1) < |\alpha^t_{ij,k}|^2 \\ 1 & otherwise \end{cases} \tag{5}$$

Where $R(.,.)$ is a uniform random number generator.

Step 4. Each binary instant x^t_{ij} is evaluated to give some measure of its objective. In this step, the fitness of all binary solutions of X^0 are evaluated and stored in B^t.

Step 5. Next the normalization is performed on B^t. The normalization is performed as:

$$B_{ij} = \frac{B_{ij} - Min}{Max - Min} \tag{6}$$

Where: $Min = minimum^S_{i,j=1}(B^t_{ij}), \quad Max = maximum^S_{i,j=1}(B^t_{ij})$

The magnetic field of each particle is normalized in the range of [0-1]. This is because the fitness values of possible solutions are problem dependent. The range of the fitness of the possible solutions can be in any range, since the amount of the magnetic field controls the movement of the particles, we normalize the amount of magnetic field.

Step 6. In this step the mass of all particles is calculated and stored in M^t:

$$M^t_{ij} = 1 + B^t_{ij} \tag{7}$$

Step 7. In this step in the "for" loop, the resultant force of all forces on each particle is calculated.

Step 8. At first the resultant force F_{ij} to particle x^t_{ij} is initialized to zero.

Step 9. In the lattice-like structure of QEA population, each particle interacts only with its neighbors i.e. each particle applies its force only to its neighbors. In this step the neighbors of x^t_{ij} are considered. The set of neighbors for particle x_{ij} can be defined as $N_{ij} = \{x_{i'j}, x_{ij'}, x_{i"j}, x_{ij"}\}$ Where: $i' = \begin{cases} i-1 & i \neq 1 \\ S & i = 1 \end{cases}$, $j' = \begin{cases} j-1 & j \neq 1 \\ S & j = 1 \end{cases}$, $i" = \begin{cases} i+1 & i \neq S \\ 1 & i = S \end{cases}$, $j" = \begin{cases} j+1 & j \neq S \\ 1 & j = S \end{cases}$ Step 10. In this step, the applied force to particle x^t_{ij} by its neighbor's x^t_{uv} $\forall x^t_{uv} \in N_{ij}$ is calculated.

Step 11. The force which is applied by x^t_{uv} to x^t_{ij} relates to the distance between two particles and the magnetic field of x^t_{uv} and is calculated as:

$$F_{ij,k} = \frac{\left(x^t_{uv,k} - (\beta^t_{ij,k})^2\right) \times B^t_{uv}}{D(x^t_{ij}, x^t_{uv})} \tag{8}$$

Here F_{ij} shows the force applied to q-individual q^t_{ij}. The part "$x^t_{uv,k} - (\beta^t_{ij,k})^2$" shows the direction which the q-individual moves and $(\beta^t_{ij,k})^2$ is the probability of q^t_{ij} representing state "1". Where $D(.,.)$ is the distance between each pair of neighboring particles and is calculated as:

$$D(x^t_{ij}) = \frac{1}{m} \sum_{k=1}^{m} \left| x^t_{ij,k} - x^t_{uv,k} \right| \tag{9}$$

Where x^t_{ij} and x^t_{uv} are $(i, j) - th$ and $(u, v) - th$ binary solutions of the population in iteration t respectively and $x^t_{ij,k}$ is the $k - th$ dimension of $(i, j) - th$ binary solution in iteration t. This step is the main step in the proposed algorithm.

Steps 12, 13, 14. In these steps the location of q-individuals are updated. Here η is the movement coefficient which controls the speed of movement.

The proposed update operator has two advantages. First according to 8 the observed binary solutions with higher fitness have bigger magnetic field B and apply more force to the q-individual, therefore the better binary solutions have more attraction force. Here unlike Q-Gate the movement of q-individuals is not constant throughout the search process and varies for various q-individuals and even various dimensions. Second in the proposed update operator even the inferior binary solutions have effect on the q-individuals but with smaller amplitude. Accordingly the interaction among possible solutions is much more than Q-Gate and the inferior binary solutions participate in the search process. It helps the algorithm escaping from local optima and if the inferior binary solutions are near an optimum, helps the q-individuals to find the optimum.

3.1 *Parameter Tuning*

As it is seen in step 13 of MQEA, the proposed algorithm has a parameter of η. This section tries to find the best parameter for the proposed update operator for several benchmark functions. The size of population for all the experiments is set to 25, and the parameter is set to η=(1,2,3,4,5,10,15,20,25,30,35,40,45,50). Figure 1 shows the parameter setting for the proposed algorithm on Knapsack problem and Generalized Schwefel's Function 2.26. The results are averaged over 30 runs. According to Figure 1, the best parameter for Knapsack problem repair type 1, the best parameter is η=5, the best parameter for Knapsack penalty type 2 is η=20 and the best parameter for Generalized Schwefel is 10. This paper finds the best parameter for the proposed update operator for several benchmark functions and the results are summarized in Table 2. As it is clear in Table 2, for all the numerical function problems the best parameter for the proposed update operator is 10.

4 Experimental Results

The proposed algorithm is compared with the original version of QEA to show the improvement on QEA. The best parameters as found in previous sections are used

Table 2 Best parameter for the proposed Update operator. The results are averaged over 30 runs

Problem	η	Problem	η	Problem	η	Problem	η
Kpck Rep 1	5	Kpck Rep 2	35	Kpck Pen 1	5	Kpck Pen 2	20
Trap	2	Schwefel [16]	10	Rastrigin [16]	10	Ackley [16]	10
Griawank [16]	10	Penalized1 [16]	10	Penalized2 [16]	10	Kennedy [15]	10
Michalewicz [15]	10	Goldberg [15]	10	Sphere [16]	10	Rosenbrock [15]	10
Schwefel 2.21 [16]	10	Dejong [15]	10	Schwefel 2.22 [16]	10		

in order to provide fair comparison between the proposed algorithm and the original version of QEA. The parameters of QEA is set to the best parameters found in [1]. The experimental results are performed on Knapsack problem Penalty type 1 and 2, Knapsack problem Repair types 1 and 2 (see Appendix), Trap problem and fourteen numerical function optimization problems, for the dimension of 100 and 250. The population size of all algorithms for all of the experiments is set to 25; termination condition is set for a maximum of 2000 iterations for Knapsack and Trap problems and 5000 iterations for the fourteen numerical functions. The parameter of QEA is set to Table 1. The parameter of the proposed update operator is set to the values found in previous section. Due to statistical nature of the optimization algorithms, all results are averaged over 30 runs.

Table 3 shows the experimental results on the proposed magnetic optimization update operator and Q-Gate update operator. According to Table 3, the proposed update operator improves the performance of QEA significantly, and in all the experimental results, the proposed algorithm reaches better results. Additionally, the standard deviation of the best reached results over 30 runs in the proposed update operator is much smaller than Q-Gate. The small STD shows better performance for the proposed algorithm.

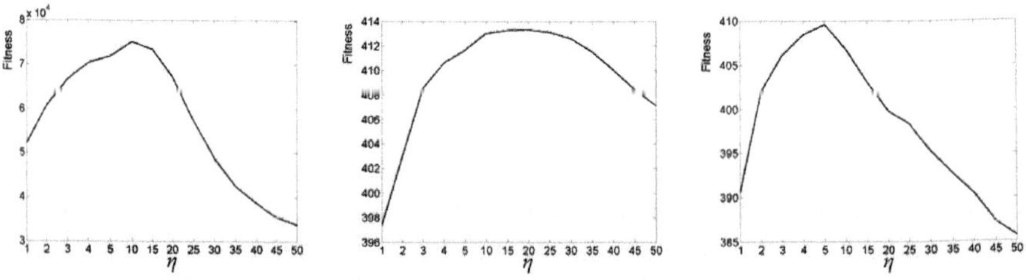

Fig. 1 The effect of the parameter η on the performance of the proposed algorithm on Generalized Schwefel's Function 2.26, Knapsack problem Penalty Type 2 and Trap problem.

Table 3 Experimental results on Knapsack problem, Trap Problem and fourteen numerical function optimization problems. The number of runs is 30. Mean and STD represent the mean of best answers and standard deviation of best answers for 30 runs respectively. *m* is the dimension of problem. The bold results are the best ones.

	m=100				*m*=250			
	Q-Gate		Magnetic		Q-Gate		Magnetic	
Problem	Mean	STD	Mean	STD	Mean	STD	Mean	STD
Kpck Rep 1	373.73	4.79	387.04	0.90	907.53	18.37	1008.50	3.11
Kpck Rep 2	438.44	6.77	456.32	0.02	887.43	16.27	991.73	1.46
Kpck Pen 1	373.73	4.79	387.04	0.90	966.76	20.96	1086.27	4.15
Kpck Pen 2	434.97	7.20	456.29	0.05	930.56	18.26	1046.00	0.59
Trap	71.50	4.65	80.33	1.18	156.67	8.15	196.90	3.13
Schwefel	4.2×10^4	5558	7.6×10^4	1304	7.9×10^4	14101	1.8×10^5	2207
Rastrigin	-1677	259	-252	35	-5858.41	720.78	-848.38	75.77
Ackley	-17.91	0.12	-5.72	1.75	-18.21	0.13	-11.86	0.81
Griewank	-32.18	5.79	-0.07	0.16	-127.65	23.17	-1.67	0.40
Penalized 1	-1.3×10^5	2.4×10^4	-33.45	288.06	-5.1×10^5	1.0×10^5	-1864	1179
Penalized 2	-3.0×10^4	7014	-107.60	69.05	-1.1×10^5	2.4×10^4	-1415	566
Michalewicz	32.99	6.33	80.97	1.372	51.79	15.07	190.24	2.86
Goldberg	46.40	3.76	89.01	1.07	93.77	12.14	212.65	2.04
Sphere Model	-3.3×10^5	8.1×10^4	-1435	1176	-1.4×10^6	2.2×10^5	-1.8×10^4	4631
Schwefel 2.22	-4.10	0.61	-0.05	0.01	-5.73	0.54	-0.19	0.02
Schwefel 2.21	-172.70	6.15	-124.64	11.57	-189.44	2.14	-161.94	4.91
Dejong	-2.2×10^7	8.3×10^6	-92603	48854	-2.6×10^8	7.3×10^7	-2.2×10^6	7.7×10^5
Rosenbrock	-7.7×10^4	2.0×10^4	-2460	1311	-3.6×10^5	9.9×10^4	-1.2×10^4	3488
Kennedy	-1.26	0.83	-0.0003	0.0003	-18.17	7.61	-0.003	0.002

5 Conclusion

This paper proposes a novel alternative for Q-Gate update operator. In the proposed update operator the q-individuals are some magnetic particles being attracted to binary solutions based on their fitness. In comparison with Q-Gate the proposed update operator has two advantages. First the movement of q-individuals is not constant throughout the search process and varies for various q-individuals and even various dimensions. Second in the proposed update operator even the inferior binary solutions have effect on the q-individuals but with smaller amplitude. Accordingly the interaction among possible solutions is much more than Q-Gate and the inferior binary solutions participate in the search process. The proposed algorithm has a parameter; this paper has also investigated the effect of the parameter for several benchmark functions. The proposed update operator is tested on several benchmark functions, and experimental results shows better performance for the proposed operator than Q-Gate. In our future works we will focus on some operators to improve the performance of the proposed algorithm and apply the proposed algorithm on some real problems.

References

1. Han, K., Kim, J.: Quantum-inspired evolutionary algorithm for a class of combinatorial optimization. IEEE Transactions on Evolutionary Computing 6(6) (2002)
2. Li, Y., Zhang, Y., Zhao, R., Jiao, L.: The immune quantum-inspired evolutionary algorithm. In: IEEE International Conference on Systems, Man and Cybernetics (2004)
3. Wang, Y., Feng, X.-Y., Huang, Y.-X., Zhou, W.-G., Liang, Y.-C., Zhou, C.-G.: A Novel Quantum Swarm Evolutionary Algorithm for Solving 0-1 Knapsack Problem. In: Wang, L., Chen, K., S. Ong, Y. (eds.) ICNC 2005. LNCS, vol. 3611, pp. 698–704. Springer, Heidelberg (2005)
4. Gao, H., Xu, G., Wang, Z.: A Novel Quantum Evolutionary Algorithm and Its Application. In: The Sixth IEEE World Congress on Intelligent Control and Automation (2006)
5. Yu, Y., Tian, Y., Yin, Z.: Hybrid Quantum Evolutionary Algorithms Based on Particle Swarm Theory. In: 1st IEEE Conference on Industrial Electronics and Applications (2006)
6. You, X., Liu, S., Shuai, D.: On Parallel Immune Quantum Evolutionary Algorithm Based on Learning Mechanism and Its Convergence. In: Jiao, L., Wang, L., Gao, X.-b., Liu, J., Wu, F. (eds.) ICNC 2006. LNCS, vol. 4221, pp. 903–912. Springer, Heidelberg (2006)
7. Cruz, D., Vellasco, A.V.A., Pacheco, M.M.B.: Quantum-Inspired Evolutionary Algorithm for Numerical Optimization. In: IEEE Congress on Evolutionary Computation (2006)
8. Feng, X.Y., Wang, Y., Ge, H.W., Zhou, C.G., Liang, Y.C.: Quantum-Inspired Evolutionary Algorithm for Travelling Salesman Problem. In: Computational Methods, pp. 1363–1367. Springer, Heidelberg (2007)
9. Platelt, M.D., Schliebs, S., Kasabov, N.: A versatile quantum-inspired evolutionary algorithm. In: IEEE Congress on Evolutionary Computation (2007)
10. Hutsell, S.R., Greenwood, G.W.: Applying evolutionary techniques to quantum computing problems. In: IEEE Congress on Evolutionary Computation (2007)
11. Tayarani, M.-H., Akbarzadeh, M.R.-T.: A Cellular Structure and Diversity Preserving operator in Quantum Evolutionary Algorithms. In: IEEE World Conference on Computational Intelligence (2008)
12. Tayarani, M.-H., Akbarzadeh, M.R.-T.: A Sinusoid Size Ring Structure Quantum Evolutionary Algorithm. In: IEEE International Conference on Cybernetics and Intelligent Systems Robotics, Automation and Mechanics (2008)
13. Tayarani, M.-H., Akbarzadeh, M.-R.-T.: Probabilistic Optimization Algorithms for Numerical Function Optimization Problems. In: IEEE International Conference on Cybernetics and Intelligent Systems Robotics, Automation and Mechanics (2008)
14. Tayarani, M.-H., Akbarzadeh, M.R.-T.: Magnetic Optimization Algorithm, A New Synthesis. In: IEEE World Conference on Computational Intelligence (2008)
15. Khorsand, A.-R., Akbarzadeh, M.-R.-T.: Quantum Gate Optimization in a Meta-Level Genetic Quantum Algorithm. In: IEEE International Conference on Systems, Man and Cybernetics (2005)
16. Zhong, W., Liu, J., Xue, M., Jiao, L.: A Multi-agent Genetic Algorithm for Global Numerical Optimization. IEEE Trans. Sys. Man and Cyber. 34, 1128–1141 (2004)
17. Koumousis, V.K., Katsaras, C.P.: A Saw-Tooth Genetic Algorithm Combining the Effects of Variable Population Size and Reinitialization to Enhance Performance. IEEE Trans. Evol. Comput. 10, 19–28 (2006)

Improved Population-Based Incremental Learning in Continuous Spaces

Sujin Bureerat

Abstract. Population-based incremental learning (PBIL) is one of the well-established evolutionary algorithms (EAs). This method, although having outstanding search performance, has been somewhat overlooked compared to other popular EAs. Since the first version of PBIL, which is based on binary search space, several real code versions of PBIL have been introduced; nevertheless, they have been less popular than their binary code counterpart. In this paper, a population-based incremental learning algorithm dealing with real design variables is proposed. The method achieves optimization search with the use of a probability matrix, which is an extension of the probability vector used in binary PBIL. Three variants of the new real code PBIL are proposed while a comparative performance is conducted. The benchmark results show that the present PBIL algorithm outperforms both its binary versions and the previously developed continuous PBIL. The new methods are also compared with well-established and newly developed EAs and it is shown that the proposed real-code PBIL can rank among the high performance EAs.

Keywords: Population-based incremental learning, evolutionary algorithms, performance comparison, meta-heuristics, continuous domains.

1 Introduction

Evolutionary algorithms (EAs) have long been regarded as alternative optimizers, apart from classical mathematical programming. This variety of optimizers is well-known and popular for their robustness, derivative – free searching and capability in tackling global optimization. However, these methods have no guarantees of convergence and search consistency. Moreover, they usually require a large number of function evaluations to achieve optimum solutions. With such advantages and disadvantages, there have been numerous evolutionary algorithms and meta-heuristic (MH) searches developed with the aim to enhance search

Sujin Bureerat
Department of Mechanical Engineering, Faculty of Engineering,
Khon Kaen University, Thailand, 40002
e-mail: sujbur@kku.ac.th

A. Gaspar-Cunha et al. (Eds.): Soft Computing in Industrial Applications, AISC 96, pp. 77–86.
springerlink.com

performance whilst maintaining their outstanding abilities e.g. global optimization capability, derivative-free strategy, and robustness.

Methods, such as genetic algorithm (GA), are said to be well-established and have been implemented on a wide range of real world applications. PBIL, on the other hand, has been proposed as an alternative binary code approach to GA. The method has been somewhat overlooked although it has an acceptable search performance when compared to some popular EAs. In this paper, population-based incremental learning dealing with real design variables is proposed. The method achieves optimization search with the use of a probability matrix, which is an extension of a probability vector used in the binary PBIL. Three variants of the new real code PBIL are proposed while a comparative performance is performed. The benchmark results show that the present PBIL algorithm outperforms its binary versions and the previously developed continuous PBIL. The new methods are also compared with well-established EAs, and some other newly developed algorithms. It is shown that the proposed real-code PBIL is considered among the best EAs.

2 PBIL Algorithms

The first version of PBIL is based upon a binary search space [1]. It was proposed as an alternative EA apart from the best known GA. The basic idea of binary PBIL is to use the so-called probability vector to represent a set of design solutions or population rather than using a set of binary design solutions as with GA. During the searching process, the probability vector is improved iteratively until termination conditions are met. This search strategy can be viewed as limiting a binary search space iteratively until reaching the optimum solution. The real-code versions of PBIL have been developed later, e.g. $PBIL_I$ and $PBIL_G$ [2-3]. Moreover, since the introduction of binary PBIL, a number of estimation distribution algorithms have been proposed, such as the bivariate marginal distribution algorithm [4], and multivariate interactions (e.g. Bayesian optimization algorithm [5]). Multiobjective versions of PBIL have also been proposed [6] and implemented on a variety of engineering applications. The most outstanding feature of PBIL, when dealing with a multiobjective problem, is its ability to provide better population diversity. This means that an efficient, approximate, non-dominated set can be expected. The real code PBIL presented in this paper is developed for single objective optimization with box constraints, which can be expressed as:

$$\begin{aligned} &\text{Min} f(\mathbf{x}) \\ &L_i \leq x_i \leq U_i;\ i = 1 \ldots n \\ &\mathbf{x} \in R^n \end{aligned} \tag{1}$$

where $\mathbf{x}$ is the vector of design variables size $n\times1$, f is an objective function, L_i are lower bounds of $\mathbf{x}$, and U_i are upper bounds of $\mathbf{x}$.

For binary PBIL, the so-called probability vector, $\mathbf{P}$, is used to create a binary population. Given that $\mathbf{x}$ is represented by a binary string, $\mathbf{b}$, size $N_b \times 1$, the probability vector will have the same size as the binary design solution. The i-th element of the probability vector determines the number of '1' bits on the row of a binary population $\{\mathbf{b}_1 \ldots \mathbf{b}_M\}$. If the population size is $N_b \times M$, the i-th row of the population will have approximately $P_i M$ elements of '1' in which their positions are randomly located.

In continuous domains, it is more complicated to represent real number design solutions with a probability vector. From the previous work of continuous PBIL, PBIL_I and PBIL_G use a probability vector, whereas the histogram PBIL_H [3] proposes the use of a probability matrix. In this work, a probability matrix similar to that used in PBIL_H, is used; however, the search strategy and probability updating scheme are totally different from PBIL_H. Given that the probability matrix is sized $n \times T$, the range $[L_i, U_i]$ is thus divided in to T equal parts leading to a set of points $\mathbf{a}_i = \{a_0 \ldots a_T\}$ where $a_0 = L_i$ and $a_T = U_i$. Each element P_{ij} indicates a number of x_i in a current population, which have values in the j-th interval of $\mathbf{a}_i$. Initially, a matrix $\mathbf{A}$, size $n \times (T+1)$ whose i-th row contains $\mathbf{a}_i$, is created and used together with the probability matrix. The procedure to generate a real code population $\{\mathbf{x}_1 \ldots \mathbf{x}_M\}$ is detailed in Fig. 1 where $\lambda \in [0,1]$ is a uniform random number. The command lines 7-9 use a bias random generation so that y_l has the possibility to be located at the bounds of the sub-interval $[a_{i,j-1}, a_{ij}]$.

The main search procedure of the proposed real code PBIL starts with a probability matrix whose elements are set to be $1/T$. A real number population is then created using the algorithm in Fig. 1. Having evaluated the objective function values of all the members in the population, the best individual, $\mathbf{x}^{\text{best}}$, is detected and will be used for updating the probability matrix. The probability matrix updating scheme is given in Fig. 2. Note that mutation in the updating process is used to prevent premature convergence of the real-code PBIL. At the i-th loop, the r-th interval of $\mathbf{a}_i$ which contains x_i^{best} is detected. The concept of updating the i-th row of $\mathbf{P}$ is to increase the value of P_{ir}. The command lines 3-6 are proposed to prevent the premature convergence of the method. The command line 10 is activated so as to fulfill the condition

$$\sum_{j=1}^{T} P_{ij} = 1 . \tag{2}$$

rand $\in [0,1]$ is a uniform random number generated every time the command line 7 is activated. In this paper, the learning rate (L_{R0}), mutation probability, and mutation shift (M_s) are set to be 0.5, 0.02, and 0.2 respectively.

The probability matrix and the best individual are updated iteratively until the termination condition is met. The procedure of the real-code PBIL is given in Fig. 3 where N_G is the maximum number of generations assigned for the termination condition.

Input: P_{ij}, n, T, M, a_{ij} (**A**)
Output: $\mathbf{X} = \{\mathbf{x}_1, \ldots, \mathbf{x}_M\}$
For $i = 1$ to n
    $l = 0$
    For $j = 1$ to T
        $N_j = round(P_{ij} \cdot M)$
        For $k = 1$ to N_j
            Set $l = l + 1$
            Assign $y_l = 0.9a_{i,j-1} + \lambda(1.1a_{ij} - 0.9a_{i,j-1})$
            If $y_l < a_{i,j-1}$; set $y_l = a_{i,j-1}$
            If $y_l > a_{ij}$; set $y_l = a_{ij}$
            If $l = M$, break
        End
    End
    Set the i-th row of $\mathbf{X}$ = randomly_permutate(y_l), $l = 1, \ldots, M$
End

Fig. 1 Creating real code design solutions from a probability matrix

Input: P_{ij}^{old}, $\mathbf{x}^{\text{best}}$, a_{ij}, L_{R0}, mutation probability, M_s
Output: P_{ij}^{new}
For $i = 1$ to n
    Search for the r-th interval of $\mathbf{a}_i$, which contains x_i^{best}
    For $j = 1$ to T
        $L_R = L_{R0} \exp(-(j - r)^2)$
        $P'_{ij} = (1 - L_R)P_{ij}^{old} + L_R$
    End
    If $rand$ < mutation probability
    Select $t \in \{1, \ldots, T\}$ randomly
    $P'_{it} = (1 - M_s)P'_{it} + round(rand)M_s$
    Normalize **P**: $P''_{ij} = \frac{1}{\sum_{j=1}^{T} P'_{ij}} P'_{ij}$
    $P_{ij}^{new} = P''_{ij}$
End

Fig. 2 Probability matrix updating scheme

Input: N_G, M, n, T, L_{R0}, mutation probability, M_s, function name (*fun*)
Output: $\mathbf{x}^{best}$, f^{best}
Initialization: $P_{ij} = 1/T$, a_{ij}, $\mathbf{x}^{best} = \{\}$

For $i = 1$ to N_G
  Generate a real code population $\mathbf{X}$ from P_{ij}, a_{ij} (**Algorithm 1**) and find $\mathbf{f} = \text{fun}(\mathbf{X})$
  Find new $\mathbf{x}^{best}$ from $\mathbf{X}$ and $\mathbf{x}^{best}$ from the previous generation
  Update P_{ij} based on the current $\mathbf{x}^{best}$ (**Algorithm 2**)
End

Fig. 3 Real-Code Population-Based Incremental Learning

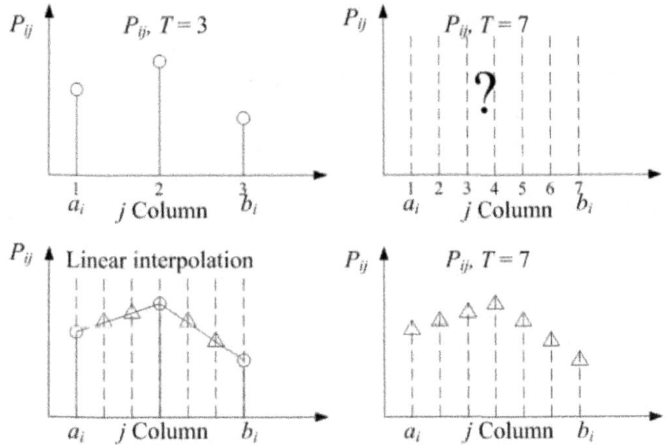

Fig. 4 Transformation of probability matrices: Δ larger T (fine), o smaller T (coarse)

One of the parameters that influence the search performance of the proposed continuous PBIL is the number of columns of the probability matrix denoted by T. The coarser probability matrix (with lower value of T) is suitable for global search whereas the finer probability matrix is suitable for local search. In this paper, three variants of real-code PBIL are proposed as follows:

RPBIL1 is a real-code PBIL using a constant size probability vector (default $T = 30$).

RPBIL2 is a real-code PBIL using two probability matrices with different sizes. The coarse probability matrix is used for global search while, at the same time, the finer probability matrix is used for local search (default $T_{\text{coarse}} = 10$, $T_{\text{fine}} = 30$). Each probability matrix is used to generate half of the solutions in a population.

RPBIL3 is a real-code PBIL using an adaptive probability vector. The strategy is to start with a coarser probability matrix at the early iterations. As the algorithm proceeds, the value of T becomes larger. The transformation from coarser to finer probability matrices can be carried out by using linear interpolation. Fig. 4

displays the transformation from a particular row of a coarser matrix ($T = 3$) to a row of a finer matrix ($T = 7$). The marker "o" represents the values of the coarser matrix elements whereas the marker "Δ" represent the values of the finer matrix elements. In this work, we use $T_{start} = 5$, and $T_{end} = 40$ where T reaches T_{end} at the half way point of an optimization run and afterwards remains constant until the search is terminated.

3 Performance Testing

Measurement of the search performance of the proposed PBILs is carried out by using a number of testing objective functions. The computation results obtained from using PBILs will be compared with those obtained from using some well-established and newly developed EAs. The implemented EAs are: binary-code genetic algorithm (GA) [7], stud-genetic algorithm (SGA) [8], real-code ant colony optimization (ACO) [9], continuous scatter search (SS) [10], particle swarm optimization (PSO) [11], charged system search (CSS) [12], fireworks algorithm (FA) [13], differential evolution (DE) [14], covariance matrix adaptation evolution strategy (CMA-ES) [15], binary PBIL [1], $PBIL_G$ [3], RPBIL1, RPBIL2, and RPBIL3. The $PBIL_H$, although being a predecessor closest to the proposed PBIL, will not be implemented because it uses all of the solutions in a population to update the probability matrix, which means it is time-consuming. Moreover, all the objective function values need to be positive or converted to be positive before entering an optimization process. This makes the algorithm difficult to use. The testing functions are detailed in Table 1. Note that function expressions may not be shown herein as we have limited paper length. The functions are taken from [9] and [10] where they are traceable from their names. For each test function, all evolutionary algorithms start with the same initial population and search for the optimum 30 simulation runs. The number of iterations is set to be $10n$ while the population size is set to be $7n$, which means the optimization procedures are terminated after $10n \times 7n$ function evaluations. It should be noted that, from observation, most of the algorithms can usually locate near optimum areas in a few early iterations. Later on, they struggle to hit the real optima. Therefore, the assigned generation number and population size used in this paper are adequate for EAs to locate near optimum regions of unconstrained optimization problems rather than to drive EAs to strict optimum points. The best results the optimizers can explore in the last generation are taken as optimum results. Several methods, such as SS and FA, due to their search mechanisms, may not use the same number of generations and population size; however, they are assigned to exploit the same number of function evaluations as the others.

Table 1 Testing functions

Function no., details	$[L,U]^n$	Function no., details	$[L,U]^n$
1, B_2 [9]	$[-50,100]^2$	18, Griewangk [9]	$[-5.12,5.12]^{10}$
2, Beale [10]	$[-4.5,4.5]^2$	19, *Perm* [10]	$[-15,15]^{15}$
3, Booth [10]	$[-10,10]^2$	20, *Perm*0 [10]	$[-15,15]^{15}$
4, Easom [9]	$[-100,100]^2$	21, Cigar [9]	$[-3,7]^{20}$
5, Goldstein & Price [9]	$[-2,2]^2$	22, Diagonal plane [9]	$[0.5,1.5]^{20}$
6, Martin & Gaddy [9]	$[-20,20]^2$	23, Dixon & Price [10]	$[-10,10]^{20}$
7, Matyas [10]	$[-5,10]^2$	24, Levy(n) [10]	$[-10,10]^{20}$
8, Penny & Linfield* [7]	$[-5,5]^2$	25, Powell(n) [10]	$[-4,5]^{20}$
9, Powersum [10]	$[0,2]^2$	26, Rastrigin(n) [9]	$[-2.56,5.12]^{20}$
10, Branin [9]	$[-5,15]^2$	27, Rosenbrock [9]	$[-5,5]^{20}$
11, Shubert [10]	$[-10,10]^2$	28, Sum Squares [10]	$[-5,10]^{20}$
12, Six Hump Camel Back [10]	$[-5,10]^2$	29, Schwefel [10]	$[-500,500]^{20}$
13, Colville [10]	$[-10,10]^4$	30, Trid(n) [10]	$[-400,400]^{20}$
14, Hartmann 3,4 [9]	$[0,4]^4$	31, Ackley(n) [10]	$[-15,30]^{30}$
15, Shekel [9]	$[0,10]^4$	32, Ellipsoid [9]	$[-3,7]^{30}$
16, Zakharov [9]	$[-5,10]^5$	33, Plane [9]	$[-0.5,1.5]^{30}$
17, Hartmann 6,4 [9]	$[0,6]^6$	34, Sphere [9]	$[-3,7]^{30}$
		35, Tablet [9]	$[-3,7]^{30}$

* $f_8(\mathbf{x}) = 0.5(x_1^4 - 16x_1^2 + 5x_1) + 0.5(x_2^4 - 16x_2^2 + 5x_2)$

4 Comparative Results

Having performed 14 optimizers to solve 35 unconstrained optimization problems, the comparative results are given in Table 2. Each value in the table is the average of 30 near optimum values each method can search for. The average values are normalized with respect to the best and the worst methods, which can be expressed as:

$$\bar{f}_i = \left| \frac{f_i - f_{\min}}{f_{\max} - f_{\min}} \right| \tag{3}$$

where $f_{\min}$ is the average near optimum value of the best method, and $f_{\max}$ is the average near optimum value of the worst method. By using Eq. 3, the best method will have $\bar{f} = 0$ whereas the worst method will have $\bar{f} = 1$.

From the overall comparison results, the five best methods are RPBIL3, ACO, FA, GA, and DE. For the functions F_{20} and F_{33}, most of the optimizers can reach the real optima. Between the top two optimizers, RPBIL3 is better when the design variables vectors have smaller size while ACO is slightly superior when dealing with design variables with larger scale. However, the method that produces the best results for most of the optimization problems is CMA-ES. Among the PBIL variants, RPBIL3 is the best while the second best is RPBIL1. The third best is the original binary PBIL whereas the worst is RPBIL2. Based on the standard deviations of the near optimum values of the various optimizers which are not shown in the paper, all of the algorithms do not have an acceptable search consistency. This could be due to the low number of generations and small population size being used.

Table 2 Comparative results by normalized function values

Fn	ACO	CSS	FA	GA	PSO	SGA	SS	DE	ES	PB	PG	PR1	PR2	PR3
1	0	0.007	0.016	0.507	0	1	0.018	0.01	0	0.429	0.032	0	0.043	0
2	0.027	1	0.395	0.189	0	0.425	0.175	0.467	0.17	0.298	0.52	0.1	0.121	0.085
3	0.004	0.048	0.021	0.367	0	1	0.135	0.159	0	0.214	0.066	0.023	0.575	0
4	0.977	1	1	0.622	0.522	0.999	1	0.993	1	0.886	0.559	1	1	0
5	0	0.712	0.036	0.358	0.181	0.92	0.527	0.242	0.333	0.295	0.084	0.11	1	0.108
6	0.001	0.023	0.02	0.174	0.002	1	0.073	0.066	0	0.589	0.018	0.004	0.315	0
7	0.034	0.067	0.085	0.375	0.02	0.585	0.515	0.228	0	1	0.926	0.073	0.146	0.001
8	0.038	0.16	0.336	0.091	1	0.193	0.092	0.031	0.545	0.128	0.032	0.71	0.007	0
9	0.009	0.009	0.063	0.191	0.086	1	0.082	0.096	0	0.571	0.073	0.009	0.456	0.009
10	0.382	0.018	0.042	0.135	0.285	1	0.15	0.368	0.603	0.319	0.288	0.019	0.366	0
11	0.459	0.295	0.103	0.194	0.265	0.199	0.506	0.416	1	0.126	0.244	0.123	0.07	0
12	0.025	0.151	0.083	0.156	0.173	0.391	0.255	0.09	0	0.196	0.094	1	0.144	0.027
13	0.077	0.062	0.118	0.083	0.052	0.075	0.078	0.211	0	0.082	1	0.045	0.13	0.056
14	0	1	0.004	0.036	0.336	0.37	0.051	0.001	0.129	0.091	0.168	0.144	0.221	0.011
15	0	0.098	0.867	0.399	0.542	1	0.186	0.687	0.439	0.962	0.822	0.661	0.342	0.027
16	0.026	0.007	0.02	0.473	0.262	1	0.188	0.232	0	0.742	0.262	0	0.635	0.005
17	0	1	0.102	0.194	0.904	0.312	0.524	0.047	0.911	0.328	0.192	0.263	0.102	0.26
18	0.502	0	0.099	0.062	0.083	0.103	0.01	0.589	1	0.044	0.317	0.058	0.715	0.028
19	0	0.447	0.585	0.032	0.8	0	0.24	0	0.201	0.059	0.341	1	0.3	0.956
20	0	0	0	0	0	0.172	0	0.001	0	0	0	0	1	0
21	0	0	0.005	0.054	0.04	0.1	0.032	0.006	0	0.186	0.737	0.01	1	0.007
22	0	0.015	0	0.001	1	0	0.057	0	0	0.022	0	0.057	0.035	0.053
23	0	0	0	0.001	0.003	0.001	0	0.001	0	0.003	0.041	0	1	0
24	0.007	0	0.31	0.013	0.16	0.112	0.012	0.042	0.005	0.119	0.209	0.085	1	0.091
25	0.003	0	0.001	0.016	0.021	0.047	0.002	0.007	0	0.057	0.152	0.006	1	0.003
26	0.886	0.259	0.371	0	0.319	0.26	0.352	0.462	0.087	0.114	1	0.615	0.678	0.424
27	0.003	0.002	0.002	0.005	0.021	0.008	0.003	0.017	0	0.017	0.088	0.014	1	0.005
28	0	0	0.006	0.052	0.218	0.308	0.04	0.012	0	0.245	1	0.012	0.783	0.01
29	0.955	0.109	0.394	0	0.716	0.031	0.789	0.402	0.63	0.107	0.832	0.603	1	0.435
30	0.124	0.289	0	0.384	1	0.512	0.619	0.071	0.153	0.455	0.897	0.442	0.644	0.329
31	0.202	0.109	0.187	0.363	0.663	0.646	0.447	0.359	0	0.543	1	0.287	0.941	0.271
32	0.002	0	0.016	0.047	0.354	0.118	0.056	0.014	0	0.172	1	0.046	0.552	0.018
33	0	0	0	0	0	0	0	0	0	0	0	0	0	0
34	0.003	0	0.004	0.045	0.405	0.457	0.079	0.026	0	0.136	1	0.019	0.702	0.011
35	0	0	0.001	0.018	0.932	0.125	0.024	0.004	0	0.053	0.448	1	0.423	0.722
total	4.746	6.887	5.292	5.637	11.36	14.46	7.317	6.357	7.206	9.588	14.44	8.538	18.44	3.952

* PB = binary PBIL, PG = PBILG, PR1 = RPBIL1, PR2 = RPBIL2, PR3 = RPBIL3, ES = CMA-ES.

Table 3 provides an alternative comparison which is based on a ranking approach. For each testing function, the 14 optimizers are ranked based on their average values in Table 2, where the better methods have smaller scores. Based on the total scores, the top five optimizers are slightly different from the first comparison, which can be ranked in order as ACO, CMA-ES, RPBIL3, CSS, and FA. CMA-ES produces the most number one rankings. Other conclusions are somewhat similar to the comparison in Table 2.

Table 3 Comparative results by ranking

Fn No.	ACO	CSS	FA	GA	PSO	SGA	SS	DE	ES	PB	PG	PR1	PR2	PR3
1	1	6	8	13	1	14	9	7	1	12	10	1	11	1
2	2	14	10	8	1	11	7	12	6	9	13	4	5	3
3	4	7	5	12	1	14	9	10	1	11	8	6	13	1
4	6	9	9	4	2	8	9	7	9	5	3	9	9	1
5	1	12	2	10	6	13	11	7	9	8	3	5	14	4
6	3	8	7	11	4	14	10	9	1	13	6	5	12	1
7	4	5	7	10	3	12	11	9	1	14	13	6	8	2
8	5	9	11	6	14	10	7	3	12	8	4	13	2	1
9	2	2	6	11	9	14	8	10	1	13	7	2	12	2
10	12	2	4	5	7	14	6	11	13	9	8	3	10	1
11	12	10	3	6	9	7	13	11	14	5	8	4	2	1
12	2	8	4	9	10	13	12	5	1	11	6	14	7	3
13	7	5	11	10	3	6	8	13	1	9	14	2	12	4
14	1	14	3	5	12	13	6	2	8	7	10	9	11	4
15	1	3	12	6	8	14	4	10	7	13	11	9	5	2
16	6	4	5	11	9	14	7	8	1	13	9	1	12	3
17	1	14	3	6	12	9	11	2	13	10	5	8	3	7
18	11	1	8	6	7	9	2	12	14	4	10	5	13	3
19	1	10	11	4	12	1	7	1	6	5	9	14	8	13
20	1	1	1	1	1	13	1	12	1	1	1	1	14	1
21	1	1	4	10	9	11	8	5	1	12	13	7	14	6
22	1	8	1	7	14	1	12	1	1	9	1	12	10	11
23	1	1	1	8	11	8	1	8	1	11	13	1	14	1
24	3	1	13	5	11	9	4	6	2	10	12	7	14	8
25	5	1	3	9	10	11	4	8	1	12	13	7	14	5
26	13	4	8	1	6	5	7	10	2	3	14	11	12	9
27	4	2	2	6	12	8	4	10	1	10	13	9	14	6
28	1	1	4	9	10	12	8	6	1	11	14	6	13	5
29	13	4	5	1	10	2	11	6	9	3	12	8	14	7
30	3	5	1	7	14	10	11	2	4	9	13	8	12	6
31	4	2	3	8	12	11	9	7	1	10	14	6	13	5
32	3	1	5	8	12	10	9	4	1	11	14	7	13	6
33	1	1	1	1	1	1	1	1	1	1	1	1	1	1
34	3	1	4	8	11	12	9	7	1	10	14	6	13	5
35	1	1	4	6	13	9	7	5	1	8	11	14	10	12
total	140	178	189	248	287	343	263	247	148	310	330	231	364	151

5 Conclusions and Discussion

Three versions of new real-code population-based incremental learning are proposed. The methods use a probability matrix to represent a set of real design variables during the search procedure. The comparative results show that the proposed real-code PBIL is comparable to high performance evolutionary algorithms. It should be noted that a more reliable comparison should be made based on a statistical approach rather than by comparing the average objective values directly. Since RPBIL3 outperforms the other PBIL versions, it can be concluded that the probability matrix should start with a low resolution at early iterations. This will help the PBIL to perform a global search. The probability matrix resolution becomes higher in the later iterations so as to make a fine search for the real optimum. For the other versions,

RPBIL1, which uses one fine probability matrix, is superior to RPBIL2, which uses both coarse and fine probability matrices at the same time. This result is due to use of the finer probability matrix tends to produce a better search performance than the use of the coarser one. Since RPBIL2 uses the coarser probability matrix to produce half of the solution members throughout the simulation run, it therefore has an inferior performance when compared to the other two RPBIL variants. The proposed real-code PBIL, with adaptive probability matrix, is an algorithm that performs among the best EAs. It is possible to enhance its search performance by hybridizing it with some other efficient evolutionary operators.

Acknowledgments. The corresponding author is grateful of the support from the Thailand Research Fund (TRF). Many thanks are also directed to my colleague, Peter Warr, for making this paper more readable.

References

1. Baluja, S.: Population-Based Incremental Learning: a Method for Integrating Genetic Search Based Function Optimization and Competitive Learning. Technical Report CMU_CS_95_163, Carnegie Mellon University (1994)
2. Sebag, M., Ducoulombier, A.: Extending Population-Based Incremental Learning to Continuous Search Spaces. In: Eiben, A.E., Bäck, T., Schoenauer, M., Schwefel, H.-P. (eds.) PPSN 1998. LNCS, vol. 1498, pp. 418–427. Springer, Heidelberg (1998)
3. Yuan, B., Gallagher, M.: Playing in Continuous Spaces: Some Analysis and Extension of Population-Based Incremental Learning. In: CEC 2003, CA, USA, pp. 443–450 (2003)
4. Pelikan, M., Muhlenbein, H.: The Bivariate Marginal Distribution Algorithm. In: Advances in Soft Computing - Engineering Design and Manufacturing, pp. 521–535 (1999)
5. Pelikan, M., Goldberg, D.E., Cantú-Paz, E.: BOA: The Bayesian Optimization Algorithm. In: Proceedings of the Genetic and Evolutionary Computation Conference (GECCO 1999), vol. I, pp. 525–532 (1999)
6. Bureerat, S., Sriworamas, K.: Population-Based Incremental Learning for Multiobjective Optimisation. AISC, vol. 39, pp. 223–232 (2007)
7. Lindfield, G., Penny, J.: Numerical Methods Using MATLAB. Ellis Horwood (1995)
8. Khatib, W., Fleming, P.: The Stud GA: A Mini-Revolution. In: 5th Int. Conf. on Parallel Problem Solving From Nature (1998)
9. Socha, K., Dorigo, M.: Ant Colony Optimization for Continuous Domains. European Journal of Operational Research 185, 1155–1173 (2008)
10. Herrera, F., Lozano, M., Molona, D.: Continuous Scatter Search: An Analysis of the Integration of Some Combination Methods and Improvement Strategies. European Journal of Operational Research 169, 450–476 (2006)
11. Reyes-Sierra, M., Coello Coello, C.A.: Multi-objective Particle Swarm Optimizers: a Survey of the State-of-the-Art. Int. J. of Computational Intelligence Research 2(3), 287–308 (2006)
12. Kaveh, A., Talatahari, S.: A Novel Heuristic Optimization Method: Charged System Search. Acta Mechanica 213(3-4), 267–289 (2010)
13. Tan, Y., Zhu, Y.: Fireworks algorithm for optimization. In: Tan, Y., Shi, Y., Tan, K.C. (eds.) ICSI 2010. LNCS, vol. 6145, pp. 355–364. Springer, Heidelberg (2010)
14. Storn, R., Price, K.: Differential Evolution – A Simple and Efficient Heuristic for Global Optimization over Continuous Spaces. Journal of Global Optimization 11, 341–359 (1997)
15. Hansen, N., Müller, S.D., Koumoutsakos, P.: Reducing the Time Complexity of the Derandomized Evolution Strategy with Covariance Matrix Adaptation (CMA-ES). Evolutionary Computation 11(1), 1–18 (2003)

Particle Swarm Optimization in the EDAs Framework

Valentino Santucci and Alfredo Milani

Abstract. Particle Swarm Optimization (PSO) is a popular optimization technique based on swarm intelligence concepts. Estimation of Distribution Algorithms (EDAs) are a relatively new class of evolutionary algorithms which build a probabilistic model of the population dynamics and use this model to sample new individuals. Recently, the hybridization of PSO and EDAs is emerged as a new research trend. In this paper, we introduce a new hybrid approach that uses a mixture of Gaussian distributions. The obtained algorithm, called PSEDA, can be seen as an implementation of the PSO behaviour in the EDAs framework. Experiments on well known benchmark functions have been held and the performances of PSEDA are compared with those of classical PSO.

1 Introduction and Related Work

Particle swarm optimization (PSO) [9] is a population-based stochastic approach mainly used for solving continuous optimization problems (although some attempts to handle combinatorial problems have been proposed [8]). Inspired by the behavioural model of bird flocking, a population of particles (the candidate solutions) move in the search space at the quest of the optimum point.

Estimation of Distribution Algorithms (EDAs) [10] are a class of evolutionary algorithms (EAs) where a population of candidate solutions is sampled from a probabilistic model learned from the previous generations. Hence, conversely of classical

Valentino Santucci
Department of Mathematics and Computer Science, University of Perugia,
via Vanvitelli 1, 06123 Perugia, Italy
e-mail: valentino.santucci@dmi.unipg.it

Alfredo Milani
Department of Mathematics and Computer Science, University of Perugia,
via Vanvitelli 1, 06123 Perugia, Italy
e-mail: milani@unipg.it

A. Gaspar-Cunha et al. (Eds.): Soft Computing in Industrial Applications, AISC 96, pp. 87–96.
springerlink.com

EAs, EDAs does not use variation operators (like crossover or mutation), but rely on purely probabilistic methods. Although mainly used for discrete problems, some continuous EDAs applications have been proposed [3].

Recently, the hybridization of PSO and EDAs is emerged as a new research trend. A significant work to mention is EDA-PSO [17] where the next generation population is built using PSO or EDA method on the basis of a continuously updated probability. Instead, in [2] the population is splitted into chunks, hence the intra-chunk update follows PSO rules, while the global update is based on an EDA scheme. Another approach is EDPSO [7], which also relies on an Ant Colony Optimization (ACO) variant for continuous domains [14]. Finally, we mention PSO_Bounds [6], which uses an EDA scheme to manipulate the allowable bounds for the PSO particles.

Unlike the various PSO-EDA hybridization proposals, our algorithm (PSEDA) can be seen as a PSO implementation in the EDAs framework. The genotype of a PSEDA individual shares various properties with that of a PSO particle. Furthermore, PSEDA and PSO base their behavioural dynamics on the same attractive points. The notable difference with respect to PSO is the absence of velocities, while the main innovation with respect to usual EDAs is the learning of an independent probability distribution model for each individual.

The rest of the paper is organized as follows. A brief background about PSO and EDAs are covered in Sections 2 and 3. The proposed PSEDA is explained in Section 4, while some experimental results about it are provided in Section 5. Conclusions are drawn in Section 6 where some considerations and future lines of research are reported.

2 Particle Swarm Optimization

PSO denotes a class of metaheuristics introduced by Kennedy and Eberhart [9] where the algorithm scheme is inspired by the collective behaviour of flocks of birds.

A PSO swarm is composed by a set of n particles $P = \{p_1, p_2, \ldots, p_n\}$ interconnected in a graph that defines a neighborhood relation among them, i.e. for each particle p_i its neighborhood set $N_i \subseteq P$ is defined (in the following we assume, as it is in general, that $p_i \in N_i$). The position of a particle represents a candidate solution of the given optimization problem represented by the objective/fitness function $f : \Theta \to \mathbb{R}$ with $\Theta \subseteq \mathbb{R}^d$ (the region of feasible solutions) to be minimized (or maximized). Each particle, as time passes, adjusts its position, i.e. it explores search space, according to its own experience as well as the experience of its neighbours. In this way PSO combines cognitive and social strategies in order to focus the search of the swarm toward the most promising areas.

At any time step t (with $t \in \mathbb{N}$) the *genotype* of a particle p_i is formed by the following d-dimensional vectors (where d is the dimensionality of the search space):

- $x_{i,t}$, i.e. the particle position,
- $v_{i,t}$, i.e. the particle velocity,

- $b_{i,t}$, i.e. the particle personal best position ever visited until time step t,
- $l_{i,t}$, i.e. the best position ever found among its neighbours until time step t.

In the more usual PSO implementation, a complete network is adopted as neighborhood graph. In this case the $l_{i,t}$ values are replaced by a global g_t that is the same for all the particles thus improving the efficiency of the algorithm. We will call this implementation *global PSO* (gPSO).

PSO starts by assigning random positions within the feasible region Θ. Furthermore, velocities are usually initialized to small random values in order to prevent particles from leaving the search space in the early iterations.

During a main loop, velocities and positions are iteratively updated until a stop criterion is met (e.g. a given amount of fitness evaluations has been performed). The update rules are:

$$v_{i,t+1} = \omega v_{i,t} + \varphi_1 r_{1,t}(b_{i,t} - x_{i,t}) + \varphi_2 r_{2,t}(l_{i,t} - x_{i,t}) \tag{1}$$

$$x_{i,t+1} = x_{i,t} + v_{i,t+1} \tag{2}$$

Weights in (1) respectively represent the inertia ω and the acceleration factors φ_1 and φ_2. Instead, $r_{1,t}$ and $r_{2,t}$ are two random numbers uniformly distributed in $[0,1]$. The three terms of the velocity update rule (1) characterize the behaviour of the particles. The first term, called *inertia* or momentum, serves as a memory of the previous flight trajectory and prevents a particle from drastically changing direction. The second term is the *cognitive component* that models the tendency of the particles to return to the previously found personal best position. Finally, the third term represents the *social component* and quantifies the velocity contribution of the neighbours (or of the entire swarm in the gPSO case).

The (personal and social) best positions are updated whenever there is an improvement in fitness. More formally (in the case of a minimization problem):

$$b_{i,t+1} = \begin{cases} x_{i,t+1} & \text{if } f(x_{i,t+1}) \leq f(b_{i,t}) \\ b_{i,t} & \text{otherwise} \end{cases} \tag{3}$$

$$l_{i,t+1} = \arg \min_{j:p_j \in N_i} f(b_{j,t+1}) \tag{4}$$

In the gPSO case the assignment (4) becomes:

$$g_{t+1} = \arg \min_{1 \leq i \leq n} f(b_{i,t+1}) \tag{5}$$

Sometimes the particles can exceed the bounds of the feasible search space Θ. There are several solutions to this problem [13], but generally the preferred one is to randomly reset an out-of-bounds component between the previous position and the bound exceeded according to:

$$x_{i,t+1,k} = \begin{cases} x_{i,t,k} + r(u_k - x_{i,t,k}) & \text{if } x_{i,t+1,k} > u_k \\ x_{i,t,k} - r(x_{i,t,k} - l_k) & \text{if } x_{i,t+1,k} < l_k \end{cases} \tag{6}$$

where r is a random number in $[0,1]$, k is the index of the exceeding component and l_k, u_k are, respectively, the lower and upper bounds for this component. Conversely of a completely random restart, in this way the original direction of the particle is somehow preserved.

Finally, note that a lot of studies on the PSO parameters tuning have been done. However, in [4] it has been proved that a good general setting, able to allow the swarm convergence without the need of bounds for the velocity components, is the following: $\omega = 0.7298$, $\varphi_1 = \varphi_2 = 1.49618$.

3 Estimation of Distribution Algorithms

Estimation of Distribution Algorithms (EDAs) [10] are a recently new scheme of evolutionary algorithms (EAs). The main difference with the other EAs is that EDAs rely on a probabilistic model and do not use variation operators such as crossover or mutation. They select some solutions from the current population and perform a learning procedure on them. In this learning phase EDAs build a *probability distribution model* (PDM) basing on some selected solutions. Successively, the solutions for the next generation are sampled from that model.

More precisely, EDAs perform the following steps:

0. Randomly generate n initial solutions.
1. Select $m \leq n$ solutions from the current population according to a selection method.
2. Build a probability model PDM basing on the m selected solutions.
3. Replace $q \leq n$ members of the current population by new solutions sampled from the probability model PDM.
4. If a stop criterion is not met go back to step 1.

The selection method in step 1 usually chooses the best m solutions basing on their fitness, although sometimes a stochastic method (like a fitness proportional selection) is preferred in order to maintain a certain degree of diversity. Instead, the replacement method in step 3 generally substitutes the q worst members, so the EDA presents an elitist behaviour[1].

The model built in step 2 generally assumes the form of a set of probability distributions, one for each dimension of the search space. For example, in the case that the genotype is a bitstring of length d, the PDM could consist in a vector of d probability values p_i (with $1 \leq i \leq d$) where each p_i is the probability that the ith bit is a 1. Note that sometimes the PDM is a multivariate probability distribution.

Finally, we report that, although EDAs have been originally introduced to tackle combinatorial optimization problems, recently, numerical applications have also been proposed [3].

[1] Note that there is no elitism if $q = n$.

4 Particle Swarm Estimation of Distribution Algorithm

Particle Swarm Estimation of Distribution Algorithm (PSEDA), more than an hybridization between PSO and EDAs, can be seen as a PSO implementation in the EDAs framework. The notable difference with respect to PSO is the absence of velocities, while the main innovation with respect to usual EDAs is the learning of an independent probability distribution model (PDM) for each individual (other than for each dimension).

In order to understand how the PSO position update process is simulated in PSEDA, we first analyze the PSO rules (1) and (2). In (2) the new particle position depends on its current position and on its velocity, which in turn is a function of the previous velocity, the personal best position, the social best position, and, again, the current position (see rule (1)). Summarizing and simplifying, we can say that $x_{i,t+1}$ depends on $x_{i,t}$, $b_{i,t}$, and $l_{i,t}$. Note that this is a coarse simplification since we have omitted the previous velocity value $v_{i,t}$, however, this component is somehow reintroduced later in the probability model.

Before explaining the PSEDA learning procedure, we note that, as usual in continuous optimization, the objective function f is provided with its feasible region $\Theta = [l_k, u_k]^d \subseteq \mathbb{R}^d$ (with $1 \leq k \leq d$).

As aforementioned, in PSEDA, a PDM is independently built for each one of the n population individuals. Let d be the dimensionality of the search space, then each PDM is modeled by d mixtures of (univariate) probability distributions (i.e. a mixture for each dimension of the problem). The probability distributions composing the mixture $M_{i,t,k}$ of the kth dimension of individual i at time step t are:

1. $TN^x_{i,t,k}(x_{i,t,k}, \sigma_t; l_k, u_k)$, i.e. a Gaussian distribution, with mean $x_{i,t,k}$ and standard deviation σ_t, truncated in $[l_k, u_k]$,
2. $TN^b_{i,t,k}(b_{i,t,k}, \sigma_t; l_k, u_k)$, i.e. a Gaussian distribution, with mean $b_{i,t,k}$ and standard deviation σ_t, truncated in $[l_k, u_k]$,
3. $TN^l_{i,t,k}(l_{i,t,k}, \sigma_t; l_k, u_k)$, i.e. a Gaussian distribution, with mean $l_{i,t,k}$ and standard deviation σ_t, truncated in $[l_k, u_k]$,
4. $U_k(l_k, u_k)$, i.e. a continuous uniform distribution in $[l_k, u_k]$,
5. $\overline{M}_{i,t-1,k}(x_{i,t-1,k}, b_{i,t-1,k}, l_{i,t-1,k}; \sigma_t; l_k, u_k)$, i.e. a relaxed version of the previous mixture distribution for the same dimension of the same particle. This mixture is composed by the three truncated (in $[l_k, u_k]$) Gaussians with means $x_{i,t-1,k}$, $b_{i,t-1,k}$, $l_{i,t-1,k}$ and standard deviation σ_t.

At the end of every generation, once built the PDMs, a new individual position is sampled from the mixture distributions that form its PDM.

In this way, the three truncated Gaussians (TN) ensure that the attractive points of PSO ($x_{i,t}$, $b_{i,t}$, $l_{i,t}$) are attractive also for PSEDA (since they are the means of the Gaussians). The uniform component of the mixture (U) is introduced in order to regulate the exploration degree of PSEDA. Finally, the last component of the mixture $\overline{M}_{i,t-1,k}$, a relaxed version of the previous mixture $M_{i,t-1,k}$, plays the role of the previous velocity in PSO rule (1).

In each one of the (truncated) Gaussian distributions, at time step t, a common standard deviation σ_t is used. This parameter, although in a different way from the uniform distribution U_k, is significant for the exploration/exploitation balance of the algorithm. In order to match the more common EAs best practices [1], σ_t should be relatively large in early generations (favoring exploration) and relatively small in later generations (favoring exploitation). For this reason we have decided to shade σ_t from a relatively great to a relatively small value with the passing of generations. After some preliminary experiments we have decided to shade σ_t from $1/10$ to $1/10^6$ of the dimension length $(u_k - l_k)$. More formally:

$$\sigma_t = \frac{u_k - l_k}{t\frac{10^6-10}{G} + 10} \tag{7}$$

where G is the maximum number of generations allowed in a computation.

As well known, the probability density function (*pdf*) of a mixture distribution is a convex combination of the *pdf*s of its component distributions [16]. Let $g_x(z)$, $g_b(z)$, $g_l(z)$, $g_u(z)$, $g_m(z)$ be the *pdf*s of the five component distributions (note that, in order to have a clear notation, we have omitted some indices), then the *pdf* $m(z)$ of the mixture distribution $M_{i,t,k}$ is:

$$m(z) = w_x g_x(z) + w_b g_b(z) + w_l g_l(z) + w_u g_u(z) + w_m g_m(z) \tag{8}$$

where the weights w_x, w_b, w_l, w_u, w_m are non-negative values[2] that sum up to 1. Obviously, these values allow to regulate the contribution of the different component distributions. Hence we have five weights against the three of PSO. However, as illustrated in Section 5, we can set them proportionally with respect to the PSO parameters (i.e.: ω, φ_1, φ_2) that we would have chosen for the given problem. In this way, the PSO concepts of inertia, cognitive acceleration and social acceleration are even more preserved.

Summarizing, the genotype of a PSEDA individual is composed, other than by the current position $x_{i,t}$, the personal best $b_{i,t}$, and the social best $l_{i,t}$, also by the same values at the previous generation, i.e. $x_{i,t-1}$, $b_{i,t-1}$, $l_{i,t-1}$. Indeed, these are needed for computing the distribution[3] $\overline{M}_{i,t-1,k}$. The complete algorithm for PSEDA is shown in Algorithm 1.

In PSEDA, exactly as in PSO, in order to have the social best values, a neighborhood relation among the individuals must be defined. In the case that a complete network is adopted as neighborhood graph, the $l_{i,t}$ values can be replaced by a global g_t thus improving the space/time efficiency of the algorithm. Similarly as in PSO, we call this implementation gPSEDA.

Finally, note that a mixture distribution can be easily sampled as explained in [16], while a truncated Gaussian can be sampled either using the accept-reject method (more suitable when σ_t is small) or the inverse cumulative function method

[2] Note that, for sampling $\overline{M}_{i,t,k}$, only the weights w_x, w_b, w_l are needed. Hence, in this case, they are temporarily normalized in order to sum up to 1.

[3] This genotype richness ensures an elitist behaviour at the algorithm.

Algorithm 1 PSEDA

procedure PSEDA
  Randomly generate n initial solutions and evaluate them
  $t \leftarrow 1$
  while $t \leq max\_t$ **do**
    for all individuals i **do**
      for all dimensions k **do**
        Learn the mixture distribution $M_{i,t,k}$
      end for
      Store previous positions $x_{i,t-1}$, $b_{i,t-1}$, $l_{i,t-1}$
      for all dimensions k **do**
        Sample $x_{i,t,k}$ from $M_{i,t,k}$
      end for
      Evaluate individual i
      Update personal best $b_{i,t}$
    end for
    Update social best $l_{i,t}$s
    Update standard deviation σ_t
    $t \leftarrow t+1$
  end while
end procedure

(more suitable when σ_t is large) [5]. With this approach, out-of-bounds individuals become no more possible, hence, repair procedures, as in PSO, are not needed.

5 Experiments

The performances of PSEDA have been evaluated on a suite of eight well known benchmark functions. These functions are non-noisy, non-translated, non-rotated versions of those proposed in [15] for the real-parameters optimization competition at CEC 2005. They are: sphere, ellipsoid, Rosenbrock, Griewank, Ackley, Rastrigin, Weierstrass, and extended Schaffer. This suite presents a variegate combination for what regards the properties of modality and separability of the composing functions.

Each benchmark is investigated with dimensionality $d = 10$. An execution is regarded convergent if $f(x) - f(x^{opt}) < \varepsilon$, on the other hand, the execution will be aborted if the number of fitness evaluations (NFES) exceeds the allowed cap of $300,000$. In Table 1, for each benchmark, its function definition, the optimization interval and the ε value used for it are reported.

The performances of a fully connected PSEDA (gPSEDA) are compared with those of a classical fully connected PSO (gPSO). The gPSO parameters are those widely used in literature and suggested in [4], i.e.: $\omega = 0.7298$, $\varphi_1 = \varphi_2 = 1.49618$. Also gPSEDA has some parameters to be set. After some preliminary tests, and inspired from the usual values for the mutation rate parameter of classical genetic algorithms [11] (it regulates the exploration degree as w_u), we have set the exploration weight w_u to a small value, i.e. $w_u = 0.05$. Instead, since the other parameters

Table 1 Experiments Setup

Function Definition	Opt. Interval	ε
$f_1(x)=\sum_{i=1}^{d} x_i^2$	$[-100,100]^d$	10^{-6}
$f_2(x)=\sum_{i=1}^{d}(10^6)^{\frac{i-1}{d-1}} x_i^2$	$[-100,100]^d$	0.5
$f_3(x)=\sum_{i=1}^{d-1}\left(100(x_i^2-x_i+1)^2+(x_i-1)^2\right)$	$[-100,100]^d$	10^{-6}
$f_4(x)=\sum_{i=1}^{d} x_i^2/4000-\prod_{i=1}^{d}\cos(x_i/\sqrt{i})+1$	$[-300,300]^d$	0.5
$f_5(x)=-20\exp\left(-0.2\sqrt{\frac{1}{d}\sum_{i=1}^{d} x_i^2}\right)-\exp\left(\frac{1}{d}\sum_{i=1}^{d}\cos(2\pi x_i)\right)+20+e$	$[-100,100]^d$	10^{-2}
$f_6(x)=\sum_{i=1}^{d}(x_i^2-10\cos(2\pi x_i)+10)$	$[-5,5]^d$	10^{-2}
$f_7(x)=\sum_{i=1}^{d}\sum_{k=0}^{20} 0.5^k\cos(2\pi\cdot 3^k\cdot(x_i+0.5))-d\sum_{k=0}^{20} 0.5^k\cos(2\pi\cdot 3^k\cdot 0.5)$	$[-0.5,0.5]^d$	0.5
$f_8(x)=\sum_{i=1}^{n-1} S(x_i,x_{i+1})+S(x_n,x_1)$, where $S(x,y)=0.5+\frac{\sin^2(\sqrt{x^2+y^2})-0.5}{(1+0.001(x^2+y^2))^2}$	$[-100,100]^d$	0.5

have a similar meaning of those of PSO, we have decided to set them proportionally with respect to those used for gPSO and reported above. Let $s=\omega+\varphi_1+\varphi_2$, then $w_x=(1-w_u)\cdot 0.5\cdot\frac{\omega}{s}=0.09313$, $w_b=(1-w_u)\cdot\frac{\varphi_1}{s}=0.38187$, $w_l=(1-w_u)\cdot\frac{\varphi_2}{s}=0.38187$, $w_m=(1-w_u)\cdot 0.5\cdot\frac{\omega}{s}=0.09313$. Other than consistent with the usual PSO parameters setting, our choice is resulted better than other settings that we have tried in some preliminary experiments (although more systematic tests are still needed).

In order to eliminate the randomness of the statistical results, for each benchmark 30 executions have been held. In each experiment, the convergence probability P_c (i.e. the number of convergent executions above the total number of executions) and the average NFES of all convergent executions C are recorded. These two indices are also synthesized in the quality measure $Q_m=C/P_c$ suggested in [15].

Table 2 Experimental Results

	gPSEDA			gPSO		
Benchmark	C	P_c	Q_m	C	P_c	Q_m
f_1	66788	1.00	66788	4227	1.00	**4227**
f_2	71651	1.00	71651	5631	0.93	**6035**
f_3	-	0.00	-	214373	0.10	**2143730**
f_4	1787	1.00	**1787**	2507	0.97	2585
f_5	5447	1.00	5447	2527	1.00	**2527**
f_6	8397	1.00	**8397**	-	0.00	-
f_7	8263	1.00	**8263**	3050	0.20	15250
f_8	6234	0.97	**6447**	-	0.00	-

In Table 2, the performance indices C, P_c, and Q_m are reported both for gPSEDA and for gPSO. The results show that gPSEDA is not as efficient as gPSO on the unimodal benchmarks (f_1, f_2) and on the Rosenbrock function (f_3), a multimodal function with a very small number of optima. However, on the other multimodal benchmarks, gPSEDA clearly outperforms gPSO both in convergence (C) than in probability of convergence (P_c) (apart on Ackley function f_5, a famous good benchmark for PSO, where the results are anyhow similar). The best results are obtained on Rastrigin function f_6 and on extended Schaffer function f_8, where our approach is able to reach a near optimum solution with a small amount of NFES and with a full convergence. Instead gPSO, on these benchmarks was never able to reach a near optimum solution before the allowed cap of NFES. Since f_6 is a separable function and f_8 is non-separable, we can conclude that PSEDA, with respect to PSO, is very versatile, other than efficient, on those problems that present a certain complexity.

6 Conclusion and Future Work

In this paper Particle Swarm Estimation of Distribution Algorithm (PSEDA), an hybridization of PSO and EDAs, has been introduced by implementing the PSO dynamics in the EDAs framework. The velocity concept of PSO has been simulated in PSEDA by a probability distribution model where the PSO attractive positions are used as peaks of probability.

Regarding PSEDA from the viewpoint of EDAs, the innovation introduced is the learning and the sampling of an independent probability distribution model for each individual and for each dimension. Furthermore, conversely of classical EDAs, elitism is maintained despite the entire population update from one generation to the next.

Experimental results show that PSEDA improves the performances with respect to classical PSO on problems that present a certain complexity on their landscapes. Further experiments are still needed to confirm the positive results on a wider range of benchmarks, such as rotated, translated and noisy functions. Furthermore, in order to provide a better understanding of the PSEDA behaviour, an analysis of the parameters setting and of the standard deviation regulation are planned for the future.

Note that, a comparison with respect to a continuous EDA scheme has been not reported since all the continuous EDAs in literature use multivariate PDMs. Indeed, for this reason the results would have been not comparable. However, our approach, through a future introduction of a multivariate PDM, could allow to incorporate an inter-dimension dependencies learning procedure in PSO.

Another future line of research is the discretization of PSEDA. Indeed, PSEDA can be easily converted to handle discrete search spaces by using mixtures of discrete probability distributions. For example, the continuous Gaussian distributions can be replaced by discrete binomial distributions in the case of ordered discrete search spaces, or by probability histograms of the form "needle in a haystack" in the case of purely combinatorial (or symbolic) search spaces (those without an ordered structure). This discretization process, once formalized and experimented, jointly with the continuous PSEDA described in this paper, could allow to handle several practical applications that generally consist of hybrid continuous/combinatorial spaces (i.e. search spaces where some dimensions are continuous and other ones are combinatorial). Imagine, for example, the automatic testing scenario where an industry, before commercializing an engineering product, must test the product in order to optimize some features (e.g. the aerodynamics of an airplane that is repetitively tested with a scale model in an automatized and robotized wind tunnel using a sort of trial-and-error method). Finally, another practical application, already approached with a similar technique [12], is the optimization of web content presentations using the users feedback as an online fitness function to optimize.

References

1. Bäck, T., Fogel, D., Michalewicz, Z.: Handbook of Evolutionary Computation. Oxford University Press, Oxford (1997)
2. Bengoetxea, E., Larrañaga, P.: EDA-PSO: A Hybrid Paradigm Combining Estimation of Distribution Algorithms and Particle Swarm Optimization. In: Dorigo, M., Birattari, M., Di Caro, G.A., Doursat, R., Engelbrecht, A.P., Floreano, D., Gambardella, L.M., Groß, R., Şahin, E., Sayama, H., Stützle, T. (eds.) ANTS 2010. LNCS, vol. 6234, pp. 416–423. Springer, Heidelberg (2010)
3. Bosman, P.A.N., Thierens, D.: Expanding from Discrete to Continuous EDAs: The IEDA. In: Proceedings of Parallel Problem Solving from Nature, vol. 6, pp. 767–776 (2000)
4. Clerc, M., Kennedy, J.: The Particle Swarm-Explosion, Stability, and Convergence in a Multidimensional Complex Space. IEEE Trans. on Evolutionary Computation 6(1), 58–73 (2002)
5. Devroye, L.: Non-Uniform Random Variate Generation. Springer, New York (1986)
6. El-Abd, M., Kamel, M.: PSO_Bounds: A New Hybridization Technique of PSO and EDAs. Foundations of Computational Intelligence 3, 509–526 (2009)
7. Iqbal, M., de Oca, M.A.M.: An Estimation of Distribution Particle Swarm Optimization Algorithm. In: Dorigo, M., Gambardella, L.M., Birattari, M., Martinoli, A., Poli, R., Stützle, T. (eds.) ANTS 2006. LNCS, vol. 4150, pp. 72–83. Springer, Heidelberg (2006)
8. Jarboui, B., Cheikh, M., Siarry, P., Rebai, A.: Combinatorial Particle Swarm Optimization (CPSO) for Partitional Clustering Problem. Applied Math. and Computation 192, 337–345 (2007)
9. Kennedy, J., Eberhart, R.: Particle Swarm Optimization. In: Proceedings of IEEE Conference on Neural Networks, vol. 4, pp. 1942–1948 (1995)
10. Larrañaga, P., Lozano, J.A.: Estimation of Distribution Algorithms: A New Tool for Evolutionary Computation. Kluwer Academic Publishers, Boston (2001)
11. Michalewicz, Z.: Genetic Algorithms + Data Structures = Evolution Programs. Springer, Berlin (1992)
12. Milani, A., Santucci, V., Leung, C.: Optimal Design of Web Information Contents for E-Commerce Applications. In: Proceedings of 25th International Symposium on Computer and Information Science. LNEE, vol. 62, pp. 339–344. Springer, Heidelberg (2010)
13. Price, K.V., Storn, R.M., Lampinen, J.A.: Differential Evolution: A Practical Approach to Global Optimization. Springer, Heidelberg (2005)
14. Socha, K., Dorigo, M.: Ant Colony Optimization for Continuous Domains. Technical Report TR/IRIDIA/2006-037, IRIDI Technical Report Series (2006)
15. Suganthan, P.N., Hansen, N., Liang, J.J., Deb, K., Chen, Y.P., Auger, A., Tiwari, S.: Problem Definitions and Evaluation Criteria for the CEC 2005 Special Session on Real-Parameter Optimization. Technical Report #2005005, Nanyang Technological University (2005)
16. Titterington, D., Smith, A., Makov, U.: Statistical Analysis of Finite Mixture Distributions. John Wiley & Sons, Chichester (1985)
17. Zhou, Y., Jin, J.: Eda-Pso - A new Hybrid Intelligent Optimization Algorithm. In: Proceedings of the Michigan University Graduate Student Symposium (2006)

Differential Evolution Based Bi-Level Programming Algorithm for Computing Normalized Nash Equilibrium

Andrew Koh

Abstract. The Generalised Nash Equilibrium Problem (GNEP) is a Nash game with the distinct feature that the feasible strategy set of a player depends on the strategies chosen by all her opponents in the game. This characteristic distinguishes the GNEP from a conventional Nash Game. These shared constraints on each player's decision space, being dependent on decisions of others in the game, increases its computational difficulty. A special solution of the GNEP is the Nash Normalized Equilibrium which can be obtained by transforming the GNEP into a bi-level program with an optimal value of zero in the upper level. In this paper, we propose a Differential Evolution based Bi-Level Programming algorithm embodying Stochastic Ranking to handle constraints (DEBLP-SR) to solve the resulting bi-level programming formulation. Numerical examples of GNEPs drawn from the literature are used to illustrate the performance of the proposed algorithm.

1 Introduction

In a game when a rational agent optimizes her welfare in the presence of others symmetrically doing the same simultaneously, game theory [23] provides a way to analyze the strategic decision variables of all players. The solution concept of such games was analyzed by Nash in [16]. A game is considered to have attained a Nash Equilibrium (NE) if no one player can unilaterally improve her payoff given the strategic decisions of all other players. While establishing that an outcome is not a NE (by showing that a player can profitably deviate) is usually not difficult, determining the NE itself is more challenging. A review of some deterministic and stochastic methodologies for determination of NE is found in [13].

Andrew Koh
Institute for Transport Studies, University of Leeds, Leeds, LS2 9JT, United Kingdom
e-mail: a.koh@its.leeds.ac.uk

A. Gaspar-Cunha et al. (Eds.): Soft Computing in Industrial Applications, AISC 96, pp. 97–106.
springerlink.com

This paper is concerned with a special class of Nash Games known as the Generalized Nash Equilibrium Problem (GNEP). In the GNEP, the players' payoffs and their strategies are continuous (and subsets of the real line) but more importantly the GNEP possesses the distinctive feature that players face constraints depending on the strategies their opponents choose. This feature is in contrast to more common Nash Game where the utility/payoff/reward the players obtain depend solely on the decisions they make and their actions are not restricted because of the strategies chosen by others. The ensuing constrained action space in GNEPs makes them more difficult to resolve than conventional Nash games discussed in monographs such as [23]. We point out in passing that the solution algorithm proposed in this paper can be easily applied to conventional Nash games (see below).

This paper is structured thus. Following this introduction, we introduce the GNEP formally along with the various game theoretic terminologies. In Sect 3 the key result emphasized is that the GNEP can be formulated as a bi-level program. Sect 4 outlines DEBLP-SR, a Differential Evolution based algorithm integrated with deterministic gradient based solvers and embodying stochastic ranking to resolve the resulting bi-level formulation. Numerical examples from the literature are discussed in Sect 5. Results of runs using the proposed DEBLP-SR are presented in Sect 6 and Sect 7 wraps up with some concluding remarks.

2 Nash Equilibrium and the GNEP

This section introduces the notation used throughout this work. The GNEP is a single shot normal form game with a set N of players indexed by $i \in \{1,2,...,n\}$ and each player can play a strategy $x_i \in X_i$ which all players are assumed to announce simultaneously. $X \subseteq \mathbb{R}^m$ is the collective action space for all players. In a standard Nash Game, $X = \prod_{i=1}^{n} X_i$. X is thus equal to the Cartesian product. In contrast, in a GNEP, the feasible strategies for player $i \in N$ depend on the strategies of all other players [1],[4],[10],[21]. We denote the feasible strategy space of each player by the point to set mapping: $K^i : X^{-i} \to X^i, \forall i \in N$ that emphasizes the ability of other players to influence the strategies available to player i [4],[7],[21]. The distinction between a conventional Nash game and a GNEP is therefore analogous to the distinction between unconstrained and constrained optimization.

To emphasize the variables chosen by player i, we write $\mathbf{x} \equiv (x_i, x_{-i})$ where x_{-i} is the combined strategies of all players in the game excluding that of player i i.e. $x_{-i} \equiv (x_1, ..., x_{(i-1)}, x_{(i+1)}, ..., x_n)$. Critically note that the notation (x_i, x_{-i}) *does not* mean that the components of $\mathbf{x}$ are reordered such that x_i becomes the first block. In addition, let $\phi_i(\mathbf{x})$ be the payoff/reward to player $i, i \in N$ if $\mathbf{x}$ is played.

Definition 1. *[21] A combined strategy profile* $\mathbf{x}^* = (x_1^*, x_2^*, ..., x_n^*) \in X$ *is a Generalized Nash Equilibrium for the game if:*

$$\phi_i(x_i^*, x_{-i}^*) \geq \phi_i(x_i, x_{-i}^*), \ \forall x_i \in K(x_{-i}^*), \forall i \in \{1,2,...,n\}.$$

At a Nash Equilibrium no player can benefit (increase individual payoffs) by unilaterally deviating from her current chosen strategy. Players are assumed not to cooperate and so each is doing the best she can given what her competitors are doing [5],[13],[23]. For a GNEP, the strategy profile $\mathbf{x}^*$ is a Generalized Nash Equilibrium (GNE) if it is feasible with respect to the mapping K^i *and* if it is a maximizer of each player's utility over the constrained feasible set [7].

2.1 Nikaido Isoda Function

The Nikaido Isoda (NI) function in Eq 1 is an useful tool used in the study of Nash Equilibrium problems eg. [3],[4],[8],[10]. Its interpretation is as follows: each summand shows the increase in payoff a player will receive by unilaterally deviating and playing a strategy $y_i \in K(x_{-i})$ while other players play according to x.

$$\Psi(\mathbf{x},\mathbf{y}) = \sum_{1}^{n} [\phi_i(y_i,x_{-i}) - \phi_i(x_i,x_{-i})] \text{ },\forall i \in \{1,2,...,n\} \tag{1}$$

The NI function is always non-negative for any combination of $\mathbf{x}$ and $\mathbf{y}$. Furthermore, this function is everywhere non-positive when either $\mathbf{x}$ or $\mathbf{y}$ is a Nash Equilibrium point by virtue of Definition 1 since at a Nash Equilibrium no player can increase their payoff by unilaterally deviating. This result is summarized in Definition 2.

Definition 2. *[10] A vector $\mathbf{x}^* \in X$ is called a Normalized Nash Equilibrium if $\Psi(\mathbf{x},\mathbf{y}) = 0$.*

2.2 Solution Approaches for the GNEP

A review of solution methods for the GNEP is discussed in the survey [4]. Deterministic (i.e. gradient-based) descent, the subject of detailed study in Von Heusinger's PhD thesis [9], is the primary solution approach for finding Normalized Nash Equilibrium (NNE). Krawczyk et al [3],[8],[14] also proposed another deterministic descent method based on minimization of the Nikaido-Isoda function. In this paper however we exploit the proof that we can find the NNE by formulating the GNEP as a special bi-level program [2],[21] as discussed in the following section.

3 A Bi-Level Programming Approach for GNEPs

The NNE solution to the GNEP can be found by solving a bi-level programming problem given by the system of equations in 2 and 3. For a proof see [2],[21].

$$\max_{(x,y)} \quad f(x,y) = -(y-x)^T(y-x) \tag{2a}$$

$$\text{subject to } x^i \in K^i(x^{-i}) \text{ },\forall i \in \{1,2,...,n\}. \tag{2b}$$

where y solves

$$\max_{(x,y)}(\phi_1(y^1,x^{-1})+\ldots+\phi_n(y^n,x^{-n})) \equiv \max_{(x,y)}\sum_{i=1}^{n}[\phi_i(y_i,x_{-i})-\phi_i(x_i,x_{-i})] \quad (3a)$$

$$\text{subject to } y^i \in K^i(x^{-i}), \forall i \in \{1,2,...,n\}. \quad (3b)$$

The upper level problem (Eq 2) is a norm minimization problem subject to strategic variable constraints (Eq.2b). The objective function of the lower level problem (Eq 3) is exactly the Nikado Isoda function (Eq 1).

Definition 3. *[21] The optimal value of $f(x,y)$ is 0 at the Normalized Nash Equilibrium.*

Definition 3 will perform the critical role of being the termination criteria of the proposed DEBLP-SR Algorithm discussed in Sect 4.

4 Differential Evolution for Bi-Level Programming

Differential Evolution (DE) for Bi-Level Programming (DEBLP) was initially proposed in [12] to solve Bi-Level Programs (posed as leader-follower games) arising in transportation systems management. It follows the Genetic Algorithms Based Approach proposed in [22] but substitutes the use of binary coded Genetic Algorithms strings with real coded DE [18] as the stochastic optimization method instead.

DEBLP integrates DE manipulation of the upper level variables with gradient based optimization of the lower level problem. The characteristic feature of GNEPs is the constraints facing the players i.e. (Eq.2b); and thus it is necessary to employ constraint handling techniques to produce solutions that satisfy the constraints. Constraint handling methods were not required for the class of Nash Games discussed in [13] and so the technique proposed here is considered more generic.

In the original DEBLP, constraints in the upper level problem were handled by degrading fitness values if constraints were not satisfied via rudimentary penalty methods [12]. In this paper, the upper level constraints in Eq 2 are handled using stochastic ranking [20]. Hence this version of DEBLP is termed DEBLP-SR.

The pseudo code of DEBLP-SR is summarized in Algorithm 1. DEBLP-SR operates thus: A population of h chromosomes is randomly initialized between the bounds of the problem and the user provides the control parameters (mutation probability and crossover factors) for the DE algorithm [18] . The evaluation of fitness is carried out in a two stage process: In the first stage (lines 5 and 13), each chromosome, representing x the upper level variable, is used as a input argument into the lower level program (Eq 3) parameterized in y. Thus given x we solve the lower level program for y using conventional gradient based optimization methods. In the second stage (lines 6 and 14), x and y are used to compute Eq 2 ($f(x)$ in line 10).

This measures how far the chromosome is from the optimal value of 0 (cf Definition 3) and thus represents the fitness of the chromosome x. In addition, the constraint violation are also output by the evaluation routine (line 7 and 15).

Stochastic ranking (SR), a robust procedure for handling constraints, uses a stochastic bubble sort procedure to rank population members taking into account both the objective function value and constraint violations. (See [20] for more details). In the first iteration (line 9) the best member of the population is the member that is assigned a rank of 1 (one) by the SR algorithm. DE operations are subsequently used to evolve child chromosomes and evaluated following the two stage process described in the foregoing.

Algorithm 1. Differential Evolution for Bi-Level Programming with Stochastic Ranking (DEBLP-SR)

1: Input: h, Max_{it}, DE Control Parameters (Mutation Probability, Crossover Factor)
2: $it \leftarrow 0$
3: Randomly initialize a population of h parent chromosomes $\mathscr{P}$
4: **for** $j = 1$ to h **do**
5:     Solve Eq 3 using deterministic optimization given chromosome $j \in \mathscr{P}$
6:     Compute Eq 2 to evaluate fitness of chromosome $j \in \mathscr{P}$
7:     Compute constraint violation of chromosome $j \in \mathscr{P}$
8: **end for**
9: Apply Stochastic Ranking to rank each member of $\mathscr{P}$ (between 1 (best member) and h)
10: **while** $it < Max_{it}$ or $f(x) \neq 0$ **do**
11:     Apply DE/best/1/bin [18] to create a child population $\mathscr{C}$
12:     **for** $j = 1$ to h **do**
13:         Solve Eq 3 using deterministic optimization given chromosome $j \in \mathscr{C}$
14:         Compute Eq 2 to evaluate fitness of chromosome $j \in \mathscr{C}$
15:         Compute constraint violation of chromosome $j \in \mathscr{C}$
16:     **end for**
17:     Pool Parents and Children Chromosomes:
18:     $\mathscr{T} \leftarrow \mathscr{P} \cup \mathscr{C}$
19:     Apply Stochastic Ranking to rank each member of $\mathscr{T}$ (between 1 (best member) and h)
20:     $\mathscr{P} \leftarrow MaxRank(\mathscr{T})$
21:     **if** $f(x) = 0$ **then**
22:         Terminate
23:     **else**
24:         $it \leftarrow it + 1$
25:     **end if**
26: **end while**
27: Output: Normalized Nash Equilibrium

To utilize the ranking information generated by SR, we modify the selection procedure used for determining whether parent or child survive into the next generation. Instead of the one to one greedy selection found in canonical DE [18], we pool the entire set of parent and child chromosomes together and then apply SR to identify the top h ranked population members which will survive (this is the set returned by the *MaxRank* procedure in line 20 of Algorithm 1). The rest of the population is discarded and such a selection procedure is reminiscent of that used in e.g. GENITOR [24]. If the best fitness attains the value of 0 *and* constraints are satisfied, then we have found the NNE and the algorithm terminates, else the iteration counter is increased and the process is repeated until Max_{it} generations are exceeded.

5 Numerical Examples

In this section, we give details of the numerical examples to which we apply DEBLP-SR and present the results of numerical experiments in Sect 6.

5.1 Problem 1

Problem 1, from [19] was originally solved using a projected gradient method. This game has 2 players with 1 decision variable each. Player 1's objective is:

$$\phi_1(x_1,x_2) = \frac{1}{2}(x_1)^2 - x_1x_2$$

Player 2's objective is:

$$\phi_2(x_1,x_2) = (x_2)^2 + x_1x_2$$

The feasible space is defined according to:

$$X = \{x \in \Re^2 | x_1 \geq 0, x_2 \geq 0, -x_1 - x_2 \leq -1\}$$

As an example, we give the NI function explicitly as:

$$\Psi(\mathbf{x},\mathbf{y}) = [(\frac{1}{2}{x_1}^2 - x_1x_2) - (\frac{1}{2}{y_1}^2 - y_1x_2)] + [({x_2}^2 + x_1x_2) - ({y_2}^2 + x_1y_2)]$$

The NNE is $x_1^* = 1, x_2^* = 0$ [9],[19].

5.2 Problem 2

Problem 2, again with 2 players and 1 decision variable each, comes from Harker [7]. Player 1's objective is:

$$\phi_1(x_1,x_2) = (x_1)^2 + \frac{8}{3}x_1x_2 - 34x_1$$

Player 2's objective is:

$$\phi_2(x_1,x_2) = (x_2)^2 + \frac{5}{4}x_1x_2 - 24.25x_2$$

The feasible space is defined according to:

$$X = \{x \in \Re^2 | x_1 \geq 0, x_2 \geq 0, x_1 + x_2 \leq 15\}$$

The NNE is $x_1^* = 5, x_2^* = 9$ [7], [9].

5.3 Problem 3

This problem describes an Environmental Pollution Control Problem known as the "River Basin Pollution Game" studied by Krawczyk and co-workers [8],[14]. There are 3 players with 1 variable each. The objective function for player j is:

$$\phi_j(\mathbf{x}) = (c_{1j} + c_{2j})x_j - (3 - 0.01(x_1 + x_2 + x_3))x_j, \forall j \in \{1,2,3\}$$

The feasible space is defined according to:

$$3.25x_1 + 1.25x_2 + 4.125x_3 \leq 100$$

$$2.2915x_1 + 1.5625x_2 + 2.8125x_3 \leq 100$$

$$x_j \geq 0, \forall j \in \{1,2,3\}$$

The NNE is $x_1^* = 21.14, x_2^* = 16.03, x_3^* = 2.927$ [9], [8], [14].

5.4 Problem 4

This problem describes an internet switching model with 10 players proposed in [11] and also studied in [9]. The cost function for player j is given by

$$\phi_j(\mathbf{x}) = -(\frac{x_j}{(x_1 + ... + x_{10})})(1 - \frac{(x_1 + ... + x_{10})}{1}), \quad \forall j \in \{1,...,10\}$$

The feasible solution space is:

$$X = \{x \in \Re^{10} | x_j \geq 0.01, \forall j \in \{1,...,10\}, \sum_{j=1}^{10} x_j \leq 1\}$$

The NNE is $x_j^* = 0.09, \quad \forall j = \{1,...,10\}$ [10].

5.5 Problems 5a and 5b

The last problem studied is a non-linear Cournot-Nash Game with 5 players proposed in [15] which we refer to as Problem 5a. Inclusion of Problem 5a serves to emphasize that the method articulated here can be applied to both standard Nash Games and GNEPs and thus demonstrate that the method in this paper is more general than that proposed in [13]. With the introduction of a production constraint in [17], it is transformed into a GNEP (and referred to as Problem 5b herein).

For both problems, each player's cost function is given as:

$$\phi_j(\mathbf{x}) = (x_j) = c_j x_j + (\frac{\beta_j}{\beta_j + 1}) L_j^{\frac{-1}{\beta_j}} x_j^{\frac{\beta_j+1}{\beta_j}} \quad - P(\mathbf{x}) x_j, \quad \forall j \in \{1,...,5\}$$

$$P(\mathbf{x}) = 5000^{\frac{1}{1.1}} (\sum_{j=1}^{5} x_j)^{-(\frac{1}{1.1})}, \forall j \in \{1,...,5\}$$

The firm dependent parameters (c_j, $\beta_j$and L_j) are found in [15],[17]. The feasible space for Problem 5a (NEP) is the positive axis as production cannot be negative. The solution of the NEP is $\mathbf{x}^* = [36.9318, 41.8175, 43.7060, 42.6588, 39.1786]^{\mathrm{T}}$ [6],[15].

The feasible space for the GNEP variant includes a joint production constraint (Problem 5b) given as follows: [17]

$$X = \{x \in \Re^5 | x_j \geq 0 \ \forall j \in \{1,...,5\}, \sum_{j=1}^{5} x_j \leq M\}$$

For the case where $M = 100$, then the NNE (for GNEP variant 5b) is $\mathbf{x}^* = [14.050, 17.798, 20.907, 23.111, 24.133]^{\mathrm{T}}$ [9].

6 Results

In numerical experiments, we carried out 30 independent runs of DEBLP-SR for each test problem allowing for a maximum of 200 iterations (Max_{it}) each run. Based on Definition 3, we terminate the algorithm when the objective function reaches a value of 0[1]. When this target value is reached *and* the maximum constraint violation is less than 0.000001, we deem a run to be "successful " and the number of such runs are reported in Table 1. All runs also utilize the DE/best/1/bin strategy [18]. The crossover and mutation factor applied to all problems are both set 0.9 without any parameter tuning. Our results illustrate that the algorithm is very useful for simpler problems but robustness (as measured by standard deviation and number of successful runs out of 30) decreases as both non-linearity (c.f. Problem 5) and number of players increases (c.f. Problem 4). However, no solution would be valid if it does not satisfy the constraints and it is evident that all constraints are satisfied for all problems since the maximum constraint violation for each run is zero.

For the purposes of comparing DEBLP-SR against others proposed in the literature, we also used PSwarm [25], which is explicitly designed to handle both bound and linear constraints, to solve our test problems. We are unable to include a comparison of DEBLP-SR with PSWARM due to space constraints but instead

[1] In practice we terminate when the best objective reached is less than or equal to 0.000001.

have made the performance comparison available at `http://goo.gl/bupz0`. For this we used the MATLAB version of PSWARM available on the world wide web at `http://www.norg.uminho.pt/aivaz/pswarm`.

Table 1 Results of Application of DEBLP-SR to Test Problems defined in Section 5

Problem Number	1	2	3	4	5a	5b
Best Objective	1E-08	0	2E-08	0	0	0
Worst Objective	1E-07	0	1E-07	3.4E-05	0	1.2E-07
Mean Objective	5.13E-08	0	5.5E-08	4.2E-06	0	5.8E-08
Median Objective	5E-08	0	5E-08	0	0	8E-08
Standard Deviation	2.57E-08	0	2.84E-08	1.1E-05	0	5.33E-08
Maximum Constraint Violation	0	0	0	0	NA	0
Minimum No. of Function Evaluations	400	420	800	4320	1920	2700
Maximum No. of Function Evaluations	660	640	1940	6000	2790	6000
Mean No. of Function Evaluations	529	535	1081	5618	2388	4935
Median No. of Function Evaluations	520	520	1040	5730	2430	5685
Population Size	20	20	20	30	30	30
No. of Successful Runs	30	30	30	24	30	30

7 Conclusions

The Generalized Nash Equilibrium Problem is a Nash Game with the characteristic that the strategic options open to each player depend on what others have chosen. One particular solution of the GNEP is the Normalized Nash Equilibrium which can be found by solving a specialized bi-level program. We have demonstrated the use of a heuristic method which integrates deterministic optimization with Differential Evolution to solve the resulting bi-level program. DEBLP-SR incorporates stochastic ranking to deal with constraints and tournament selection to select survivors when comparing parent and child chromosomes. We illustrated the performance of DEBLP-SR with numerical examples drawn from the literature and evidence suggests that DEBLP-SR is a viable algorithm for this class of Nash games.

Acknowledgements. The research reported here is funded by the Engineering and Physical Sciences Research Council of the UK under the "Competitive Cities" Grant EP/H021345/1. The author thanks Dr Simon Shepherd and colleagues for comments on an earlier draft.

References

1. Aussel, D., Dutta, J.: Generalized Nash equilibrium problem, variational inequality and quasiconvexity. Oper. Res. Lett. 36(4), 461–464 (2008)
2. Bouza Allende, G.B.: On the calculation of Nash Equilibrium points with the aid of the smoothing approach. Revista Investigación Operacional 29(1), 71–76 (2008)
3. Contreras, J., Klusch, M., Krawzyck, J.B.: Numerical solutions to Nash-Cournot equilibrium in electricity Markets. IEEE T. Power Syst. 19(1), 195–206 (2004)

4. Facchinei, F., Kanzow, C.: Generalized Nash equilibrium problems. Q.J. Oper. Res. 5(3), 173–210 (2007)
5. Gabay, D., Moulin, H.: On the uniqueness and stability of Nash-equilibria in non cooperative games. In: Bensoussan, A., et al. (eds.) Applied Stochastic Control in Econometrics and Management Science, pp. 271–293. North Holland, Amsterdam (1980)
6. Harker, P.T.: A variational inequality approach for the determination of Oligopolistic Market Equilibrium. Math. Program 30(1), 105–111 (1984)
7. Harker, P.T.: Generalized Nash games and quasi-variational inequalities. E. J. Oper. Res. 54(1), 81–94 (1991)
8. Haurie, A., Krawczyk, J.: Optimal charges on river effluent from lumped and distributed sources. Environ. Model Assess 2(3), 177–189 (1997)
9. von Heusinger, A.: Numerical Methods for the Solution of the Generalized Nash Equilibrium Problem. PhD Thesis, Institute of Mathematics, University of Wüzburg, Würzburg (2009)
10. von Heusinger, A., Kanzow, C.: Relaxation Methods for Generalized Nash Equilibrium Problems with Inexact Line Search. J. Optimiz. Theory App. 143(1), 159–183 (2009)
11. Kesselman, A., Leonardi, S., Bonifaci, V.: Game-Theoretic Analysis of Internet Switching with Selfish Users. In: Deng, X., Ye, Y. (eds.) WINE 2005. LNCS, vol. 3828, pp. 236–245. Springer, Heidelberg (2005)
12. Koh, A.: Solving transportation bi-level programs with Differential Evolution. In: Proc IEEE CEC, Singapore, September 25–28, pp. 2243–2250 (2007)
13. Koh, A.: Nash Dominance with Applications to Equilibrium Problems with Equilibrium Constraints. In: Gao, X., et al. (eds.) Soft Computing in Industrial Applications: Algorithms, Integration, and Success Stories. Advances in Soft Computing, vol. 75, pp. 71–79. Springer, Heidelberg (2010)
14. Krawczyk, J., Uryasev, S.: Relaxation algorithms to find Nash equilibria with economic applications. Environ. Model Assess 5(1), 63–73 (2000)
15. Murphy, F.H., Sherali, H.D., Soyster, A.L.: A mathematical programming approach for determining oligopolistic market equilibrium. Math. Program 24(1), 92–106 (1982)
16. Nash, J.: Equilibrium points in N-person games. Proc. Natl. Acad. Sci. USA 36(1), 48–49 (1950)
17. Outrata, J.V., Kočvara, M., Zowe, J.: Nonsmooth Approach to Optimization Problems with Equilibrium Constraints. In: Nonconvex Optimization and its Applications, vol. 28. Kluwer, Dordrecht (1998)
18. Price, K., Storn, R., Lampinen, J.: Differential evolution: a practical approach to global optimization. Springer, Berlin (2005)
19. Rosen, J.B.: Existence and Uniqueness of Equilibrium Points for Concave N-person Games. Econometrica 33(3), 520–534 (1965)
20. Runarsson, T.P., Yao, X.: Search biases in constrained evolutionary optimization. IEEE T. Syst. Man. Cy. C 35(2), 233–243 (2005)
21. Sun, L.: Equivalent Bilevel Programming Form for the Generalized Nash Equilibrium Problem. Journal of Mathematics Research 2(1), 8–13 (2010)
22. Yin, Y.: Genetic Algorithm based approach for bilevel programming models. J. Transp. Eng-ASCE 126(2), 115–120 (2000)
23. Webb, J.N.: Game theory: decisions, interaction and Evolution. Springer, London (2007)
24. Whitley, D.: The GENITOR Algorithm and Selection Pressure: Why Rank-Based Allocation of Reproductive Trials is Best. In: Proc ICGA3, Fairfax, Virginia, pp. 116–121 (1989)
25. Vaz, A.I.F., Vicente, L.N.: PSwarm: A hybrid solver for linearly constrained global derivative-free optimization. Optim. Method Softw. 24(4-5), 669–685 (2009)

Part II: Fuzzy Control and Neuro-Fuzzy Systems

Estimating CO Conversion Values in the Fischer-Tropsch Synthesis Using LoLiMoT Algorithm

Vahideh Keikha, Sophia Bazzi, Mahdi Aliyari Shoorehdeli, and Mostafa Noruzi Nashalji

Abstract. In this paper, a new method for estimation of CO conversion in a range of temperatures, pressures and H2/CO molar ratios in the Fischer-Tropsch (FT) synthesis based on Locally Liner Model Tree (LoLiMoT) has been introduced. LoLiMoT is an incremental tree-construction algorithm that partitions the input space by axis-orthogonal splits. In each iteration two new local models as the result of splitting the worst local model has been inserted into the previous structure and result decreasing the total error. The system has been evaluated through two methods and results show estimated CO conversion values by LoLiMoT are in good agreement with experimental data.

1 Introduction

FT synthesis is an important chemical process for the production of liquid fuels and olefins [1, 2]. The exothermic FT reaction is the polymerization of methylene groups [-*(CH2)*-] forming mainly linear alkanes and alkenes, ranging from methane to high molecular weight waxes [3]:

Vahideh Keikha
Department of Computer Science, University of Sistan and Baluchestan, Zahedan, Iran
e-mail: va.keikha@yahoo.com

Sophia Bazzi
Department of Chemistry, Faculty of Sciences, University of Sistan and Baluchestan, Zahedan, Iran
e-mail: Sophia.bazzi@yahoo.com

Mahdi Aliyari Shoorehdeli
Faculty Member of Mechatronics Group, Electrical and Computer Engineering Faculty, K. N. Toosi University of Technology, Tehran, Iran
e-mail: m_aliyari@eetd.kntu.ac.ir

Mostafa Noruzi Nashalji
Islamic Azad University, South Tehran Branch. Young Researchers Club, Tehran, Iran. Research ID: C-7636-2009
e-mail: mostafanoruzi@hotmail.com

A. Gaspar-Cunha et al. (Eds.): Soft Computing in Industrial Applications, AISC 96, pp. 109–119.
springerlink.com

$$nCO + 2nH_2 \rightarrow (-CH_2-)_n + nH_2O \quad (1)$$

The kinetic study of the FT reaction is a very important task for the industrial practice, being a prerequisite for the industrial FT process design, optimization and simulation [4]. In general, the rate of synthesis gas conversion depends on the partial pressure of the feed constituents, as well as temperature. Understanding of the mechanism of a chemical reaction makes it possible to develop physically-realistic mathematical models based on these mechanisms [5]. A number of mechanistic schemes have been developed for FT synthesis over iron [6].

Most studies support a carbene mechanism involving dissociation of CO to C and O atoms followed by stepwise hydrogenation of carbon and oxygen. Wang et al. developed a kinetic model for the rate of CO consumption on $CO/Fe/SiO_2$ catalyst [7]. CO conversion values have been achieved in various temperatures, pressures and H_2/CO molar ratios. If the used experimental data is enough, a kinetic study will be reliable; in the kinetic study, the mechanism's accuracy of chemical reactions increases as the number of experimental data increases. Provision of the numerous data using the fixed bed micro-reactor is a time-consuming and costly procedure. Thus a mathematical model that can estimate the major experimental data will be very valuable and interesting in industry.

System modeling plays a very important role in many applications such as control, communication and industrial [8-12]. Process control devices utilizing fuzzy algorithms have been used for some times in chemical/biochemical engineering. Industrial and academic research chemists will now encounter in their laboratories both hardware programmed with fuzzy algorithms and software functioning via fuzzy rules. Neuro-fuzzy methods are popular solutions for nonlinear system identification that has been successfully applied in system modeling. Neuro-fuzzy approach possesses all advantages of the pure neural and fuzzy methods, such that the low level learning and computational power of neural networks into fuzzy systems and provision of the high-level human-like thinking are the reasons of predominance of fuzzy systems into neural networks [13]. Also, these methods solve the problems such finding good neural network architecture, high number of hidden units and long computational time of every epoch in the learning process in neural networks. Thus, neuro-fuzzy approach is a good selection for many real world problem such estimation *of* CO conversion values in FT synthesis. Also, the use of computer simulation as a powerful tool in industrial process, model predictive control, static and dynamic simulation and to process analysis is now well established. Many of these applications contain complex nonlinearities and bioinspired solutions that are based on a linearized model of the process. However, if the process is highly nonlinear and subject to large frequent disturbances, a nonlinear model will be necessary. An alternative approach is to design a nonlinear model consists of several linear models that major output derives from combination of these linear models. Also, many training algorithms and structures for solving such problems have been suggested such as LoLiMoT, adaptive network based fuzzy inference system (ANFIS), Takagi-Sugeno (TS) and piecewise linear networks (PLN) [14-17]. Likewise, LoLiMoT has been greatly applied for training neuro-fuzzy network and chemical process control because of its rapid and accurate operation in computation [11-13]. The kinetic study of FT reaction would be more reliable if we had CO conversion values in the various experimental conditions which are a time-consuming and costly process. Thus, a complementary method is a requirement for the kinetic study of FT synthesis. The development of attempts to build satisfactory models for the kinetic study and the mechanism of FT synthesis are the theme of the present work. In the present study, the CO conversion values which have been achieved in various temperatures, pressures and H_2/CO molar ratios using LoLiMoT algorithm, are compared with the experimentally achieved ones. The estimated values have an acceptable accuracy for use in the kinetic study of FT synthesis.

2 Experimental Studies

2.1 Catalyst Preparation

The tested catalysts in this study were prepared using co-precipitation method. An aqueous solution of $Co(NO_3)_2.6H_2O$ (0.5 mol/l) and $Fe(NO_3)_3.9H_2O$ (0.5 mol/l) with 15 wt% of SiO_2 support, has been added with nominal composition of $40\%Fe/60\%Co$ on molar basis. Then the resulting solution was heated to 70°C in a round- bottomed flask fitted with a condenser. Aqueous Na_2CO_3 (0.5 mol/l) was added to the mixed nitrate solution with stirring while the temperature was maintained at 70°C until pH 7.0 was achieved. The precipitate resulting was then left in this medium for times ranging from 2 h. The aged suspension was then filtered, washed several times with warm distilled water until no further Na^+ was observed in the washings tested by flame atomic absorption. The precipitate was then dried in the oven (110°C, 18 h) to give a material denoted as the catalyst precursor, which was subsequently calcined in static air in the furnace (600°C, 6 h) to give the final catalyst.

2.2 Catalyst Testing

The $Co/Fe/SiO_2$ catalyst tests were carried out in a fixed bed stainless steel micro reactor at various operating conditions (Fig. 1). All gas lines to the reactor bed were made from 1/4" stainless steel tubing. Three mass flow controllers (Brooks, Model 5850E) equipped with a four-channel control panel (Brooks 0154) were used to adjust the flow rate of the inlet gases (CO, H_2, and N_2 with purity of 99.99%) automatically. The mixed gases passed into the reactor tube, which was placed inside a tubular furnace (Atbin, Model ATU 150-15) capable of producing temperature up to 1300ºC and controlled by a digital programmable controller (DPC). The catalyst bed is located at the middle of the reactor. The reaction temperature was controlled by a computer, based on the temperature measurement obtained by a thermocouple inserted into the catalyst bed. The meshed catalyst (1.0 g) was held in the middle of the reactor with 110 cm length using quartz wool. The studied system consists of an electronic back pressure regulator which can control the total pressure of the desired process. The catalyst was pre-reduced in situ atmospheric pressure in a flowing H_2 stream (flow rate=30 ml/min) at 400ºC for 16 h before synthesis gas exposure. Analysis of the reactants and products was by means of a GC (THERMO ONIX; LNICAM Pro

GC^+) equipped with a thermal conductivity detector. The CO conversion (%) was calculated according to the normalization method:

$$CO\ conversion(\%) = \frac{(Moles\ CO_{in}) - (Moles\ CO_{out})}{Moles\ CO_{in}} \times 100 \qquad (2)$$

2.3 Kinetic Experimental Data

Kinetic experiments were carried out in a fixed bed micro-reactor containing 1g of $Co/Fe/SiO_2$ catalyst. CO, N_2 and H_2 were fed to the reactor. The partial pressures of CO and H_2 were varied. Prior to reaction the catalyst was pre-treated in pure H_2 at 400°C for 16 hours. Experimental conditions were varied in the following ranges: P=1–12 bars, T=250–300°C, GHSV=3600 ml/h/g_{cat}, and H_2/CO feed ratio=1–2.5. Each experiment was replicated three times in order to verify the experimental data accuracy and reproducibility.

3 Modeling Study

3.1 Local Linear Neuro-Fuzzy Network

The Local Linear Neuro-Fuzzy (LLNF) network is depicted in Fig. 2. A local linear modeling approach is based on a dived-and-conquer strategy. A complex modeling problem is divided into a number of smaller and thus simpler sub problems, which could be solved almost independently by simple models. The most important factor for the success of such an approach is the division strategy for the original complex problems [18-20]. The main objective of learning from data extracts a family of rules that cover these data explicitly. As data used in learning contain generally vagueness and ambiguity, there is a need of developing algorithms for learning from imprecise data [13]; LoLiMoT algorithm provides a simple, fast and deterministic model which has low number of trial-and-error steps for system identification.

3.2 Locally Linear Model Tree

LoLiMoT has been introduced as a local linear modeling algorithm [11, 16-20] LoLiMoT can be regarded as a radial basis function that the output layer weights are replaced with a linear function of the network inputs (Fig. 2). Each neuron realizes a local linear model (LLM) and an associated validity function that determines the region of validity of the LLM [18]. It is an incremental tree-construction algorithm that partitions the input space by axis-orthogonal splits. In each iteration, a new rule or LLM is added to the model. Thus, LoLiMoT belongs to the class of incremental or growing algorithms [13]. In each iteration the validity functions with correspond to the actual partitioning of the input space are compared and the corresponding rule consequents are optimized by the local weighted least square technique [18, 21]. The model construction algorithm has an outer loop (upper level) that determines the parameters for nonlinear partitioning of the input space (structure) an inner loop (lower level) that estimates the parameters of those local linear models [18, 19, 22]. The validity functions, which are similar to basis functions in RBF and could be Gaussians, are normalized such that for any input u, $\sum_{i=1}^{M}\phi_i(\underline{u})=1$, and he output of this model is computed as:

$$\hat{y}=\sum_{i=1}^{M}(w_{i0}+w_{i1}u_1+w_{i2}u_2+w_{i3}u_3+\ldots+w_{ip}u_p)\phi_i(\underline{u}) \quad (3)$$

Where the local linear models and the validity functions depend on $\underline{u}=[u_1u_2\cdots u_p]^T$ as input of the model and p is number of the dimension of the model input.

This network simply interpolates linear hyper-planes, which are used to approximate the functions locally, by nonlinear neurons called validity function. A choice for validity function is normalized Gaussians. If these Gaussians are furthermore axis- orthogonal the validity functions are

$$\begin{aligned}\phi_i(\underline{u})&=\frac{\mu_i(\underline{u})}{\sum_{j=1}^{M}\mu_j(\underline{u})}\\ \mu_i(\underline{u})&=\exp\left(\frac{(u_1-c_{i1})^2}{-2\sigma_{i1}^2}\right)\times\ldots\times\exp\left(\frac{(u_p-c_{ip})^2}{-2\sigma_{ip}^2}\right)\end{aligned} \quad (4)$$

The center coordinates c_{ij} and standard deviations σ_{ij} as the hidden layer parameters of the neural network are nonlinear network parameters and each weight w_{ij} as the *j-th* local weight of the linear system *i* is a linear parameter. Each Gaussian function in (4) performs as a membership function with input vector $\underline{u}$ for the locally linear model. The global parameter vector contains M(p+1) elements:

$$w=[w_{10}w_{11}\cdots w_{1p}w_{20}w_{21}\cdots w_{M0}w_{Mp}] \quad (5)$$

And the associated regression matrix X for N measured data samples are:

$$X=[X_1X_2\ldots X_M] \quad (6)$$

$$x_i=\begin{bmatrix}\phi_i(\underline{u}(1)) & u_1(1)\phi_i(\underline{u}(1)) & \ldots & u_p(1)\phi_i(\underline{u}(1))\\ \phi_i(\underline{u}(2)) & u_2(2)\phi_i(\underline{u}(2)) & \ldots & u_p(2)\phi_i(\underline{u}(2))\\ \vdots & \vdots & \vdots & \vdots\\ \phi_i(\underline{u}(N)) & u_1(N)\phi_i(\underline{u}(N)) & \ldots & u_p(N)\phi_i(\underline{u}(N))\end{bmatrix} \quad (7)$$

Therefore:

$$\hat{y}=X\hat{w}\quad,\quad \hat{w}=(X^TX)^{-1}X^Ty \quad (8)$$

The input space is decomposed in axis orthogonal style yielding hyper-rectangles which centers of Gaussian membership functions $\mu_i(u)$ are placed. The standard division of these Gaussians is set to 0.159 of the length of their rectangles in each dimension.

$$\sigma_{ij} = k_{\sigma}.\Delta_{ij} \tag{9}$$

$$k_{\sigma} = 0.159 \tag{10}$$

Where $\Delta_{i,j}$ denotes the extension of the hyper rectangle of local model i in dimension u_j [18]. k_σ has been found by trial and error and 0.001 perturbation on it make a big difference in estimated data and reduce the accuracy of the result.

The LoLiMoT algorithm is classified as follow:

1. Start with an initial model: start with a single neuron, which is a global linear model over the whole input space with $\phi_1(u)=1$ and set M=1. If there is a priori input space partitioning it can be used as the initial structure.
2. Find the worst locally linear model: Calculate a local loss function e.g. Sum Square Error (SSE) for each of the i =1,..., M LLMs, and find the worst performing neuron.
3. Check all divisions: The worst locally linear model is considered for further refinement. The hyper rectangle of this LLM is split into two halves with an axis orthogonal split. Divisions in all dimensions are tried, and for each of the p divisions the following steps are carried out:
   a) Construction of the multi-dimensional validity functions for both generated hyper rectangles;
   b) Construction of all validity functions.
   c) Estimation of the rule consequent parameters for newly generated LLMs.
   d) Calculations of the loss function for the current overall model.
4. Find the best division: The best of the p alternatives was checked in step 3 is selected, and the related validity functions and LLMs are constructed. The number of LLM neurons is incremented (M = M + 1).
5. Test for convergence: If the termination condition is met, then stop, else go to step 2.

4 Results and Discussion

The basis form of LoLiMoT algorithm has been used to estimate the CO conversion in the specified range of temperatures, pressures and H_2/CO molar ratios. The standard division calculated based on trial and error is equal to 0.159. The algorithm was implemented in MATLAB 2009. The efficiency of algorithm has been evaluated through two methods. During the first method, the available data have been split into two non-overlapped parts, test set and train set. The test set data (eight points) is held out and not looked at during training; fifty four points are used as train set. The maximum number of neurons is restricted to the number of experimental train data that is 54, although we reach to the accuracy of 10^{-12} by 34 neurons in 2.1606 seconds and after this, just the amount of error has been decreased. Fig.4 shows the error reducing process in the procedure of increasing number of neurons. The *CO* conversion values are estimated in the pressures of 1, 3, 6 and 12 bar and approximation by LoLiMoT has been done in 3 and 6 bars by different temperatures and H_2/CO molar ratios (see Table 2). The kinetics of the gas–solid Fischer–Tropsch synthesis over a $Co–Fe–K–SiO_2$ catalyst was studied in a fixed bed micro-reactor. Experimental conditions were varied as follows: p=1–12 bar, T=250– 300 °C, GHSV=3600 ml/h/g_{cat} and H_2/CO feed ratio=1–2.5. The experimental results obtained in this investigation at different temperatures and pressures are given in Table 1. The comparison of estimated and test data for FTS in fixed bed reactor has been shown in Table 3. Also, Fig.3 shows that estimated *CO* conversion values are in relatively good accordance with the real ones. During the second method, the system has been evaluated trough 8-fold cross-validation. The data are partitioned into 8 nearly equally sized segments. Eight iterations of training and validation are performed, during each iteration a different fold of the data is held-out for validation while the remaining 7 folds are used for learning. The values of mean error and the variance are 0.0270 and 0.0287, respectively.

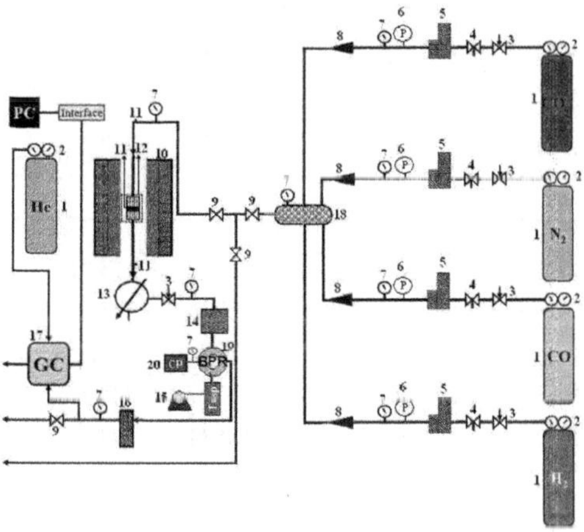

Fig. 1 Fixed bed stainless steel micro reactor

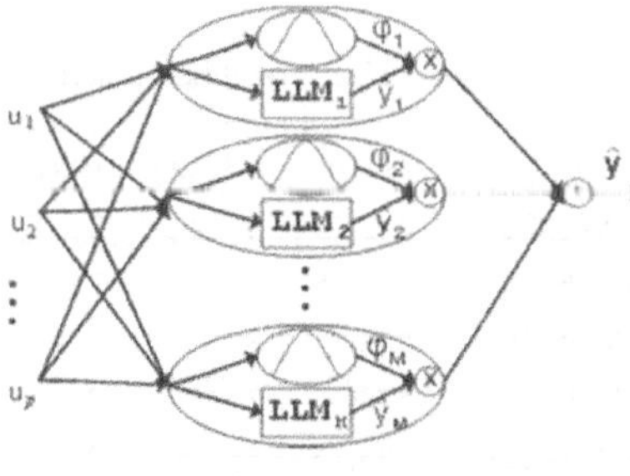

Fig. 2 The network structure of a local linear neuro-fuzzy model with M neurons and p inputs

Table 1 The experimental data for FTS in fixed bed reactor

CO conv.	T(K)	Pt[Bar]	H2/CO	CO conv.	T(K)	Pt[Bar]	H2/CO
0.042521	523.15	1	1	0.097355	533.15	12	1.5
0.047098	533.15	1	1	0.116366	543.15	12	1.5
0.051209	543.15	1	1	0.129036	563.15	12	1.5
0.067585	553.15	1	1	0.136978	573.15	12	1.5
0.100756	563.15	1	1	0.14257	583.15	12	1.5
0.182049	583.15	1	1	0.087076	543.15	12	2
0.089412	513.15	1	1.5	0.081119	553.15	12	2
0.066127	513.15	1	2	0.057969	533.15	3	1
0.120704	513.15	1	1	0.055105	543.15	3	1
0.076497	513.15	1	2.5	0.075608	553.15	3	1
0.12247	523.15	1	1.5	0.088885	563.15	3	1
0.115699	523.15	1	2	0.102707	573.15	3	1
0.102454	523.15	1	2.5	0.082219	583.15	3	1
0.089412	513.15	1	1.5	0.104912	533.15	3	1.5
0.109588	533.15	1	1.5	0.108637	543.15	3	1.5
0.113235	543.15	1	1.5	0.11434	553.15	3	1.5
0.096039	553.15	1	1.5	0.11952	563.15	3	1.5
0.106931	563.15	1	1.5	0.118202	573.15	3	1.5
0.045188	533.15	1	2	0.123707	583.15	3	1.5
0.08588	543.15	1	2	0.076736	533.15	3	2
0.102756	553.15	1	2	0.05738	583.15	6	2
0.127657	563.15	1	2	0.113717	533.15	6	1
0.06159	543.15	12	1	0.113641	543.15	6	1
0.113173	553.15	12	1	0.114307	553.15	6	1
0.09375	563.15	12	1	0.10713	563.15	6	1
0.102616	573.15	12	1	0.113876	573.15	6	1
0.126951	583.15	12	1	0.110089	583.15	6	1

Table 2 The experimental test data for FTS in fixed bed reactor

CO conv	T(K)	Pt[Bar]	H2/CO
0.088400523	543.15	3	2
0.082965138	553.15	3	2
0.080143368	563.15	3	2
0.077996684	573.15	3	2
0.094118521	583.15	3	2
0.128986569	543.15	6	1.5
0.171944000	583.15	6	1.5
0.068357485	533.15	6	2

Table 3 Comparison of estimated and test experimental data for FTS in fixed bed reactor

Estimated	0.0861	0.0988	0.0841	0.0721	0.0848	0.1017	0.1453	0.0702
Real	0.0884	0.0830	0.0801	0.0780	0.0941	0.1290	0.1719	0.0684

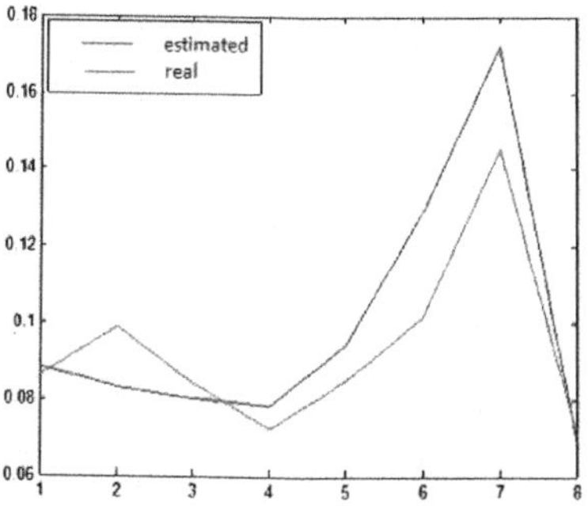

Fig. 3 Real and estimated output in comparing LoLiMoT and experimental test data

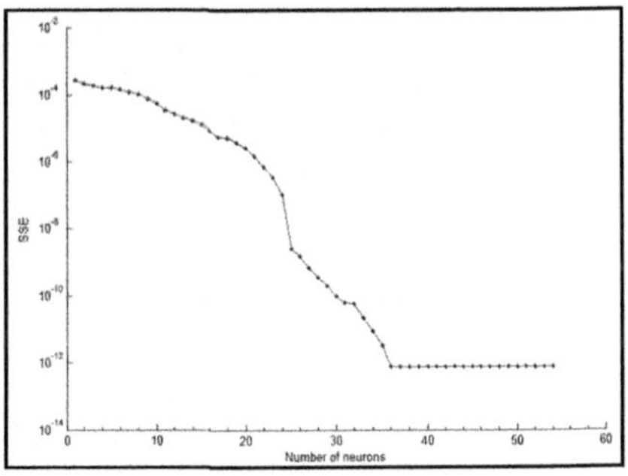

Fig. 4 Convergence curve in learning process by training data

5 Conclusions

In this study, a new method for estimation of *CO* conversion values in the FT synthesis based on LoLiMoT algorithm has been presented. LoLiMoT has been used to approximate *CO* conversion values in a range of temperatures, pressures and H_2/CO molar ratios. Experimental conditions were varied in the following ranges: P=1-12 bar, T=250–310 °C, H_2/CO feed ratio=1–2 mol/mol, GHSV=3600 cm^3 (STP)/h/g_{cat}. The system has been evaluated through two methods, hold-out exercise and 8-fold cross validation. The results show that estimated *CO* conversion values have good accordance with the real ones and accuracy of estimated *CO* conversion values is acceptable for use in the kinetic study of FT reaction, also with few discrete training data. In addition, this algorithm can be used in order to save time and to reduce the costs of kinetic study of FT reaction. Therefore it can be used as a complement in the kinetics modeling and similar systems. Since it is a novel approach for approximation of *CO* conversion values, there is not any results, obtained by other techniques in the previous works, to be compared with our results.

References

1. Mirzaei, A., Habibpour, R., Faizi, M., Kashi, E.: Characterization of iron-cobalt oxide catalysts: Effect of different supports and promoters upon the structure and morphology of precursors and catalysts. Applied Catalysis A: General 301(2), 272–283 (2006)
2. Cheng, J., Gong, X., Hu, P., Lok, C., Ellis, P., French, S.: A quantitative determination of reaction mechanisms from density functional theory calculations: Fischer–Tropsch synthesis on flat and stepped cobalt surfaces. Journal of Catalysis 254(2), 285–295 (2008)

3. Yates, I., Charles, N.: Intrinsic Kinetics of the Fischer-Tropsch Synthesis on a Cobalt Catalyst. Energy & Fuels 5(1), 168–173 (1991)
4. Visconti, C., Tronconi, E., Lietti, L., Zennaro, R., Forzatti, P.: Development of a complete kinetic model for the Fischer–Tropsch synthesis over Co/Al2O3 catalysts. Chemical Engineering Science 62(18-20), 5338–5343 (2007)
5. Critchfield, B.: Statistical methods for kinetic modeling of Fischer-Tropsch synthesis on a supported Iron catalyst. Ph.D thesis, Brigham Young University (2006)
6. Bartholomew, C.H., Farrauto, R.J.: Fundamentals of Industrial Catalytic Processes. Wiley-Interscience, Chichester (2006)
7. Wang, Y., Ma, W., Lu, Y., Yang, J., Xu, Y., Xiang, H., Li, Y., Zhao, Y., Zhang, B.: Kinetics modelling of Fischer–Tropsch synthesis over an industrial Fe–Cu–K catalyst. Fuel 82(2), 195–213 (2003)
8. Mohammadzaman, I., Jamab, A.: Adaptive Predictive Control of an Electromagnetic Suspension System with LOLIMOT Identifier. In: 14th Mediterranean Conference on Control and Automation (MED 2006), pp. 1–6. IEEE, Los Alamitos (2007)
9. Schmid, M., Nelles, O.: Filtering and Differentiating Noisy Signals using Neural Networks. In: Procedings of the American Control Conference 1998, vol. 5, pp. 2730–2731. IEEE, Los Alamitos (2002)
10. Vahabie, A., Yousefi, M., Araabi, B., Lucas, C., Barghinia, S., Ansarimehr, P.: Mutual Information Based Input Selection in Neuro-Fuzzy Modeling for Short Term Load Forecasting of Iran National Power System. In: IEEE International Conference on Control and Automation ICCA 2007, pp. 2710–2715. IEEE, Los Alamitos (2007)
11. Jalili-Kharaajoo, M.: Predictive control of a solution copolymerization reactor using locally linear identifier and evolutionary programming optimizer. In: Proceedings of 2nd International IEEE Conference on Intelligent Systems, vol. 3, pp. 102–106. IEEE, Los Alamitos (2004)
12. Robat, A., Salmasi, F.: State of charge estimation for batteries in HEV using locally linear model tree (LOLIMOT). In: International Conference on Electrical Machines and Systems ICEMS 2007, pp. 2041–2045. IEEE, Los Alamitos (2007)
13. Jamab, A., Araabi, B.: A Learning Algorithm for Local Linear Neuro-fuzzy Models with Self-construction through Merge & Split. In: IEEE Conference on Cybernetics and Intelligent Systems, pp. 1–6. IEEE, Los Alamitos (2006)
14. Bolognese Fernandes, P., Trierweiler, J.: Local Thermodynamic Models Networks for Dynamic Process Simulation. Industrial & Engineering Chemistry Research 48(18), 8529–8541 (2009)
15. Ashoori, A., Moshiri, B., Khaki-Sedigh, A., Bakhtiari, M.: Control of a nonlinear fed-batch fermentation process using model predictive approach. Journal of Process Control 19(7), 1162–1173 (2009)
16. Rouhani, H., Jalili, M., Araabi, B., Eppler, W., Lucas, C.: Emotional learning based intelligent controller applied to neurofuzzy model of micro-heat exchanger. Expert Systems with Applications 32(3), 911–918 (2007)
17. Rezaie, J., Moshiri, B., Araabi, B.: Distributed Estimation Fusion with Global Track Feedback Using a Modified LOLIMOT Algorithm. In: Annual Conference on SICE 2007, pp. 2966–2973. IEEE, Los Alamitos (2008)
18. Nelles, O.: Nonlinear system identification. Measurement Science and Technology 13, 646 (2002)
19. Mehran, R., Fatehi, A., Lucas, C., Araabi, B.: Particle swarm extension to LOLIMOT. In: Sixth International Conference on Intelligent Systems Design and Applications ISDA 2006, vol. 2, pp. 969–974. IEEE, Los Alamitos (2006)

20. Castellano, G., Fanelli, A.: A self-organizing neural fuzzy inference network. In: Proceedings of the IEEE-INNS-ENNS International Joint Conference on Neural Networks, IJCNN 2000, vol. 5, pp. 14–19. IEEE, Los Alamitos (2002)
21. Nelles, O., Fink, A., Isermann, R.: Local linear model trees (LOLIMOT) toolbox for nonli-near system identification. In: 12th IFAC Symposium on System Identification (SYSID), Santa Barbara, USA, pp. 845–850 (2000)
22. Holzmann, H., Nelles, O., Halfmann, C., Isermann, R.: Vehicle dynamics simulation based on hybrid modeling. In: Proceedings of IEEE/ASME International Conference on Advanced Intelligent Mechatronics, pp. 1014–1019. IEEE, Los Alamitos (2002)

[illegible] In: Proceedings of the [illegible] [illegible], pp. [illegible] (20[illegible])
[illegible]
[illegible]

Global Optimization Using Space-Filling Curves and Measure-Preserving Transformations

Hime A. e Oliveira Jr. and Antonio Petraglia

Abstract. This work proposes a multi-start global optimization algorithm that uses dimensional reduction techniques based upon approximations of space-filling curves and simulated annealing, aiming to find global minima of real-valued (possibly multimodal) functions that are not necessarily well behaved, that is, are not required to be differentiable or continuous. Given a real-valued function with a multidimensional and compact domain, the method builds an equivalent, one-dimensional problem by composing it with a space-filling curve (SFC), searches for a small group of candidates and returns to the original higher-dimensional domain, this time with a small set of "promising" starting points. Finally, these points serve as seeds to the algorithm known as Fuzzy Adaptive Simulated Annealing, aiming to find the global optima of the original cost functions. New SFCs are built with basis on the well-known Sierpiński SFC, a subtle modification of a theorem by Hugo Steinhaus and several results of ergodic theory.

1 Introduction

A significant number of techniques for global optimization of numerical functions based on space-filling curves or its approximations have been proposed [2, 6]. One common characteristic shared by these contributions is that the functions under study have certain regularity properties, such as being of Lipschitz type or differentiable. If those properties are not satisfied, or we cannot prove whenever they are, the problem is outside the scope of the corresponding method. It is thus of interest

Hime A. e Oliveira Jr
National Cinema Agency, Rio de Janeiro, Brazil
e-mail: `hime@engineer.com`

Antonio Petraglia
Department of Electrical Engineering, Federal University of Rio de Janeiro, Rio de Janeiro, Brazil
e-mail: `antonio@pads.ufrj.br`

A. Gaspar-Cunha et al. (Eds.): Soft Computing in Industrial Applications, AISC 96, pp. 121–130.
springerlink.com

to devise a way of handling such difficulties, as well. Another situation that occurs frequently is related to the poor precision attained by some methods whenever the domain dimension gets higher - such a fact is related to the difficulty in minimizing the resulting one-dimensional auxiliary functions that exhibit a large number of local minima and very noisy graphs, fractal-like indeed. Considering that the great majority, if not all, of global minimization methods fail in such extreme situations, such an approach is in some cases considered to be ineffective. The reasons for this are related to the way the multidimensional domain is filled up and to the quality of the one-dimensional minimization process applied to the resulting auxiliary function.

To be successful in practice, the global optimization approach must devise a good approximation of a space-filling curve (SFC) whose image is, or contains, the compact domain of the function under study, and to use a one-dimensional global minimization algorithm that can find precise approximations to optimizing points of the composed map, whose domain is a closed interval, say [0,1], and that assumes real values, possessing the same extremes as the original function, among which the desired global optimum is included. Unfortunately, such conditions are not easily satisfied and past efforts were only partially successful in finding good results. In [2], for example, an interesting and promising paradigm is presented, but the results focus on low dimensional spaces. Another issue is related to the adequacy of the chosen way for filling up the original multidimensional domain - by projecting the image of an approximation of a particular SFC onto the space generated by vectors corresponding to directions that have larger variance (by means of principal component analysis), we can show whether there are poorly visited regions, inside which extreme points could be located.

To find SFCs capable of overcoming such obstacles, measure-preserving transformations, key theorems of general topology and ergodic theory were taken as inestimable tools [1, 3, 4]. In this paper, we assume, without loss of generality, that all optimization problems are related to unconstrained global minimization of real functions and all SFCs have the unitary interval [0,1] as their domain.

It's worth to highlight that an important qualitative property of SFCs is their ability to "sweep" deterministically high-dimensional domains, so as to improve the likelihood of finding good "seeds" for complementary, posterior global optimization stages, taking into account the existence of several methods whose final results depend strongly upon their starting points [7]. In this fashion, it is of interest to investigate new ways of finding adequate starters, particularly those located in attraction basins of global optima.

Despite of the existence of many good global optimization methods that could be used in such a posterior stage, we have chosen the fuzzy adaptive simulated annealing algorithm, taking into account its excellent performance in difficult optimizations tasks [9] and the maturity of the adaptive simulated annealing paradigm itself [7, 10]. Nevertheless, it's possible to replace it by any other method, especially those ones depending on good starting points.

2 Auxiliary Theoretical Results

In this work, we define a SFC as a surjective and continuous function from a real interval, say [0,1], to a compact subset of a finite-dimensional vector space, which can be identified to $\mathbb{R}^n$, the n-dimensional Euclidean space. The SFCs were well studied in the past and there are many theoretical results stating necessary conditions for their existence [11]. Besides, long before the invention of digital computers, a number of great mathematicians have proposed constructive examples and established several interesting properties [11] of SFCs. More recently, researchers have found ways to apply previous knowledge about SFCs to various relevant areas, including global optimization of numerical functions. In this paper, the fundamental idea is to compose a given objective function with a SFC corresponding to a compact superset of the respective domain. In such a way we can always reduce a multivariate problem to a univariate one. Hence, at least in theory, it would be possible, by solving the auxiliary one-dimensional problem, to go back to the n-dimensional domain and find the desired optimum point. Unfortunately, when such ideas are brought to the digital computer realm, some complications arise, particularly in high-dimensional domains.

The main drawback concerning implementation issues is that virtually all curves idealized in the far past did not take into account the finite word length of digital computers (one good reason for that is that digital computers were invented long after their synthesis). Typically, the first proposed SFCs were based on infinite expansions and used, for instance, the property that elements in [0,1] can be represented as $0.t_1t_2t_3t_4t_5...$ in a given basis B, where each t_i is an integer between 0 and $B-1$ (extremes included). It is thus necessary to find adequate approximations of SFCs if we want to pursue this kind of approach. Initially, a reasonable alternative seems to be the use of (approximations of) Sierpiński SFCs, taking into account the availability of their precise defining formulas, as follows [11]:

$$x(t) = f(t), \;\; y(t) = f(t-1/4), \;\; t \in [0,1] \tag{1}$$

where f is a bounded, even and continuous real function whose expression is given by

$$f(t) = \frac{\Theta(t)}{2} - \frac{\Theta(t)\Theta(\tau_1(t))}{4} + \frac{\Theta(t)\Theta(\tau_1(t))\Theta(\tau_2(t))}{8} - \dots \tag{2}$$

The 1-periodic functions $\Theta(t)$ and $\tau_k(t)$ are defined in [11], so that the resulting curve $(x(t), y(t))$ is a 2-dimensional SFC and shows to be well-suited to numerical calculations. To build higher-dimensional SFCs starting from this one, some results were crucial, as stated by the following theorems, whose proofs can be found in [11]. First, however, we need to present some definitions.

Definition 1. A function $\varphi : [0,1] \to \mathbb{R}$ is uniformly distributed with respect to the Lebesgue measure if, for any (Lebesgue) measurable set $A \subset \mathbb{R}$, we have

$$\Lambda_1(\varphi^{-1}(A)) = \Lambda_1(A) \tag{3}$$

where Λ_1 is the Lebesgue measure in the real line.

Definition 2. n measurable functions $\varphi_1, \varphi_2, ..., \varphi_n : [0,1] \to \mathbb{R}$ are stochastically independent with respect to Lebesgue measure if, for any n measurable sets $A_1, A_2, ..., A_n \subset \mathbb{R}$,

$$\Lambda_1(\bigcap_{j=1}^{n} \varphi_j^{-1}(A_j)) = \prod_{j=1}^{n} \Lambda_1(\varphi_j^{-1}(A_j)) \tag{4}$$

Theorem 1. *(H. Steinhaus) If $\varphi_1, \varphi_2, ..., \varphi_n : [0,1] \to \mathbb{R}$ are continuous, non-constant and stochastically independent with respect to the Lebesgue measure, then*

$$f = (\varphi_1, \varphi_2, ..., \varphi_n) : [0,1] \to \varphi_1([0,1]) \times \varphi_2([0,1]) \times ... \times \varphi_n([0,1]) \tag{5}$$

is a SFC.

Theorem 2. *If $f = (\varphi, \psi) : [0,1] \to [0,1] \times [0,1]$ is (Lebesgue) measure-preserving and onto, then its coordinate functions φ, ψ are uniformly distributed and stochastically independent.*

Taking into account that the Sierpiński SFC is measure-preserving ([11], page 111), we conclude that its coordinate functions are uniformly distributed and stochastically independent, and can be used to synthesize higher-dimensional SFCs with coordinates

$$\begin{aligned} x_1(t) &= \varphi(t) \\ x_2(t) &= \varphi \circ \psi(t) \\ &\dots\dots\dots\dots\dots \\ x_n(t) &= \varphi \circ \psi \circ \psi \circ \ldots \circ \psi(t) \\ t &\in [0,1] \end{aligned} \tag{6}$$

that are non-constant, continuous and stochastically independent. Unfortunately, after a few experiments it was clear that for higher-dimensional domains (around 8 dimensions), approximations of such "pure" Sierpiński based SFCs failed to fill up adequately the compact domains of interest, as will be shown in the sequel (Fig. 1(a)). This fact is due to distortions caused by numerical approximations, despite the theoretical curve being really a SFC one. Aiming to find a better curve, we composed the original Sierpiński function with an invertible (Lebesgue) measure-preserving transformation that is a natural extension of a particular Generalized Lüroth Series transformation [4], mapping [0,1)x[0,1) onto itself. That new mapping was found through a new partition, that we called the bisection partition, as defined in (7). To generate a new SFC mapping [0,1] onto [-1,1]x[-1,1], it was necessary to use linear homeomorphisms from [-1,1]x[-1,1] to [0,1]x[0,1] and vice-versa. Let us denote such a transformation as τ, and derive its expression as shown below

$$D = \{1,2,3,...\} = \mathbb{N}, \;\; I_k = [l_k, r_k) = [\frac{1}{2^k}, \frac{1}{2^{k-1}}), \; k \in D$$

$$
\begin{aligned}
s(x) &= \frac{1}{r_k - l_k} = \frac{1}{\frac{1}{2^{(k-1)}} - \frac{1}{2^k}} = 2^k,\ x \in I_k \\
h(x) &= \frac{l_k}{r_k - l_k} = \frac{1}{2^k} 2^k = 1,\ \ x \in I_k \\
&\left\{ \begin{array}{ll} s_1(x) = s(x) = 2^k & \\ h_1(x) = h(x) = 1 & ,\ x \in I_k \end{array} \right\} \\
T(x) &= x.s(x) - h(x) = 2^k x - 1,\ x \in I_k \\
\tau(x,y) &= (T(x), \frac{h_1(x) + y}{s_1(x)}) = \\
(2^k x - 1, \frac{1}{2^k} + \frac{y}{2^k}) &= (2^k x - 1, 2^{-k}(1+y)),\ x \in I_k
\end{aligned}
\tag{7}
$$

Here, the $\{I_k : k \in \mathbb{N}\}$ form the bisection partition.

The proposed SFC is given by the following sequence of operations:

$$
[0,1] \rightarrow [-1,1] \times [-1,1] \rightarrow [0,1] \times [0,1] \rightarrow [0,1] \times [0,1] \rightarrow [-1,1] \times [-1,1] \tag{8}
$$

Sierpiński Homeomorphism τ Homeomorphism

It should also be observed that although τ was initially defined only in [0,1)x[0,1), we extended it to [0,1]x[0,1] in an obvious way, so that the composite path can reach all regions of the desired domain. To assess the filling degree of the curves relatively to the compact set $[-1,1]^n$, we present in Figs. 1(a) and 1(b) the plots produced by PCA projections of the corresponding multidimensional curves onto the 2 maximum variance directions. Qualitatively, it can be stated that the more filled the graph area is, the more adequate the filling curve will be. The generated paths consisted of 20,000 8-dimensional points of each kind of curve, and conventional PCA was carried out through a customized computer program. The parameterizing domain was chosen as the closed interval [0,1] in all cases. As can be seen from the plot in Fig. 1(a), the projected points of "pure" Sierpiński curves did not fill adequately the maximum variance region. The composed transformation, in Fig. 1(b), presented substantially better performance, showing denser and more uniform covering. Let

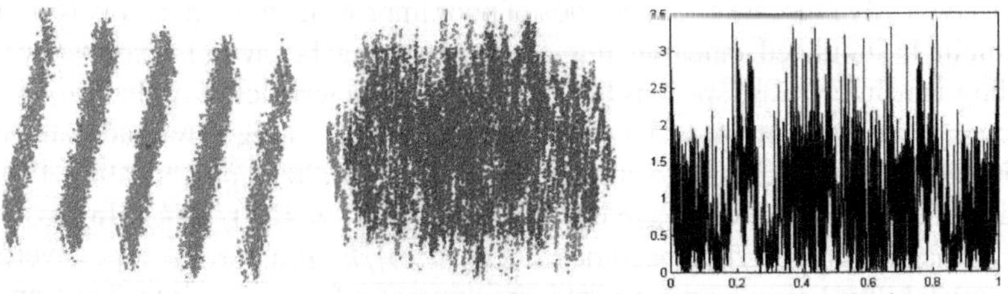

(a) PCA (Sierpiński) (b) PCA (proposed) (c) Reduced function

Fig. 1 Domain projections and example of dimensional reduction (2 → 1)

us thus analyze what happens to the graph of the very well behaved bi-dimensional function $f(x_1,x_2) = (x_1 - 0.3)^2 + (x_2 - 0.3)^2$, restricted to the square [-1,1]x[-1,1], when it is composed with the proposed SFC, yielding a univariate, real-valued function with compact domain [0,1]. The result of uniformly sampling the interval [0,1] by 10000 points is displayed in Fig. 1(c), from which we conclude that a simple dimensional reduction from 2 to 1 transforms a smooth surface into the very rough line $g(t) = f(\varphi_1(t), \varphi_2(t))$. This phenomenon worsens as the original number of dimensions increases. On the other side, by carefully observing Fig. 1(c), we notice that there are regions where the maximum (3.38) and minimum (0) values of f in [-1,1]x[-1,1] are visibly approximated, and could serve as hints to start more effectively preexisting and efficient global optimization algorithms, be them stochastic or deterministic.

Returning to the above example function we find that for t = 0.2908 the one-dimensional function attains the value 0.000119, which is the minimum of the 10000 points used in the illustrative discretization and an approximation for the actual 0 value, corresponding to (0.3, 0.3). Going back to the two-dimensional domain, we find (0.299, 0.289) as the associated bi-dimensional point, which is reasonably close to the actual minimizer. It is worth noting that although Steinhaus theorem is stated only for continuous functions, it is also true for surjective (over [0,1]), piecewise continuous coordinate functions, as well, as is the case for the components of the proposed SFC. In fact, we could state the following theorem, whose proof follows from the corresponding one of Steinhaus theorem in [11].

Theorem 3. *(modified Steinhaus) If* $\varphi_1, \varphi_2, \ldots, \varphi_n : [0,1] \to [0,1]$ *are piecewise continuous, surjective, non-constant and stochastically independent with respect to the Lebesgue measure, then f (defined below) is a SFC.*

$$f \equiv (\varphi_1, \varphi_2, ..., \varphi_n) : [0,1] \to \varphi_1([0,1]) \times \varphi_2([0,1]) \times \ldots \times \varphi_n([0,1]) \quad (9)$$

3 Fuzzy Adaptive Simulated Annealing

Adaptive simulated annealing (ASA) [7] is a sophisticated and rather effective global optimization method. The ASA technique is particularly well suited to applications involving neuro-fuzzy systems and neural network training [10], thanks to its superior performance and simplicity. Unfortunately, stochastic global optimization algorithms typically present certain periods of poor improvement in their way to a global optimum. In simulated annealing implementations, that behavior is mainly due to the cooling schedule, whose speed is limited by the characteristics of probability density functions, which are employed with the purpose of generating new candidate points. In this fashion, if we choose to employ the so-called Boltzmann annealing, the temperature has to be lowered at a maximum rate of $T(k) = T(0)/ln(k)$. In the case of fast annealing, the schedule becomes $T(k) = T(0)/k$, if assurance of convergence with probability 1 is to be maintained, resulting in a faster schedule. The approach based on ASA has an even better default scheme, given by

$$T_i(k) = T_i(0)\exp(-C_i k^{1/D}) \tag{10}$$

because of its improved generating distribution. The constant C_i is a user-defined parameter, and D is the number of independent variables of the function under minimization (dimension of the domain). Notice that subscripts indicate independent evolution of temperatures for each parameter dimension. In addition, it is possible to take advantage of simulated quenching, that is,

$$T_i(k) = T_i(0)\exp(-C_i k^{Q_i/D}) \tag{11}$$

where Q_i is termed the quenching parameter. By attributing to Q_i values greater than 1 we obtain a gain in speed, but the convergence to a global optimum is no longer assured [7]. Such a procedure could be used for higher-dimensional parameter spaces, whenever computational resources are scarce. The internal structure of a successful approach to accelerate the ASA algorithm, using a simple fuzzy controller that dynamically adjusts certain user's parameters related to quenching, is described in [9] - the so-called fuzzy ASA algorithm. As in any other method aiming at global optimization of arbitrary numerical functions, ASA and fuzzy ASA techniques could benefit from the choice of good starting points. Accordingly, the main point of the present work is to find a small set of good seeds able to avoid convergence to suboptimal regions.

4 Proposed Algorithm

We propose the following algorithm to find a global minimum of a given function $f: C \to \mathbb{R}$, where C is a compact subset of some n-dimensional Euclidean space $\mathbb{R}^n$. No condition of regularity, such as differentiability or even continuity, is imposed on f, and C is usually a hyper-rectangle. If it is not, we can always find one hyper-rectangle that contains it, taking into account its compactness. So, from this point on we assume that $C = [a_1, b_1] \times [a_2, b_2] \times \ldots \times [a_n, b_n]$. The algorithm is:

(i) Using (6) and the sequence of operations shown in (8), find an SFC $\varphi = (\varphi_1, \varphi_2, ..., \varphi_n) : [0,1] \to [-1,1] \times [-1,1] \times ... \times [-1,1]$;
(ii) Transform the original, multidimensional minimization problem into a unidimensional one by composing φ and a linear isomorphism $\Psi : [-1,1] \times [-1,1] \times \ldots \times [-1,1] \to [a_1,b_1] \times [a_2,b_2] \times ... \times [a_n,b_n]$, defining g as the composition of φ, Ψ and f, from [0,1] onto $[a_1,b_1] \times [a_2,b_2] \times \ldots \times [a_n,b_n]$, sharing with f the same extreme values;
(iii) Submit g to a one-dimensional global minimization process and find a finite subset of best candidates to global minimizers of f, say $\{t_1, t_2, t_3, ..., t_N\}$, contained in [0,1]. In this work, this set has N=3 elements, but such a number could be easily reconfigured, if necessary;

(iv) Compute the set $\phi = \{\Psi(\varphi(t_1)), \Psi(\varphi(t_2)), \ldots, \Psi(\varphi(t_N))\}$, jumping back to the original domain;
(v) Use the elements of ϕ as starting points for the fuzzy ASA algorithm;
(vi) Choose the best point (corresponding to the minimum value of f) as the final output of the algorithm.

The unidimensional minimization process in (iii) is to be chosen by the implementer. In this work we used intentionally a simple scheme in our experiments (uniform sequential search), aiming to highlight the filling ability of the proposed SFC.

5 Experiments

As noticed in [8], it is usual in the literature to employ certain sets of test functions for evaluation of optimization methods. However, the chosen problems may not be the best ones for testing global optimization algorithms, as the functions belonging to them are relatively simple and regions of attraction of the global minimizers could be easily caught, despite their complicated appearances. Consequently, it is argued that they do not present sufficient difficulty to stress the minimization ability of new optimization approaches. Hence, it is necessary to idealize more sophisticated and systematic tests to verify their performances. To assess the effectiveness of the proposed method, a scheme similar to the one used in [8] was adopted, by employing a certain class of test functions, specified in Table 1, produced by the GKLS generator [5], which allows us to evaluate algorithms in a more complete way. It generates classes of 100 test functions having the same number of local minima plus one global minimum, supplying complete parametric information about each of the functions, such as their dimensions, the values of local minimizers and respective coordinates, placement and sizes of attraction regions of the global minimizer, which are described by parameters r_g (radius of the approximate attraction region of the global minimizer) and d (distance from the global minimizer to the paraboloid vertex). We refer the reader to [5] for more details. In what follows, we consider a global minimum found when candidate points reach a ball B_i of radius $\rho = 0.01\sqrt{N}$, where N is the Euclidean dimension of the function domain, that is,

$$B_i = \{y \in \mathbb{R}^N : ||y - y_i^G|| \leq \rho\} \tag{12}$$

where y_i^G is the global minimizer of the i-th function in a given test class and $i \in \{1, ..., 100\}$. Unlike in the original tests, just one function class (shown in Table 1) was used in the experiments. It should also be noted that all functions are non-differentiable, for (expected) greater difficulty. The present method was tested against the best one presented in [8], therein termed ALI. The authors proposed 4 new global optimization methods denoted as AG, AGI, AL and ALI. Figure 2 displays the number of function evaluations and respective global minima found by the proposed technique, in a similar manner as those reported in [8] to compare AG and ALI for functions of class 5. As indicated in Table 2, the proposed algorithm produced better performance.

Table 1 Parameters pertaining to the function class used in the experiments

Function class	Domain dimension	No. of local minima	Global minimum value	d	r_g	Function type
5	4	10	-1	0.66	0.33	ND

Table 2 ALI and the proposed method minimizing 100 ND-type class 5 functions

Method	Average number of function evaluations	Maximum number of function evaluations in individual minimization operations
ALI	14910	48210
Proposed	**10716.5**	**17350**

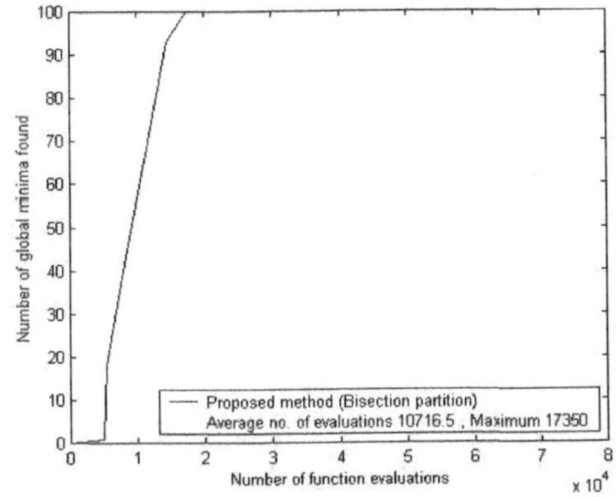

Fig. 2 Results for 100 class 5 functions - average 10716.5 and maximum 17350 evaluations

6 Conclusions

This work presented a multi-start approach for global minimization of multidimensional functions using dimensional reduction and a nonlinear stochastic method. When compared to previously published techniques, the method showed superior performance and adequacy for use in difficult cost functions. This is mainly due to the ability of space-filling curves in serializing the space, thereby allowing us to reduce the search domain to real line intervals. As an important and desirable byproduct, the method is able to find points located in the attraction basins of global minimum points that will be used as seeds when jumping back to the original, multidimensional domain. That ability is invaluable in functions presenting large planar regions, where the great majority of methods get trapped, due mainly to the lack of differential indications. Although the reduced function typically presents a fractal-like graph even whenever the original one is smooth or "well-behaved", existing one-dimensional optimization techniques can obtain good estimates of near optimal points.

Finally, it should be said that although the method could be tested against many other classes of multimodal functions, in this work we have focused on (which we consider) the hardest test used in [8] (cited in section 4.5 and shown in Figure 12 of that reference), for the sake of exact comparison of our findings to the best performing method therein shown, namely ALI (in truth, all benchmarks presented in [8] were privately done and our method performed well, but, for lack of space, we are not presenting them here).

References

1. Boyarsky, A., Góra, P.: Laws of Chaos: Invariant Measures and Dynamical Systems in One Dimension. Birkhäuser, Boston (1997)
2. Cherruault, Y., Mora, G.: Optimisation Globale - Theorie des courbes alpha-denses. Economica, Paris (2005)
3. Choe, G.H.: Computational Ergodic Theory. Springer, Berlin (2005)
4. Dajani, K., Kraaikamp, C.: Ergodic Theory of Numbers. The Mathematical Association of America, Washington DC (2002)
5. Gaviano, M., Kvasov, D.E., Lera, D., Sergeyev, Y.D.: Software for generation of classes of test functions with known local and global minima for global optimization. ACM TOMS 29(4), 469–480 (2003)
6. Goertzel, B.: Global Optimization with Space-Filling Curves. Applied Mathematics Letters 12, 133–135 (1999)
7. Ingber, L.: Adaptive simulated annealing (ASA): Lessons learned. Control and Cybernetics 25(1), 33–54 (1996)
8. Lera, D., Sergeyev, Y.D.: Lipschitz and Hölder global optimization using space-filling curves. Applied Numerical Mathematics 60, 115–129 (2010)
9. Oliveira Jr., H.A.: Fuzzy control of stochastic global optimization algorithms and VFSR. Naval Research Magazine 16, 103–113 (2003)
10. Rosen, B.: Function optimization based on advanced simulated annealing, `http://www.inger.com`
11. Sagan, H.: Space-Filling Curves. Springer, New York (1994)

Modelling Copper Omega Type Coriolis Mass Flow Sensor with an Aid of ANFIS Tool

Patil Pravin, Sharma Satish, and Jain Satish

Abstract. For a variety of practical uses, modelling techniques are being building up with the endeavor of reducing the expenditure and time related with the improvement of new Coriolis mass flow sensors [CMFS]. In this paper the phase shift which is linearly proportional to mass flow rate is modeled using an ANFIS. This technique is competent of understanding an immense diversity of non-linear correlations of substantial intricacy. The experimental data obtained from experimentation on indigenously developed Copper CMFS test rig is used for training the Anfis model then this model is accessible to the network in the structure of input-output pairs, thus the best possible correlation is found between the phase shift and influential important parameters. The training data is having phase shift at changeable input factors like sensor location, drive frequency and mass flow rate. Further, the multilayer feed forward neural network (MFNN) model is developed and compared with the ANFIS model results. These results reveal that ANFIS models could be effectively used in the expansion of Copper Coriolis mass flow sensors.

1 Introduction

A range of Coriolis mass flow sensors are recognized and commercially existing. In such devices, one or two fluid flows are normally pass on through rotating or oscillating tubes, usually driven into oscillation by one or more electromagnetic

Patil Pravin
Department of Mechanical and Industrial Engineering, Indian Institute of Technology, Roorkee, India-247667
e-mail: pravinppatil2004@gmail.com

Sharma Satish
Department of Mechanical and Industrial Engineering, Indian Institute of Technology, Roorkee, India-247667
e-mail: sshmefme@iitr.ernet.in

Jain Satish
Department of Mechanical and Industrial Engineering, Indian Institute of Technology, Roorkee, India-247667
e-mail: sjainfme@ iitr.ernet.in

A. Gaspar-Cunha et al. (Eds.): Soft Computing in Industrial Applications, AISC 96, pp. 131–140.
springerlink.com

oscillators acting at a resonant frequency of the system. This generates a Coriolis acceleration acting on the flowing fluid, and results in a Coriolis force directed at right angles to the flow path and in decisive opposing directions as between two legs of each tubes. This causes a sinusoidal time-varying twisting motion of the tube which can be sensed by conventional motion sensors to produce corresponding analog sinusoidal outputs of measurable amplitude, frequency, and phase relative to a selected reference. By determining a phase difference between such sinusoidal outputs from two sensors, each sensing a motion at a different predetermined location on the tube carrying the flow; it is possible to determine the mass flow rate of the fluid flow through the tube. Coriolis mass flow sensor are very popular for the next generation mass flow measurement techniques and this is attributed due to sensing of the true mass flow rate directly, unlike some other instruments that measure the volumetric flow rate. Physical models play a most important role in assisting the understanding of the various stages for design and optimization of the mass flow sensors. Option to the experiments, which are costly and in some cases not easy to carry out, a well validated physical model can give constructive information about the performance prediction of the mass flow sensor for diverse set situations [1].

However, the major harms with the physical models are the difficulties connected with their creation and restricted accuracies due to the multifaceted nature of the physical designs. As substitute approaches, ANN and ANFIS take out the preferred information straightforwardly from experimental data, and necessitate not take into account the comprehensive information of structures and interactions in the systems and get better the prediction accurateness compared to the conventional models and they have been used extensively. Fuzzy logic reduces the probable complicacies in modelling and analysis of intricate data and also, it is suitable for integrating the qualitative aspects of human understanding within its mapping rules, which are to provide a way of communicable information. Artificial neural networks (ANNs) have also been used to recognize models of difficult systems because of their high computational rates, strength and capability to learn. For the same purpose Neuro- fuzzy systems are fuzzy systems which use ANNs theory in order to determine their properties (fuzzy sets and fuzzy rules) by processing data samples. A specific approach in neuro-fuzzy is the adaptive neuro-fuzzy inference system (ANFIS) that is one of the first integrated hybrid neuro-fuzzy models [2]. An ANFIS gives the mapping relation between the input and output data by using hybrid learning method to determine the optimal allocation of membership functions [3]. Both artificial neural network (ANN) and fuzzy logic (FL) are used in ANFIS structural design [4]. Such structure makes the ANFIS modeling more organized and not as much of dependent on specialist acquaintance [5].

In this work, in order to show the applicability of ANFIS for prediction of phase shift of a copper omega type mass flow sensor a hybrid grid partitioning ANFIS was used. The parameters like sensor location, drive frequency and mass flow rate were considered input features. All these parameters have been observed influential in performance evaluation of such type of sensors in authors' previous experimental study [6]. The model was trained and tested for the region where the experimental data is available. Further the model is well validated with the testing data as well as with the multilayer feed forward neural network modeling.

2 Adaptive Network Based Fuzzy Inference System (ANFIS)

The qualitative aspects of human knowledge and reasoning processes without employing accurate quantitative analyses can be model by a fuzzy inference system which is make use of fuzzy if–then rules. Fuzzy logic modeling techniques can be classified into three categories, that is to say the linguistic (Mamdani- type) [7], the relational equation, and the Takagi–Sugeno–Kang (TSK) [8]. In linguistic models, both the antecedent and the consequence are fuzzy sets while in the TSK model the antecedent consists of fuzzy sets but the consequence is made up of linear equations. Fuzzy relational equation models aim at building the fuzzy relation matrices according to the input–output process data. Jang, [2] has been introduced an adaptive network based fuzzy inference system (ANFIS) based on the TSK model. ANFIS is fuzzy inference system implemented as neural network. Each layer in the network corresponds to a part of the fuzzy inference system (FIS) specifically input fuzzification, rule inference and fire strength computation, and output defuzzification. The main advantage of this kind of illustration is that the FIS parameters are encoded as weights in the neural network and, thus, can be optimized via influential well recognized neural net learning methods. This model is frequently appropriate to the modeling of nonlinear systems. It combines the recursive least-square estimation and the steepest descent algorithms for calibrating both premise and consequent parameters iteratively.

Fig. 1 illustrates an example of a simple FIS structure in an ANFIS network. In ANFIS architecture, a FIS is explained in a layered, feed-forward network structure, where some of the parameters are correspond to by adjustable nodes (represented as rectangular entities in Fig. 1) and the others as fixed nodes (represented as spherical entities in Fig. 1). The raw inputs are fed into the nodes of layer 1 that represent the membership functions. The parameters in this layer are called premise parameters and they are adjustable. The second layer represents the T-norm operators that merge the potential input membership grades in order to calculate the firing strength of the rule. The third layer implements a normalization function to the firing strengths producing normalized firing strengths. The fourth layer represents the consequent parameters that are adjustable. The fifth layer represents the aggregation of the outputs performed by weighted summation. It is not adjustable.

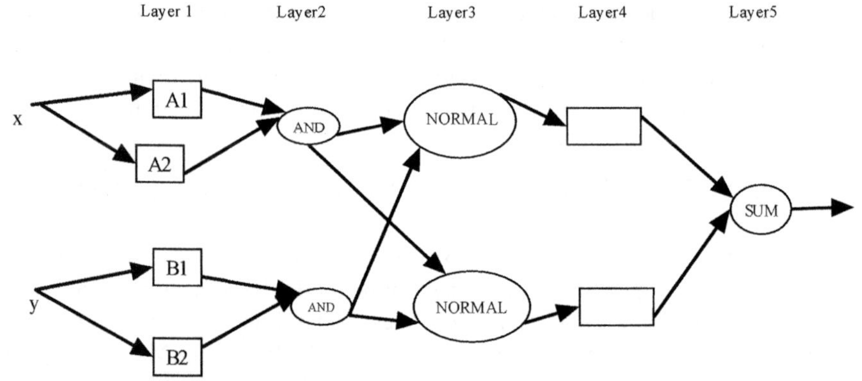

Fig. 1 An ANFIS network structure for a simple FIS.

3 ANFIS/Neural Network Modeling of Phase Shift

ANFIS is dominant model in solving multifaceted problems. Since ANFIS has the impending of solving nonlinear problem and can easily accomplish the input–output mapping, it is perfect to be used for solving the prediction problem. In this work, the ANFIS model on the basis of grid partitioning algorithm with three inputs (sensor location, drive frequency , mass flow rate and one output (phase shift) was designed for prediction modeling of CMFS in a wide range of all input parameters. In the grid partitioning method, the domain of each antecedent variable is partitioned into equidistant and identically shaped membership functions. The Gaussian membership function used in the ANFIS model. Hybrid learning rule is used to train the model according to input/ output data pairs. A hybrid algorithm can be divided to forward pass and a backward pass. The forward pass of the learning algorithm stop at nodes at layer 4 and the consequent parameters are identified by least squares method. In the backward pass, the error signals propagate backward and the premise parameters are updated by gradient descent. It has been demonstrated that this hybrid algorithm is extremely capable in training the ANFIS [2]. Further multilayer feed forward neural network model (MFNN) is developed using the backpropagation network training function: Levenberg-Marquardt and backpropagation weight/bias learning function: Gradient descent having number of hidden layers 8 and epochs 100. In order to evaluate the predicting accuracy of models, the phase shift prediction uses the root mean square error (RMSE) to measure the difference between the predicted and measured values.

$$RMSE = \sqrt{\frac{1}{n}\sum_{i=1}^{n}(phaseshift_{\exp t} - phaseshift_{pred})^2}$$

where n is the total number of data considered. All ANFIS and neural network results were derived from the code developed in MATLAB 7 [9].

4 Experimental Test Conditions

The experimental studies were performed on the copper omega type Coriolis mass flow sensor. Vibrating tube of copper material was used having 12.7 mm outer diameter and 10.9 mm inner diameter. Water was used as flowing fluid through vibrating tube. Tube was having height of 300mm and width of 500 mm. The Experimental set up used in the present study has been designed on Pro Engineer Wildfire modeling software and later manufactured at the Instrumentation project laboratory of Mechanical and Industrial Engineering Department, IIT, Roorkee. The photographic view of the experimental setup has been shown in figure 2, which consists of the several functional elements such as: Hydraulic bench for providing regulated water supply to the flow sensor. Test bench for supporting the tubes of the mass flow sensor. Excitation system for providing mechanical excitation to the mass flow sensor, consists of an Electrodynamics shaker, control unit, accelerometer and vibration meter. Virtual instrumentation comprising of non-contact displacement laser sensors, and a signal conditioning unit for extraction of phase shift. Various settings of sensor location, drive frequency and mass flow rate were used in the experiments. A response surface three level factorial experimental design with 32 runs was used to perform the experiments. The factors and level of each factor are illustrated in Table 1.

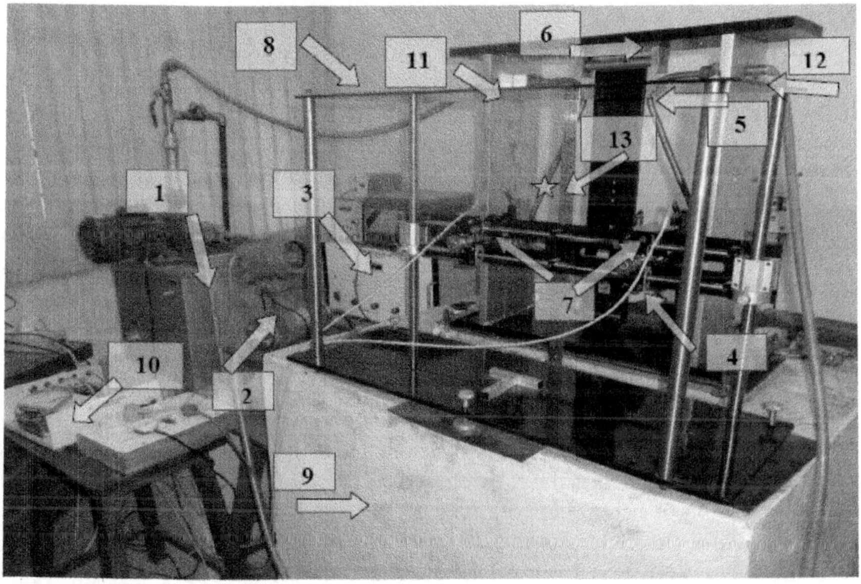

Fig. 2 actual photographic view of experimental setup

Various design components as follows:

1. Hydraulic bench
2. Electromagnetic flow meter
3. Vibration Control unit
4. Vibration driver
5. Omega tube
6. Test Bench
7. Laser sensors
8. Sensor holding stand
9. Foundation
10. Data Acquisition box
11. Inlet pipe
12. Outlet pipe
13. Sensor locations

Table 1 CMFS parameters and their levels

Input parameters	Level 1	Level 2	Level 3
Sensor Location (cm)	6	10	14
Drive frequency (Hz)	18	19	20
Mass flow rate (kg/s)	0.1	0.2	0.3

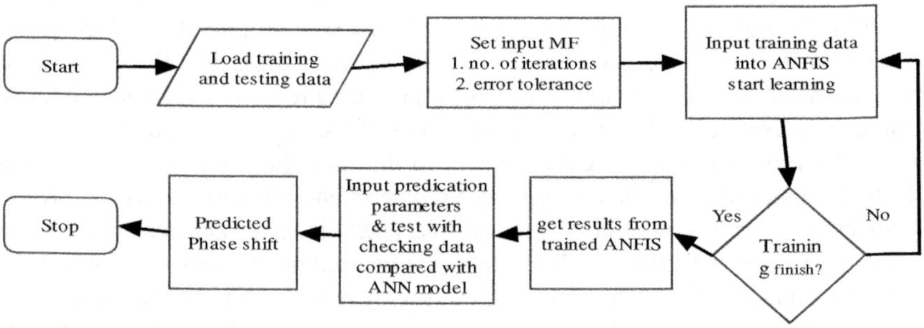

Fig. 3 Flow chart of ANFIS model

5 Results and Discussions

In this study, an ANFIS model based on both ANN and FL has been developed to predict phase shift in CMFS. Three input parameters specifically sensor location, drive frequency and mass flow rate were taken as input features. Further multi-layer feed forward neural network model (MFNN) is also developed for comparison with the Anfis model. The experimental data for phase shift is collected from the experimentation carried out on the indigenously developed copper CMFS test rig. The experiments were divided into two group for training (the first 16 experiment) and testing (remaining) of ANFIS and neural network model. For this purpose, computer simulation results were carried out and further results were validated with the testing data in terms of root mean square error (RMSE) for determining the performance of the proposed methodology. According to the simulation results, the proposed method is efficient for estimating of the phase shift in CMFS. Figs. 3 and 4 depict the flow chart of procedure followed for Anfis modelling; comparison of neural network and ANFIS results for the phase shift respectively. Similarly fig. 5 depicts the RMSE for both the model developed and it is found that the ANIFS is having less compared to ANN. Therefore, it may be proved that the ANFIS method used in this paper is realistic and well improved over neural network results and could be used to predict the phase shift for coriolis mass flow sensor. The comparised lines seem to be close to each other indicating with good agreement. Fig. 6 - 8 shows the consequential surface plot identifying the correlation between preferred variables. It can be observed from the surface plot that the identified correlation by ANFIS methodology is non-linear in nature for sensor location as well as drive frequency and linear for mass flow rate.

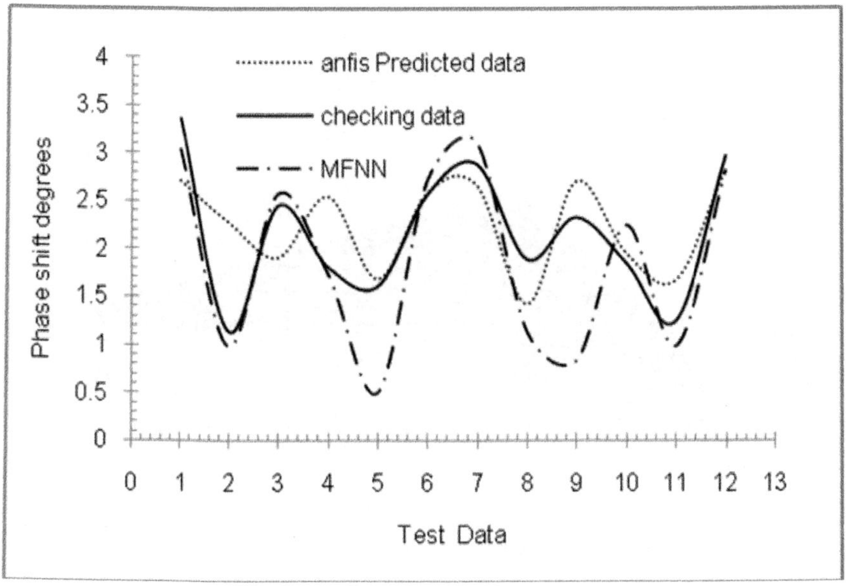

Fig. 4 Comparison of predicted results for ANFIS/ MFNN with actual values for phase shift

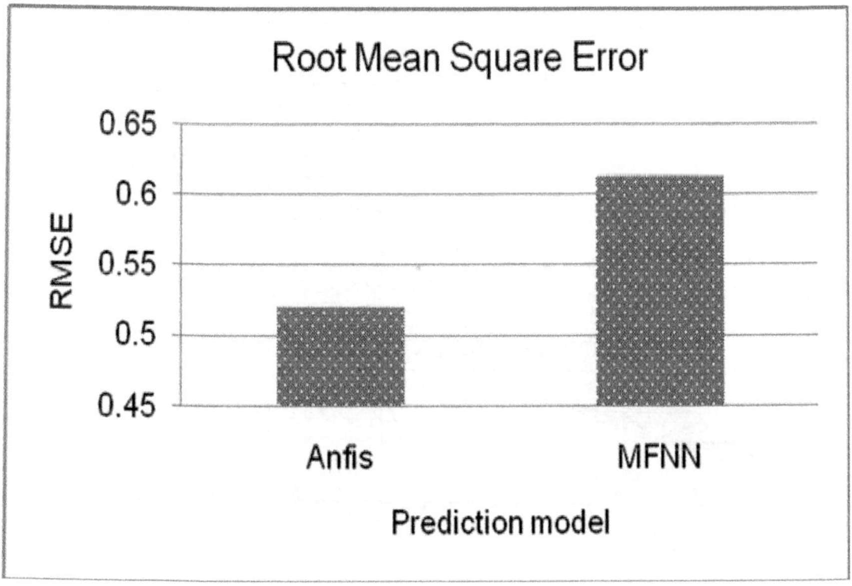

Fig. 5 RMSE for ANFIS/MFNN model

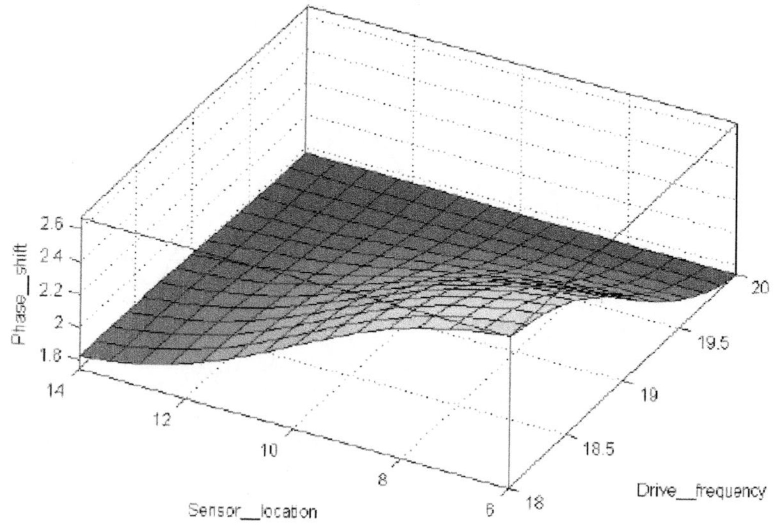

Fig. 6 overall input output surface of sensor location, drive frequency and phase shift

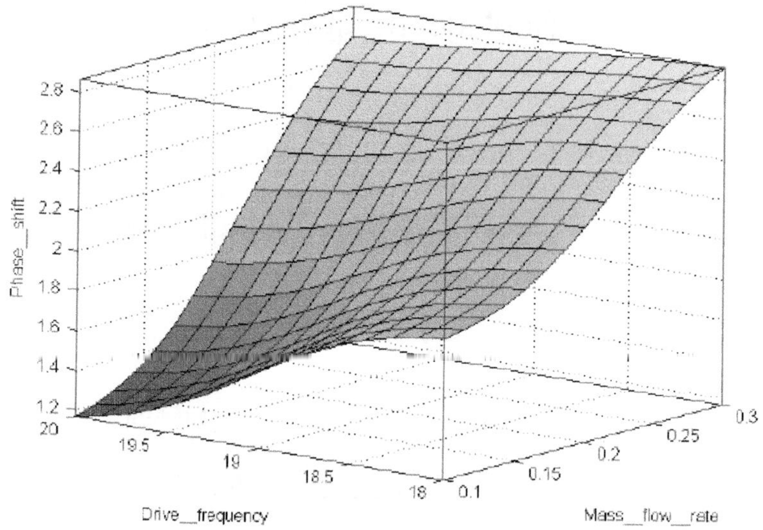

Fig. 7 overall input output surface of drive frequency, mass flow rate and phase shift

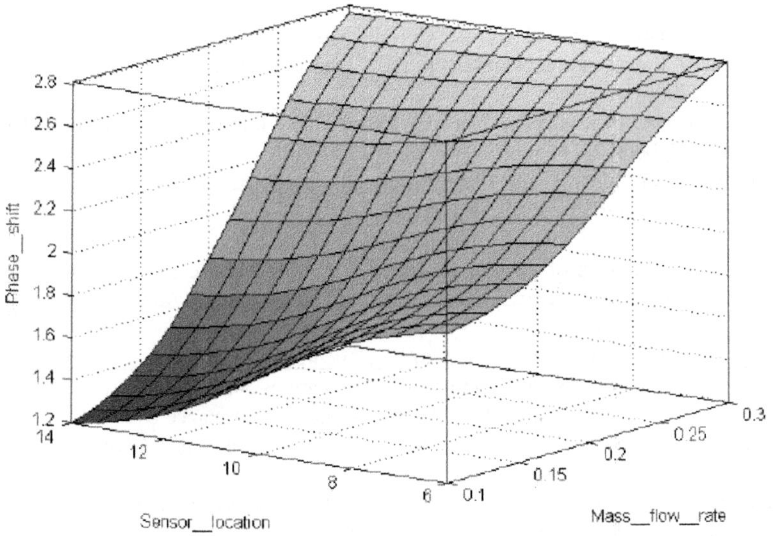

Fig. 8 overall input output surface of sensor location, mass flow rate and phase shift

6 Conclusion

The paper has illustrated the use of the adaptive neuro-fuzzy inference system method for predicting phase shift of copper Coriolis mass flow sensor. ANFIS tests were performed to predict the preferred performances which are competent of realizing a immense multiplicity of non-linear interaction of significant intricacy. The experimental data obtained from experimentation on indigenously developed Copper CMFS test rig is used for training the Anfis model then this model is accessible to the network in the structure of input-output pairs, thus the best possible correlation is found between the phase shift and influential important parameters. The training data is having phase shift at changeable input factors like sensor location, drive frequency and mass flow rate. Further, the multilayer feed forward neural network (MFNN) model is developed and compared with the ANFIS model results. These results reveal that ANFIS models could be effectively used in the expansion of Copper Coriolis mass flow sensors.

Acknowledgement

The authors would like to thank the Department of Science and Technology (DST) Government of India for providing the necessary funding to carry out this research work.

References

[1] Anklin, M., Drahm, W., Rieder, A.: Coriolis mass flow meters: Overview of the current state of the art and latest research. Flow Measurement and Instrumentation 17(6), 317–323 (2006)
[2] Jang, J.S.R.: ANFIS: adaptive Network-Based Fuzzy Inference System. IEEE Trans. Syst. Man Cybern. 23, 665–685 (1993)
[3] Ying, L.C., Pan, M.C.: Using adaptive network based fuzzy inference system to forecast regional electricity loads. Energy Conversation and Management 49, 205–211 (2008)
[4] Avci, E.: Comparison of wavelet families for texture classification by using wavelet packet entropy adaptive network based fuzzy inference system. Applied Soft Computing 8, 225–231 (2008)
[5] Sengur, A.: Wavelet transform and adaptive neuro-fuzzy inference system for color texture classification. Expert Systems with Applications 34, 2120–2128 (2008a)
[6] Sharma, S.C., Patil, P.P., Vasudev, M.A., Jain, S.C.: Performance Evaluation of an Indigenously Designed Copper (U) tube Coriolis Mass sensors. Measurement 43(9), 1165–1172 (2010)
[7] Mamdani, E.H., Assilian, S.: An experiment in linguistic synthesis with a fuzzy logic controller. International Journal of Man-Machine Studies 7, 1 (1975)
[8] Sugeno, M.: Industrial applications of fuzzy control. Elsevier, Amsterdam (1985)
[9] Fuzzy Toolbox User's Guide of MATLAB 7.2. The Math Works Company, Natick, MA (2006)

Gravitational Search Algorithm-Based Tuning of Fuzzy Control Systems with a Reduced Parametric Sensitivity

Radu-Emil Precup, Radu-Codruţ David, Emil M. Petriu, Stefan Preitl, and Adrian Sebastian Paul

Abstract. This paper proposes the tuning of a class of fuzzy control systems to ensure a reduced parametric sensitivity on the basis of a new Gravitational Search Algorithm (GSA). The GSA is employed to solve the optimization problems characterized by the minimization of objective functions defined as integral quadratic performance indices. The performance indices depend on the control error and on the squared output sensitivity functions of the sensitivity models with respect to the parametric variations of the controlled process. The controlled processes in the fuzzy control systems are benchmarks modeled by second-order linearized systems with an integral component and Takagi-Sugeno proportional-integral fuzzy controllers are designed and tuned for these processes.

Radu-Emil Precup
Department of Automation and Applied Informatics, "Politehnica" University of Timisoara, Bd. V. Parvan 2, 300223 Timisoara, Romania
e-mail: radu.precup@aut.upt.ro

Radu-Codruţ David
Department of Automation and Applied Informatics, "Politehnica" University of Timisoara, Bd. V. Parvan 2, 300223 Timisoara, Romania
e-mail: davidradu@gmail.com

Emil M. Petriu
School of Information Technology and Engineering, University of Ottawa, 800 King Edward, Ottawa, ON, K1N 6N5 Canada
e-mail: petriu@site.uottawa.ca

Stefan Preitl
Department of Automation and Applied Informatics, "Politehnica" University of Timisoara, Bd. V. Parvan 2, 300223 Timisoara, Romania
e-mail: stefan.preitl@aut.upt.ro

Adrian Sebastian Paul
Department of Automation and Applied Informatics, "Politehnica" University of Timisoara, Bd. V. Parvan 2, 300223 Timisoara, Romania
e-mail: adi11p@yahoo.com

A. Gaspar-Cunha et al. (Eds.): Soft Computing in Industrial Applications, AISC 96, pp. 141–150.
springerlink.com

1 Introduction

The uncontrollable variations of the parameters of the processes lead to undesirable behaviors of the control systems. The parametric sensitivity of the control systems is studied in the time domain [23, 29] and in the frequency domain [20, 24, 33]. Some optimal control applications employing sensitivity models in the objective functions are dealing with objective functions as extended quadratic performance indices used in the design and tuning of Takagi-Sugeno PI-fuzzy controllers [23], Bellman-Zadeh's approach applied to decision making in fuzzy environments in multi-criteria optimization problems [10], augmented state feedback tracking guaranteed cost control [24], optimal human arm movement control [5] or Iterative Learning Control [7]. Attractive applications are given in [18, 35].

Solving the optimization problems for the usually non-convex objective functions used in many control systems is not a trivial task as it can lead to several local minima. Different solutions including derivative-free optimization algorithms are proposed in the literature [4, 6, 8, 9, 11, 14, 17, 19, 22, 32,] to solve the optimization problems in control systems with objective functions that can have several local minima. The Gravitational Search Algorithm (GSA) [26] is such an algorithm inspired by Newton's law of gravity to solve the optimization problems with non-convex objective functions which eventually have several local minima.

The main contributions of this paper are: the application of the GSA to the optimal tuning of Takagi-Sugeno proportional-integral (PI)-fuzzy controllers, a new GSA which is based on an additional constraint regarding system's overshoot and on the modification of the depreciation equation of the gravitational constant with the advance of the algorithm's iterations, and the simple implementation of the GSA by the application of the Extended Symmetrical Optimum (ESO) method [25] which uses of a single design parameter in the tuning conditions of the PI controller parameters which are next mapped onto the parameters of the PI-fuzzy controller in terms of the modal equivalence principle, and the number of parameters of the tuning parameters of the PI-fuzzy controllers (viz. the variables of the objective functions) is much reduced.

This paper treats the following topics. Section 2 gives the description of the optimization problems in terms of definition and GSA-based solving. Section 3 is focused on the case study which deals with the optimal tuning of Takagi-Sugeno PI-fuzzy controllers for a class of second-order processes with integral component. A discussion on the results is included. Section 4 outlines the conclusions.

2 Optimization Problems: Definition and GSA-Based Solving

The control system structure is presented in Fig. 1, where C is the controller, P is the controlled process, r is the reference input, d_{inp} is the disturbance input, y is the controlled output, u is the control signal, and e is the control error, $e = r - y$,

$\boldsymbol{\alpha} = [\alpha_1 \ \ \alpha_2 \ \ \dots \ \ \alpha_m]^T \in R^m$ is the parameter vector containing the parameters α_a, $a = \overline{1,m}$, of the controlled process, and $\boldsymbol{\rho} = [\rho_1 \ \ \rho_2 \ \ \dots \ \ \rho_q]^T \in R^q$ is the parameter vector containing the tuning parameters ρ_l, $l = \overline{1,q}$, of the controller.

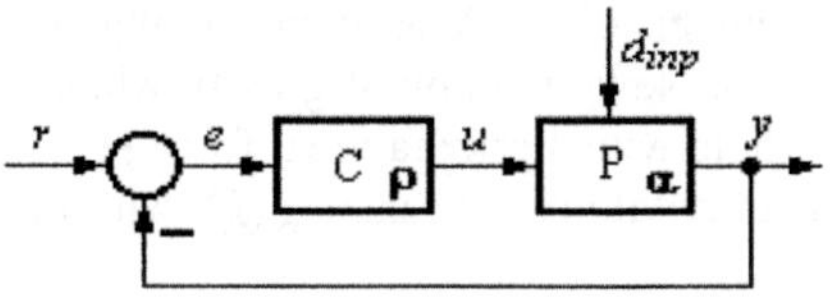

Fig. 1 Control system structure.

Accepting that the state vector of the controlled process $\mathbf{x}_P = [x_{P,1} \ \ x_{P,2} \ \ \dots \ \ x_{P,n}]^T \in R^n$ and that the state vector of the controller $\mathbf{x}_C = [x_{C,1} \ \ x_{C,2} \ \ \dots \ \ x_{C,p}]^T \in R^p$ (the superscript T indicates the matrix transposition) are grouped in the state vector of the control system $\mathbf{x}$

$$\mathbf{x} = [\mathbf{x}_P^T \ \ \mathbf{x}_C^T]^T = [x_1 \ \ x_2 \ \ \dots \ \ x_{n+p}]^T \in R^{n+p},$$
$$x_b = \begin{cases} x_{P,i} & \text{if } b = \overline{1,n} \\ x_{C,i-n} & \text{otherwise} \end{cases}, \ b = \overline{1,n+p}, \tag{1}$$

and that the state-space model of the control system is differentiable with respect to α_a, $a = \overline{1,m}$, the state sensitivity functions $\lambda_b^{\alpha_a}$, $b = \overline{1,n+p}$, and the output sensitivity function σ^{α_a} are

$$\lambda_b^{\alpha_a} = \left[\frac{\partial x_b}{\partial \alpha_a}\right]_{\alpha_a,0}, \ \sigma^{\alpha_a} = \left[\frac{\partial y}{\partial \alpha_a}\right]_{\alpha_a,0}, \ b = \overline{1,n+p}, \ a = \overline{1,m}, \tag{2}$$

where the subscript 0 indicates the nominal value of the appropriate parameter which is subjected to variations. The following discrete-time objective functions are defined to ensure the sensitivity reduction with respect to that parameter:

$$I_{ISE}^{\alpha_a}(\boldsymbol{\rho}) = \sum_{t=0}^{\infty} \{e^2(t) + (\gamma^{\alpha_a})^2[\sigma^{\alpha_a}(t)]^2\}, \ a = \overline{1,m}, \tag{3}$$

where t, $t \in IN$, is the time variable, γ^{α_a}, $a = \overline{1,m}$, are the weighting parameters, all variables in the sum depend on $\boldsymbol{\rho}$, and ISE is the Integral of Squared Error.

The objective functions defined in (3) can be viewed as extended ISE criteria. Their minimization aiming the sensitivity reduction is expressed in terms of the following optimization problems:

$$\boldsymbol{\rho}^* = \arg\min_{\boldsymbol{\rho}\in Do} I_{ISE}^{\alpha_a}(\boldsymbol{\rho}),\ a = \overline{1,m}, \tag{4}$$

where $\boldsymbol{\rho}^*$ is the optimal value of the vector $\boldsymbol{\rho}$ and Do is the feasible domain of $\boldsymbol{\rho}$. The stability of the control system should be first of all taken into consideration when setting the domain Do.

The operating mechanism of GSA used in solving the optimization problem defined in (4) comes from Newton's law of gravity which states that each particle attracts every other particle with a gravitational force [15]. We suggest the following depreciation of the gravitational constant $g(k)$ with the increase in the GSA's iterations number k:

$$g(k) = \delta\, k / k_{\max}, \tag{5}$$

where $k_{\max}$ is the maximum number of iterations, and $\delta > 0$ is constant set such that to ensure the GSA's convergence and to influence the search accuracy.

Considering N masses (agents) and the q-dimensional search space, the position of the ith agent is defined by the vector

$$\mathbf{X}_i = [x_i^1 \ \ \dots \ \ x_i^d \ \ \dots \ \ x_i^q]^T,\ i = \overline{1,N}, \tag{6}$$

where x_i^d is the position of the ith agent in the dth dimension, $d = \overline{1,q}$. The total force acting on the mass i is

$$F_i^d(k) = \sum_{j=1, i\neq j}^{N} \rho_j g(k) \frac{m_i(k) m_j(k)}{r_{ij}(k)+\varepsilon}[x_j^d(k) - x_i^d(k)], \tag{7}$$

where ρ_j, $0 \le \rho_j \le 1$, is a random generated number, $m_i(k)$ and $m_j(k)$ are the related masses of the ith and j^{th} agent, respectively, $\varepsilon > 0$ is a small constant, and $r_{ij}(k)$ is the Euclidian distance between the ith and the jth agent. The distance is used in (7) instead of the square distance to reduce the computational complexity according to the GSA presented in [26]. The law of motion results in the acceleration of the ith agent at the iteration index k in the dth dimension

$$a_i^d(k) = F_i^d(k) / m_{Ii}(k), \tag{8}$$

where $m_{Ii}(k)$ is the inertia mass related to the ith agent. The next velocity of an agent, $v_i^d(k+1)$, is considered as a fraction of its current velocity added to its acceleration. Therefore, the position and velocity of an agent are updated in terms of the following state-space equations [27]:

$$v_i^d(k+1) = \rho_i v_i^d(k) + a_i^d(k),\ x_i^d(k+1) = x_i^d(k) + v_i^d(k+1), \tag{9}$$

where ρ_i, $0 \le \rho_i \le 1$, is a uniform random variable.

The gravitational and inertial masses are calculated according to [26, 27]:

$$n_i(k) = [f_i(k) - w(k)]/[b(k) - w(k)], \; m_i(k) = n_i(k) / \sum_{j=1}^{N} n_j(k), \tag{10}$$

where $f_i(k)$ is the fitness value of the ith agent at the iteration index k, and $b(k)$ (corresponding to the best agent) and $w(k)$ (corresponding to the worst agent) are defined as follows for minimization problems as those defined in (4):

$$b(k) = \min_{j=1,n} f_j(k), w(k) = \max_{j=1,n} f_j(k). \tag{11}$$

The relations between the fitness function f and value $f_i(k)$ in the GSA and the objective functions defined in (3), and between the position of the ith agent of the GSA defined in (6) and the parameter vector are

$$f_j(k) = I_{ISE}^{\alpha_a}(\boldsymbol{\rho}), \; a = \overline{1,m}, \; j = \overline{1,N}, \; \mathbf{X}_i = \boldsymbol{\rho}, \; i = \overline{1,N}. \tag{12}$$

Our GSA consists of the following steps:

1. Initialize the q-dimensional search space, the number of agents N and randomly generate the agents' position vector $\mathbf{X}_i$.
2. Evaluate the agents' fitness using the equations (3) and (12) involving simulations and/or experiments conducted with the fuzzy control system.
3. Update $g(k)$, $b(k)$, $w(k)$ and $m_i(k)$ using (5), (10) and (11) for $i = \overline{1,N}$.
4. Calculate the total force in different directions using (7), and update the agents' velocities and positions using (8) and (9).
5. Validate the obtained vector solution $\mathbf{X}_i(k)$ in terms of checking the following inequality-type constraint which guarantees that the fuzzy control system with the obtained controller tuning parameters $\boldsymbol{\rho} = \mathbf{X}_i(k)$ ensures the convergence of the objective function:

$$| y(t_f) - r(t_f) | \le 0.001 | r(t_f) - r(0) |, \tag{13}$$

   where t_f is the final time moment (theoretically ∞ according to (3)).
6. Increment k and go to step 2 until the maximum number of iterations is reached, i.e. $k = k_{\max}$.

3 Case Study and Discussion of Results

The controlled process with the following transfer function is considered in order to validate the application of our GSA to the optimal tuning of fuzzy controllers with a reduced parametric sensitivity:

$$P(s) = k_P / [s(1 + T_\Sigma s)], \tag{14}$$

where k_P is the controlled process gain and T_Σ is the small time constant or the sum of parasitic time constants, $k_P = 139.88$ and $T_\Sigma = 0.92\,\text{s}$. The accepted sampling period is $T_s = 0.05\,\text{s}$.

The transfer function defined in (14) includes the actuator and measuring element dynamics. Such transfer functions are simplified linearized models of processes used as servo systems in various applications [1, 2, 12, 13, 21, 28, 30]. As shown in [25] the PI controllers can be tuned by the ESO method to guarantee a good compromise to the desired / imposed control performance indices using a single design parameter referred to as β, and the Takagi-Sugeno PI-fuzzy controllers (Fig. 2) are designed and tuned to improve the system's performance.

Fig. 2 Takagi-Sugeno PI-fuzzy controller structure.

The Two Inputs-Single Output fuzzy controller (TISO-FC) block in Fig. 2 is characterized by the fuzzification according to Fig. 3, the weighted average method is employed for defuzzification, and the SUM and PROD operators are used in the inference engine.

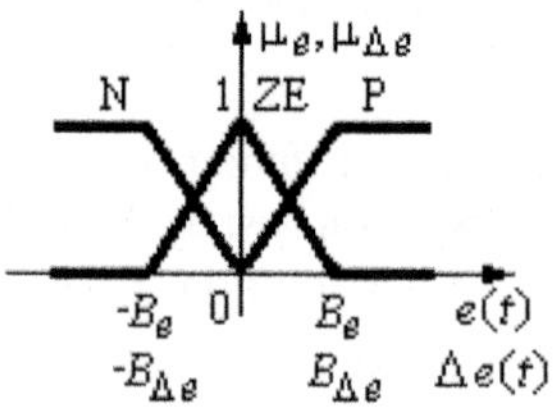

Fig. 3 Input membership functions of Takagi-Sugeno PI-fuzzy controller.

Tustin's method applied to the discretization of the linear PI controller with the transfer function

$$C(s) = k_c(1 + sT_i)/s = k_C[1 + 1/(sT_i)],\ k_C = k_c T_i, \tag{15}$$

with the controller gain k_c and integral time constant T_i leads to the discrete-time incremental PI controller with the recurrent equation and parameters

$$\Delta u(t) = K_P[\Delta e(t) + \mu\, e(t)],\ K_P = k_c(T_i - T_s/2),\ \mu = 2T_s/(2T_i - T_s). \tag{16}$$

The rule base of TISO-FC is presented in Table 1, where the consequents of the rules are characterized by

$$f_1(t) = K_P[\Delta e(t) + \mu\, e(t)],\ f_2(t) = \eta\, f_1(t). \tag{17}$$

Table 1 Decision table of TISO-FC

$\Delta e(t)$	$e(t)$		
	N	ZE	P
P	$\Delta u(t) = f_1(t)$	$\Delta u(t) = f_1(t)$	$\Delta u(t) = f_2(t)$
ZE	$\Delta u(t) = f_1(t)$	$\Delta u(t) = f_1(t)$	$\Delta u(t) = f_1(t)$
N	$\Delta u(t) = f_2(t)$	$\Delta u(t) = f_1(t)$	$\Delta u(t) = f_1(t)$

The parameter η is introduced to alleviate the overshoot of when $e(t)$ and $\Delta e(t)$ have the same sign, and the modal equivalence principle results in

$$B_{\Delta e} = \mu B_e \,. \tag{18}$$

The PI tuning conditions specific to the ESO method and the reference filter transfer function (for performance improvement) are

$$k_c = 1/(\beta \sqrt{\beta} T_\Sigma^2 k_P),\ T_i = \beta T_\Sigma,\ F(s) = 1/(1+T_i s)\,. \tag{19}$$

The parameter vector of the controller $\boldsymbol{\rho}$ ($q = 3$) and the parameter vector of the controlled process $\boldsymbol{\alpha}$ ($m = 2$) obtain the following particular expressions:

$$\begin{aligned} \boldsymbol{\rho} &= [\rho_1 = \beta \quad \rho_2 - B_e \quad \rho_3 = \eta]^T \in R^3, \\ \boldsymbol{\alpha} &= [\alpha_1 = k_P \quad \alpha_2 = T_\Sigma]^T \in R^2. \end{aligned} \tag{20}$$

The evaluation of the agents and the update of the worst and best masses are conducted using the Simulink-based digital simulation of the fuzzy control system behavior with respect to the step-type modification of the reference input. The GSAs are implemented by means of $N \in \{10,20,50\}$ masses generated randomly. The weighting parameter values were set to $(\gamma^{k_P})^2 \in \{0,1000,10000,100000\}$ and $(\gamma^{T_\Sigma})^2 \in \{0,0.05,0.5,5\}$. Each particle was a $q = 3$-dimensional vector, and each dimension was initialized using $22.5 \le B_e \le 40,\ 0.55 \le \eta \le 1,\ 4 \le \beta \le 16$, and these boundaries stand for inequality-type constraints and they define the domain *Do* as well. The maximum number of iterations used as stop criterion in the step 6 of the GSA was set to {75, 100, 150}. The parameter δ in (5) was set to 0.5 in order to ensure acceptable GSA convergence and search accuracy as well.

For the sake of simplicity just the analysis of the effect of different combinations of the number of agents and of the maximum number of iterations on the optimal values of the controller tuning parameters and on the minimum values of the objective function $I_{ISE}^{k_P}$ is presented here. The results are synthesized in Tables 2 and 3, where the superscript * shows the optimum value of a certain parameter.

The performance index $StDev(I_{ISE}^{k_P})\%$ is introduced to evaluate the convergence of the algorithms. This index is defined as the percentage represented by the standard deviation $StDev(I_{ISE}^{k_P})$ compared to the average value $Avg(I_{ISE}^{k_P})$ of the objective function obtained after all simulations.

Table 2 Parameters and objective function for 20 agents and maximum number of iterations of 75 for $I_{ISE}^{k_P}$

$(\gamma^{k_P})^2$	η^*	B_e^*	k_C^*	T_i^*	$I_{ISE}^{k_P}$
0	0.225813	32.82812	0.00276	7.292998	9328.354
1000	0.213898	30.91186	0.002767	7.258782	10136.25
10000	0.221227	32.1809	0.002759	7.298408	18721.98
100000	0.360172	35.25418	0.00329	5.427772	99524.9

Table 3 Results for 20 agents and maximum number of iterations of 75 for $I_{ISE}^{k_P}$

$(\gamma^{k_P})^2$	$Avg(I_{ISE}^{k_P})$	$StDev(I_{ISE}^{k_P})$	$StDev(I_{ISE}^{k_P})\%$
0	9328.354	248.3207	2.661999
1000	10136.25	265.8096	2.622365
10000	18721.98	410.8041	2.194234
100000	99524.9	3228.985	3.244399

4 Conclusions

The paper has proposed the GSA-based tuning of a class of Takagi-Sugeno fuzzy control systems such that to obtain a reduced sensitivity with respect to the parametric variations of the controlled process. Details concerning the implementation of a new GSA are given.

The future research will extend the optimal tuning to other classes of fuzzy control systems with applications to other processes and controller structures [3, 16, 31]. The strong reduction of the number of simulations is aimed in order to replace the digital simulations in the step 2 of the GSA by real-time experiments.

Another will deal with the comparison of the performance of different evolutionary-based algorithms from the point of view of convergence speed and computational complexity, and with the comparison with classical non-optimized fuzzy controllers and with PI controllers. System's robustness analysis will be tackled.

Acknowledgments. This work was supported by the CNMP and CNCSIS of Romania. The support from the cooperation between the "Politehnica" University of Timisoara, Romania, the Óbuda University, Budapest, Hungary, and the University of Ljubljana, Slovenia, in the framework of two Intergovernmental Science & Technology Cooperation Programs is acknowledged.

References

1. Abonyi, J.: Fuzzy Model Identification for Control. Birkhäuser, Boston (2003)
2. Bodenhofer, U., Klawonn, F.: Robust rank correlation coefficients on the basis of fuzzy orderings: Initial steps. Mathware and Soft Computing 15, 5–20 (2008)
3. Blažič, S., Škrjanc, I., Gerkšič, S., Dolanc, G., Strmčnik, S., Hadjiski, M.B., Stathaki, A.: Online fuzzy identification for an intelligent controller based on a simple platform. Engineering Applications of Artificial Intelligence 22, 628–638 (2009)

4. Boucher, X., Bonjour, E., Grabot, B.: Formalisation and use of competencies for industrial performance optimisation: A survey. Computers in Industry 58, 98–117 (2007)
5. Campos, F.M.M.O., Calado, J.M.F.: Approaches to human arm movement control - A review. Annual Reviews in Control 33, 69–77 (2009)
6. Carrano, E.G., Takahashi, R.H.C., Fonseca, C.M., Neto, O.M.: Non-linear network optimization - An embedding vector space approach. IEEE Transactions on Evolutionary Computation 14, 206–226 (2010)
7. Chen, J., Kong, C.K.: Performance assessment for iterative learning control of batch units. Journal of Process Control 19, 1043–1053 (2009)
8. David, R.C., Rădac, M.-B., Preitl, S., Tar, J.K.: Particle swarm optimization-based design of control systems with reduced sensitivity. In: Proceedings of 5th International Symposium on Applied Computational Intelligence and Informatics, Timisoara, Romania, pp. 491–496 (2009)
9. Deb, K., Gupta, S., Daum, D., Branke, J., Mall, A.K., Padmanabhan, D.: Reliability-based optimization using evolutionary algorithms. IEEE Transactions on Evolutionary Computation 13, 1054–1074 (2009)
10. Ekel, P.Y., Menezes, M., Schuffner Neto, F.H.: Decision making in a fuzzy environment and its application to multicriteria power engineering problems. Nonlinear Analysis: Hybrid Systems 1, 527–536 (2007)
11. Ferreira, J.C., Fonseca, C.M., Gaspar-Cunha, A.: Assessing the quality of the relation between scalarizing function parameters and solutions in multiobjective optimization. In: Proceedings of IEEE Congress on Evolutionary Computation, Trondheim, Norway, pp. 1131–1136 (2009)
12. Haber, R.E., Haber-Haber, R., Jiménez, A., Galán, R.: An optimal fuzzy control system in a network environment based on simulated annealing. An application to a drilling process. Applied Soft Computing 9, 889–895 (2009)
13. Hermann, G., Tar, J.K., Kozłowski, K.R.: Design of a planar high precision motion stage. In: Kozłowski, K.R. (ed.) Robot Motion and Control 2009. LNCIS, vol. 396, pp. 371–379. Springer, Heidelberg (2009)
14. Hidalgo, D., Melin, P., Castillo, O.: Optimal design of type-2 fuzzy membership functions using genetic algorithms in a partitioned search space. In: Proceedings of IEEE Conference on Granular Computing, Silicon Valley, CA, USA, pp. 212–216 (2010)
15. Holliday, D., Resnick, R., Walker, J.: Fundamentals of Physics, 7th edn. John Wiley & Sons, Hoboken (2005)
16. Johanyák, Z.C., Kovács, S.: Sparse fuzzy system generation by rule base extension. In: Proceedings of 11th International Conference on Intelligent Engineering Systems, Budapest, Hungary, pp. 99–104 (2007)
17. Ko, M., Tiwari, A., Mehnen, J.: A review of soft computing applications in supply chain management. Applied Soft Computing 10, 661–674 (2010)
18. Köppen, M.: Light-weight evolutionary computation for complex image-processing applications. In: Proceedings of 6th International Conference on Hybrid Intelligent Systems, Auckland, New Zealand, pp. 3–3 (2006)
19. López-Ibáñez, M., Stützle, T.: The impact of design choices of multiobjective antcolony optimization algorithms on performance: An experimental study on the biobjective TSP. In: Proceedings of Genetic and Evolutionary Computation Conference, Portland, OR, USA, pp. 71–78 (2010)
20. Navarro-López, E.M., Licéaga-Castro, E.: Combining passivity and classical frequency-domain methods: An insight into decentralised control. Applied Mathematics and Computation 215, 4426–4438 (2010)

21. Nouyan, S., Gross, R., Bonani, M., Mondada, F., Dorigo, M.: Teamwork in self-organized robot colonies. IEEE Transactions on Evolutionary Computation 13, 695–711 (2009)
22. Oliveira, C., Henggeler Antunes, C.: Multiple objective linear programming models with interval coefficients – an illustrated overview. European Journal of Operational Research 181, 1434–1463 (2007)
23. Precup, R.E., Preitl, S.: Optimisation criteria in development of fuzzy controllers with dynamics. Engineering Applications of Artificial Intelligence 17, 661–674 (2004)
24. Precup, R.E., Preitl, S., Korondi, P.: Fuzzy controllers with maximum sensitivity for servosystems. IEEE Transactions on Industrial Electronics 54, 1298–1310 (2007)
25. Preitl, S., Precup, R.E.: An extension of tuning relations after symmetrical optimum method for PI and PID controllers. Automatica 35, 1731–1736 (1999)
26. Rashedi, E., Nezamabadi-pour, H., Saryazdi, S.: GSA: A gravitational search algorithm. Information Sciences 179, 2232–2248 (2009)
27. Rashedi, E., Nezamabadi-pour, H., Saryazdi, S.: BGSA: binary gravitational search algorithm. Natural Computing 9, 727–745 (2010)
28. Ridluan, A., Manic, M., Tokuhiro, A.: EBaLM-THP – artificial neural network thermo-hydraulic prediction tool for an advanced nuclear components. Nuclear Engineering and Design 239, 308–319 (2009)
29. Rosenwasser, E., Yusupov, R.: Sensitivity of Automatic Control Systems. CRC Press, Boca Raton (2000)
30. Saxena, A., Saad, A.: Evolving an artificial neural network classifier for condition monitoring of rotating mechanical systems. Applied Soft Computing 7, 441–454 (2007)
31. Vaščák, J.: Fuzzy cognitive maps in path planning. Acta Technica Jaurinensis, Series Intelligentia Computatorica 1, 467–479 (2008)
32. Wang, X., Gao, X.Z., Ovaska, S.J.: Fusion of clonal selection algorithm and harmony search method in optimisation of fuzzy classification systems. International Journal of Bio-Inspired Computation 1, 80–88 (2009)
33. Wang, Y.G., Xu, X.M.: PID tuning for unstable processes with sensitivity specification. In: Proceedings of Chinese Control and Decision Conference, Guilin, China, pp. 3460–3464 (2009)
34. Xu, J., Wu, H., Wang, Y.: Unpower aerocraft augmented state feedback tracking guaranteed cost control. Journal of Systems Engineering and Electronics 19, 125–130 (2008)
35. Zhou, H., Schaefer, G., Shi, C.: A mean shift based fuzzy c-means algorithm for image segmentation. In: Proceedings of 30th Annual International Conference of the IEEE Engineering in Medicine and Biology Society, Vancouver, BC, Canada, pp. 3091–3094 (2008)

Application of Fuzzy Logic in Preference Management for Detailed Feedbacks

Zhengping Wu and Hao Wu

Abstract. In consumer-to-consumer (C2C) e-commerce environments, the magnitude of products and the diversity of vendors have caused confusion and difficulty for consumers to choose the right product from a trustworthy vendor. Feedback system is a widely used solution to help consumers evaluate vendors' reputations. Some C2C environments have started to provide detailed feedback besides the overall rating system to help consumers distinguish individual vendors from multiple aspects. However, the increase in detailed feedback may add to consumer confusion and increase the time needed to consider all aspects for a reputation evaluation decision. This paper analyzes a typical feedback and reputation system for the e-commerce environment and proposes a novel, perception-based reputation model for individual vendors.

Keywords: Detailed feedback, perception-based, fuzzy logic, preference management, eBay.

1 Introduction

E-commerce, especially consumer-to-consumer (C2C) e-commerce platform such as eBay and Amazon Marketplace has become a very popular type of e-commerce sites in recent years due to its convenience and flexibility. Meanwhile, researchers have begun to study the trustworthiness of e-commerce environments [1] and have built many trust models for this type of e-commerce environment [4][9]. From the shopper's point of view, vendor trustworthiness involves many factors [7]. Personalized recommendation is very difficult to develop because of the diversity and uncertainty of these factors. In e-commerce environments, the management of uncertainty in various factors enables us to increase confidence and prevent untrustworthy vendors or products from conducting business. From the shopper's point of view, an online shopping decision is difficult to make because of these uncertain factors in reputation evaluation plus variable price factors. In this paper, we

Zhengping Wu · Hao Wu
Department of Computer Science and Engineering, University of Bridgeport
221 University Avenue, Bridgeport, CT 06604, USA
e-mail: {zhengpiw,wuhao}@bridgeport.edu

A. Gaspar-Cunha et al. (Eds.): Soft Computing in Industrial Applications, AISC 96, pp. 151–161.
springerlink.com

present a perception-based personalized decision-making system for e-commerce applications, which incorporates all possible factors, especially detailed feedbacks for shopping decisions. These detailed feedbacks can provide extra information to recommend the right product from a trustworthy vendor, following the shopper's own rules. This paper aims to model detail feedbacks for increasing the discriminations of different vendors with similar general feedbacks. Two sub-models, derived from this perception-based reputation model for different practical situations in C2C e-commerce environments, are also discussed. An adjustable feedback index that is generated from detailed feedback information on the eBay platform is used in the reasoning mechanism for shopping decision-making to maximize the usage of shoppers' own preferences. An application of this system demonstrates that this novel framework can provide convenient and accurate recommendations for shoppers in e-commerce environments. A comparison with other extant reputation models and the two different sub-models discussed in this paper also offers a clear view of the merits and drawbacks of the entire reputation modeling.

2 Related Work

As one of the most popular C2C e-commerce platforms, eBay provides not only a flexible and convenient shopping environment but also, a complete feedback mechanism [2] that helps shoppers evaluate vendors and review vendor reputations when choosing products. The eBay feedback system includes a 3 degree overall rating mechanism plus a 5-scale rating mechanism from five different detailed aspects. eBay-like C2C systems have been popping up all over the world and have been widely adopted by online retailers in recent years. Researchers chose the eBay system as a typical C2C model for their analysis of trust and reputation in this type of system [9][8]. We attempt to incorporate detailed feedbacks into the analysis of trustworthiness in the eBay environment and propose a novel, perception-based reputation model for eBay, eBay-like systems and general C2C systems, using available information as much as we can in these systems to provide a more accurate trust and reputation evaluation in shopping decision–making.

Fuzzy reputation models have also been studied recently. In [4], authors surveyed several existing reputation models and proposed a fuzzy computational model ("fuzzy beta model") for trust and reputation systems. They focused on centralized computation, which collects ratings from all community members and presents a beta fuzzy formula suitable for any rating system. In their model, trust is defined as a subjective expectation a user has about another's future behavior based on the history of their encounters, and reputation is defined as belief about a person's or thing's character or standing. By introducing fuzzy logic into reputation and trust computation, they built their own fuzzy reputation and trust systems. As an application, they proposed a recommendation system that utilizes the results from fuzzy reputation and fuzzy trust systems.

3 Fuzzy Model and Uncertainty

To manage a collection of reputation-oriented activities in e-commerce applications, we need to understand reputation itself. Reputation of a vendor itself does not mean anything, but it can be reflected in the trust toward that vendor, as felt by shoppers. Research in [3] indicates that trust has three properties: transitivity, asymmetry and personalization. According to the properties of trust and the categorization described by Beth et al. [6], we categorize trust into four classes - direct trust, indirect trust, subjective trust and objective trust. Indirect trust is derived from direct trust. Indirect trust is a function of direct trust. It may add value to direct or indirect trust from a trusted party. Added values are uncertain at some level and tend to be fuzzy. Both subjective trust and objective trust can be derived from indirect or direct trust. However, subjective trust and indirect trust may vary greatly when different sources of information are considered. Due to this variation, subjective trust and indirect trust are uncertain at some level. Furthermore, the objective judgment from the feedback system is a combination of a huge number of other people's subjective judgments, which are also uncertain at some level. Therefore, we need special representations and enforcement processes to handle this uncertain aspect of reputation management and reputation-based decision-making for e-commerce applications.

L. Zadeh proposed an extended perception-based fuzzy logic [5], which aims to lay the groundwork for a radical enlargement of the role for natural languages in probability theory and its applications, especially in the realm of decision analysis. We believe it has the capability to operate on perception-based information and preferences. By introducing perception-based fuzzy logic into the research of decision-making, we aim to solve the issues associated with uncertainty in reputation management and then extend it to personalized shopping decision-making or recommendations for e-commerce applications.

As discussed above, an e-commerce activity may contain several types of uncertainty. When shopping on eBay, buyers may consider various aspects to determine a vendor's reputation. And it is hard for buyers to quantify all the factors. However it is much easier to describe the degree of each factor using linguistic. Thus, we use a collection of linguistic rules to describe uncertainties in words.

After we define a linguistic rule set, we use fuzzy logic to represent these rules. Following the definition in [9], we represent the set of subject as X in this paper. All the fuzzy sets on X are represented as $P(X)$. Then we can use a group of fuzzy sets from $P(X)$ to group all the elements of X into several sets with different levels of uncertainty. Here, we represent one of the factors which is feedback score as a set F and use P^* to represent the probability that a vendor's reputation is high. Then the linguistic rule set can be described as:

"If F is high then P^* is high."
"If F is normal then P^* is medium."
"If F is low then P^* is low."

The probability of "the vendor's reputation is high" can be represented as an F-granular distribution [5] (Figure 1) and written as:

$$P^* = \text{high*high} + \text{normal*medium} + \text{low*low}$$

Then we use Z(*F*) to represent the fuzzy set in *F* domain and $P(P^*)$ to represent fuzzy set in P^* domain. So the words "high", "normal", "low" belong to Z(*F*), and the words "high", "medium", "low" belong to $P(P^*)$. Then, the linguistic rule set can be represented as:

"If F is Z_i then P^* is P_j", where $Z_i \in Z(F)$, $P_j \in P(P^*)$.

"if… than…" rule is the most widely used but not the only format to present linguistic rule. As long as uncertainties need to be contained in rules, our proposed fuzzy term description can be embedded into any rule format.

Using f-granular to describe P^* (as illustrated in Figure 2), P^* can be written as:

$$P^* = \sum_{i,j} Z_i \times P_j$$

We apply this general form of fuzzy rule descriptions to reputation modeling and management for e-commerce applications in order to express uncertainty in real-life trust and decision-making with human linguistics. Since multiple uncertain factors co-exist in the e-commerce environment that influence a shopper's decision-making process regarding the trustworthiness of a vendor, and each individual factor is related to others, the final decision is made by co-activation of all related factors. General rules should have the capacity to involve all possible factors whether they are uncertain factors or normal factors. And we extended the definition of general rules described in [9] to complete linguistic rules.

Definition 3.1 Complete linguistic rule set
For subject set X, $X_i \in X$, and $Z_i(X_i)$ is a fuzzy set of X_i. Let P be the probability distribution of an action or a decision, which is determined by X, and P(P*) is a fuzzy set of P*.*

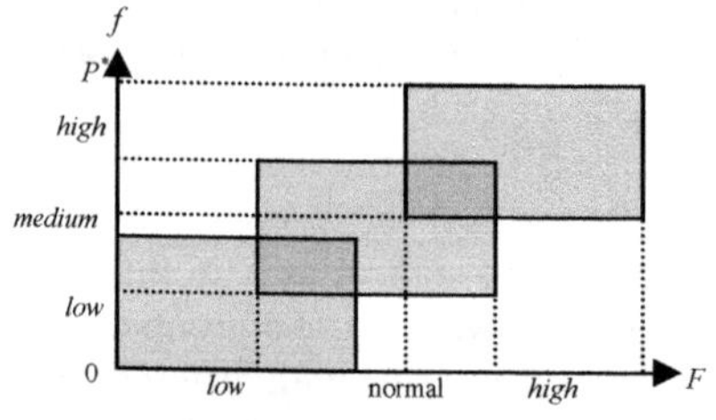

Fig. 1 F-granular

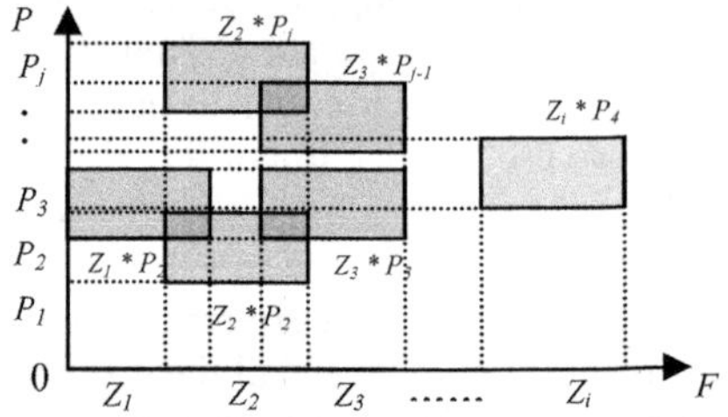

Fig. 2 F-granular of P^*

4 Modeling Detailed Feedback

Determining a vendor's reputation in an e-commerce environment is quite difficult due to various types of relevant factors in its decision-making process, which include both uncertain and certain information and the implicit relationships between factors. We use the eBay feedback system to illustrate several concrete examples hereafter, but the proposed models can be applied to any e-commerce environments and applications. Since an eBay vendor's reputation can be summarized in a feedback score, positive feedback percentage, recent performance index and detailed feedback ratings. However, most shoppers only use the feedback score, positive feedback percentage and recent performance index to derive a reputation decision for their shopping activities. Detailed feedback ratings are supplemental information to support a vendor's reputation. But we cannot ignore the significant contribution that detailed feedback can provide to evaluate the reputation of a vendor, especially to distinguish vendors with similar feedback scores and positive feedback percentages.

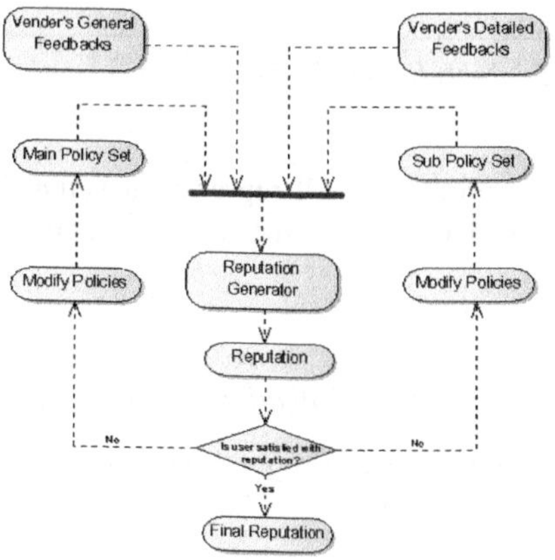

Fig. 3 Parallel Model

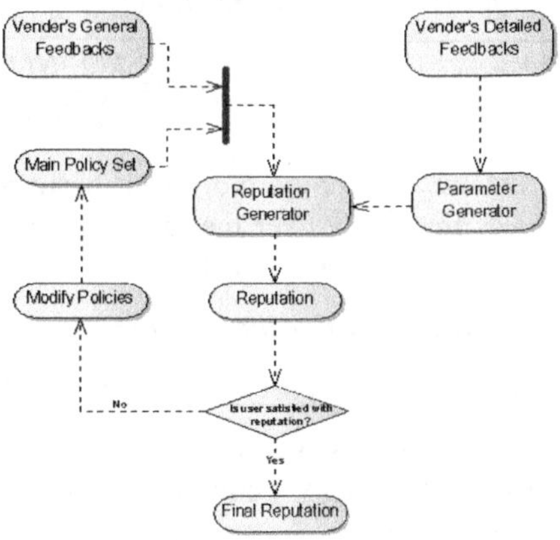

Fig. 4 Feedback Adj. Model

First, we propose a parallel model as shown in Figure 3, treating detailed feedbacks as a sub set of all relevant factors, in which detailed feedbacks are as important as general factors so that the reputation can be calculated, depending not only upon overall performance factors but also consisting of user's preferences on different aspects within detailed feedbacks. Shoppers can define main policy set for general factors and sub policy set for detailed feedbacks. And these two sets are co-actively processed by reputation generator to calculate a reputation for a vendor. On the other hand, we propose another model - the feedback adjustable model (Figure 4) – in which detailed feedbacks are only used to augment general performance factors, which use available detailed feedback ratings to adjust the decision-making process in order to meet shoppers' perceptions toward a vendor. Unlike the parallel model, the feedback adjustable model uses a parameter generator to map detailed feedback ratings into one parameter that has the range from 0 to 1([0,1]). Then this parameter is used in the reputation generation process in order to adjust the calculation of the overall reputation using general feedback information.

In the feedback adjustable model, we use detailed feedbacks to generate a parameter that is called the feedback index, represented as μ, $\mu \in [0, 1]$. We can define the set for each detailed feedback aspect as follows.

Definition 4.1 Set of detailed feedback aspect
The set of detailed feedback aspect is the set containing all the aspects to evaluate a vendor's reputation according to a 5-point average rating and the number of total ratings. In this paper, we use FA to represent the set of detailed feedback aspect.

For each element f_i in FA, $R(f_i)$ represents the average rating of f_i, and $N(f_i)$ represents the rating number of f_i. Then we can define feedback index μ as follows.

Definition 4.2 Feedback index

$$\mu = \frac{\sum_i R(f_i)N(f_i)}{5 * \sum_i N(f_i)}, \text{ in which } R(f_i) \in [0,5].$$

And we also propose a recent performance index that will represent a vendor's sales frequency over the last 12 months, in order to overcome the inaccuracy of the current feedback score.

Definition 4.3 Feedback Score

$$Feedback\ Score = Total\ Positive\ Feedbacks - Total\ Negative\ Feedbacks$$

The feedback score is defined to indicate a user's general reputation. It is widely accepted by many e-commerce systems including eBay.

Definition 4.4 Recent performance index (RP index)

$$\text{RP}Index = \frac{PositiveFe\ edbacksIn\ 12Months - NegativeFe\ edbacksIn\ 12Months}{\frac{FeedbackSc\ ore}{ActiveMont\ h} * N},$$

in which if active month<12, N=active month; If active month>=12, N=12.

The higher the recent performance index value, the more active a vendor's sales activity has been within the last year (or active period).

5 Experiments

To examine the performance and adaptability of the system, we select an unlocked Nokia N900 cell phone as the target product for shopping. Then we run the system with the eBay environment. Hundreds of vendors who sell this cell phone (with the "buy it now" option) are compared in the experiments. And according to the percentage of vendors with/without detailed feedbacks (two categories), we proportionally choose first 7 and 4 vendors from the raw result returned by eBay representing both categories for the comparison. Detailed vendors' information is shown in table 1. Then we use different models to calculate the reputation of each vendor respectively. Then we add price information into decision-making and provide users a final recommendation following users' preferences on different policy sets and different models.

Table 1 Vendors' information

Vendor	Price	Positive feedback (%)	Feedback score	12 month feedback	Active Months	Item as des. (rating/# of rating)	Communication(rating/# of rating)	Shipping time (rating/# of rating)	Shipping handle (rating/# of rating)
eBay1	465.00	50	4	6	48	N/A	N/A	N/A	N/A
eBay2	538.00	100	98	83	39	N/A	N/A	N/A	N/A
eBay3	480.99	83.3	32	2	26	N/A	N/A	N/A	N/A
eBay4	499.95	100	144	150	23	N/A	N/A	N/A	N/A
eBay5	649.66	100	57	1	117	4.4/34	4.5/35	4.9/38	4.4/35
eBay6	449.95	99.6	2616	2894	78	4.9/1712	4.9/1703	4.8/1703	4.9/1703
eBay7	529.99	99	684	551	46	4.8/415	4.8/412	4.8/418	4.8/410
eBay8	540.00	98.4	3974	67	106	4.7/52	4.5/51	4.4/51	4.7/52
eBay9	589.99	98.1	743	775	36	4.8/635	4.7/631	4.8/629	4.8/632
eBay10	599.00	100	141	91	17	4.5/43	4.7/43	4.7/43	4.5/43
eBay11	575.00	98.9	105	95	16	4.8/62	4.8/62	4.6/62	4.8/62
eBay12	499.95	100	132	137	15	5.0/61	5.0/61	4.9/60	4.9/61
eBay13	538.00	99.4	1310	188	50	4.9/142	4.8/140	4.9/141	5.0/136
eBay14	538.99	99.7	1708	316	114	4.9/229	4.6/229	4.9/226	4.9/227
eBay15	490.00	100	1439	1119	97	4.9/905	4.9/902	4.9/899	4.9/899

5.1 Comparisons of Reputation Models

In order to compare the reputations generated from different models, we first defined the main policy set, which is used in both the parallel model and the feedback adjustable model. The main policy set is defined following most shoppers' common sense, which gives a vendor a higher reputation when he obtains a higher feedback score, more positive feedback percentage and being active recently. We also defined two different sub policy sets for the parallel model in order to incorporate detailed feedbacks into the decision-making process. The first one considers all four aspects in detailed feedbacks. The second one, however, only prefers better performance on shipping time. We also set up a policy set that includes only the feedback score and the percentage of positive feedbacks, and run the system using this policy set to generate a set of reputations for vendors as the baseline for comparison. We also build a system following the fuzzy beta reputation model described in [4] to calculate the reputation scores for these 15 eBay vendors. Since the fuzzy beta reputation model does not take recent activity into consideration of vendors' reputation calculation, we also introduce RP index into the fuzzy beta reputation model as a weight to re-calculate the reputation, and view the impact of the RP index.

5.1.1 Comparisons

Figure 5 illustrates the comparisons of reputations generated by different models. RP1 to PR7 represents the general model without detailed feedback, parallel model with sub policy set1, parallel model with sub policy set2, feedback adjustable model, general model without RP index and detailed feedbacks, fuzzy beta reputation model and fuzzy beta reputation model with RP index respectively. Our system provides reputations at three different levels: low, normal and high. The confidence of each level will be represented by a percentage following that level. In order to perform the comparison, we normalize our reputations into a reputation score ([0,1]).

As mentioned before, we apply the RP index as a weight factor into the fuzzy beta reputation model. Compared RP6 with RP7, we can see its capability to increase reputation scores for recent active vendors such as eBay6 and eBay9. We can also clearly find its capability to downgrade the reputations of not-so-active vendors. We can see from the diagram that our preference-based system results are very close to the fuzzy beta reputation model's result on RP5. We believe that if we tune this policy set to reflect the implicit preferences in the fuzzy beta reputation model, the difference will be even smaller. So our system is more flexible and adaptable for reputation calculations in different types of e-commerce applications and even other types of applications as long as implicit or explicit preferences can be clearly expressed by policy rules.

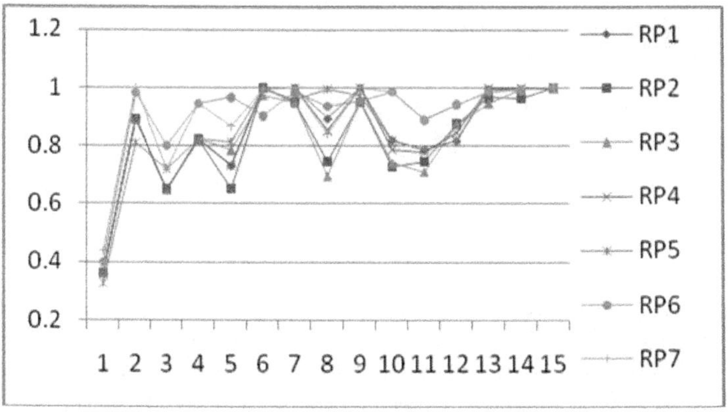

Fig. 5 Comparisons of different reputation models

When we incorporate detailed feedback as supplemental and extra support into the reputation calculation (RP2), most vendors' reputations will decrease (compared to RP1), because the proposed parallel model uses all four aspects of detailed feedback as equally weighted factors in the reputation calculation process. When we change to the second sub policy set, the reputations for those with good shipping time feedbacks are increased (such as eBay5 and eBay14) and the reputations for those with bad shipping time feedbacks are decreased (such as eBay6, eBay8 and eBay11). RP4 shows the same trend as the parallel model for final reputations, which pulls the final reputations in the same direction as the parallel model. The only difference is that the parallel model generates more accurate reputations. From the above results, we can see that incorporating detailed feedbacks into reputation calculation can offer users more accurate reputation evaluations. Compared to the feedback adjustable model, the parallel model provides users with a more scalable mechanism for users to get not only accurate reputations but also to get more focused reputations based on their own preferences. However, the feedback adjustable model can provide the same reputation evaluation adjustment trend as the parallel model does, and it is more convenient and efficient for users to use, since users do not have to define an additional policy set for various aspects of detailed feedback. Besides, the feedback adjustable model can be easily adapted into different ecommerce environments no matter detailed feedbacks are provided or not.

5.2 Comparison of Shopping Recommendations

As a direct application, the models and mechanisms we proposed in this paper can help shoppers choose a suitable product from a reputable vendor following shoppers' own preferences. We add product price to the policy definition and decision-making process for the final recommendation. We also define two different main policy sets in order to illustrate the accuracy and flexibility of our system. Then we run the system under these policy sets with different models separately. We set the sub policy set 1 used in Section 5.1 for the parallel model as our default sub policy set. The first main policy set is defined following human common sense, which always tries to select a low-priced product from a reputable vendor. The second main policy set is defined as an extreme case, which always prefers an expensive product from a reputable vendor.

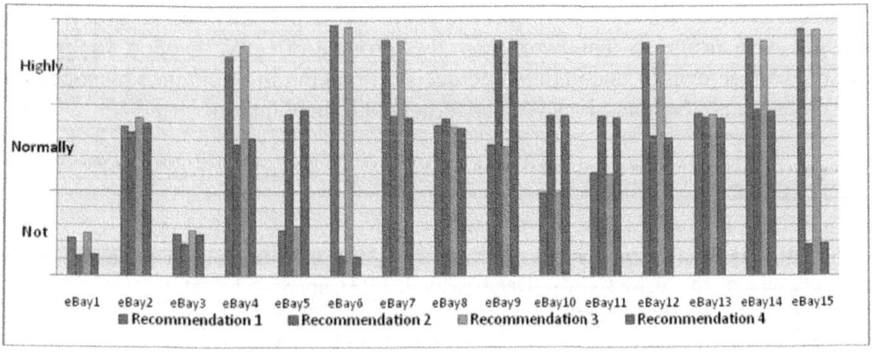

Fig. 6 Comparison of recommendations

Figure 6 indicates the huge differences between two policy sets. Recommendations of those vendors selling at a lower price dramatically drop to a very low level regardless the vendors' reputations. Especially eBay6 and eBay15, who have top reputations among all 15 vendors, suddenly drop to the same recommendation level as eBay1 and eBay3 or even lower. However, this situation is compatible with the policy set we define here, which does not recommend vendors selling at very low prices. The only highly recommended vendor for the second policy set is eBay9, who has an outstanding reputation (85.7% high) and sells at a very high price. For other vendors such as eBay5, eBay10 and eBay11, as long as they sell at high prices and receive average reputations, the recommendation levels will climb increasingly to a high level compared with their recommendations received under the first policy set. There are also three vendors normally recommended under both the first and the second policy sets, because both policy sets recommend at a normal level for average reputable vendors with average prices.

6 Conclusions and Future Work

This paper analyzed reputation modeling and management of a typical C2C environment – eBay – and proposed a fuzzy-logic-based uncertainty modeling for detailed feedback. Two sub-models derived from the general model were also proposed in order to deal with different situations. The introduction of perception-based fuzzy logic into the modeling and decision-making system can help shoppers handle uncertainty and fuzziness in personalized recommendation for e-commerce applications involving reputation, price and other factors. As illustrated in Section 5, a comparison of reputation evaluation between two sub-models gives users a clear view of the advantages and disadvantages of each model. An application of this model in the eBay environment can help shoppers make better decisions that follow their own intentions and preferences for their online shopping activities. The experiments also show the flexibility and adaptability of the system. We will extend our models to different types of e-commerce environments, such as business-to-business (B2B) and business-to-consumer (B2C) environments. A user survey asking users to evaluate the usability and accuracy of our system is in progress. This work can also provide effective support for policy-based management of uncertainty and decision-making in preference management for other types of applications with uncertain and fuzzy factors.

References

[1] Gefen, D.: Reflections on the dimensions of trust and trustworthiness among online consumers. In: ACM SIGMIS Database, vol. 33(3), pp. 38–53 (2002)

[2] eBay help. All about Feedback, `http://pages.ebay.com/help/feedback/allaboutfeedback.html`

[3] Golbeck, J., Hendler, J.: Inferring Binary Trust Relationships in Web-Based Social Networks. ACM Transactions on Internet Technology 6(4), 497–529 (2006)

[4] Bharadwaj, K.K., Al-Shamri, M.Y.H.: Fuzzy computational models for trust and reputation systems. Electronic Commerce Research and Applications 8, 37–47 (2009)
[5] Zadeh, L.A.: Toward a perception-based theory of probabilistic reasoning with imprecise probabilities. In: Intelligent Systems for Information Processing, pp. 3–34 (2003)
[6] Beth, T., Borcherding, M., Klein, B.: Valuation of Trust in Open Networks. In: Proceedings of the European Symposium on Research in Security, pp. 3–18 (1994)
[7] Chen, S.C., Dhillon, G.S.: Interpreting Dimensions of Consumer Trust in E-Commerce. Journal of Information Technology and Management 4(2-3), 303–318 (2004)
[8] Rubin, S., Christodorescu, M., Ganapathy, V., Giffin, J.T., Kruger, L., Wang, H., Kidd, N.: An auctioning reputation system based on anomaly. In: Proceedings of the 12th ACM Conference on Computer and Communications Security, pp. 270–279 (2005)
[9] Wu, Z., Wu, H.: An Agent-Based Fuzzy Recommendation System Using Shoppers' Preferences for E-Commerce Applications. International Journal of Uncertainty, Fuzziness and Knowledge-Based Systems 18(4), 471–492 (2010)

Negative Biofeedback for Enhancing Proprioception Training on Wobble Boards

Alpha Agape Gopalai and S.M.N. Arosha Senanayake

Abstract. Biofeedback has been identified to improve postural control and stability. A biofeedback system communicates with the humans' Central Nervous System through many available modalities, such as vibrotactile. The vibrotactile nature of feedback is presented in a simple and realistic manner, making the presentation of signals safe and easy to decipher. This work presents a wobble board training routine for rehabilitation combined with real-time biofeedback. The biofeedback was stimulated using a fuzzy inference system. The fuzzy system had two inputs and one output. Measurements to test this rehabilitation approach was taken in Eyes Open and Eyes Close states, with and without biofeedback while subjects stood on the wobble board. An independent T-test was conducted on the readings obtained to test for statistical significance. The goal of this work was to determine the feasibility of implementing a negative close-loop biofeedback system to assist in proprioceptor training utilizing wobble boards.

1 Introduction

Postural control, a fundamental to the maintenance of balance, is essential to carry out all Activities of Daily Living (ADL) [12]. An effective postural control mechanism requires the use and integration of several sensory inputs namely: (1) visual, (2) vestibular, and (3) somatosensory. Studies have proven that, in the absence or degradation of these sensory inputs, a poor postural control is exhibited [13]. A potential solution to this scenario is to use biofeedback to provide additional sensory or augmented information via sensory modalities.

Biofeedback augments the extrinsic information about task success or failure provided to the performer. These systems are designed to improve postural control [1]. Biofeedback systems typically consists: (1) sensory device, (2) restitution device

Alpha Agape Gopalai · S.M.N. Arosha Senanayake
Monash University Sunway campus, Jalan Lagoon Selatan, 46150 Bandar Sunway, Malaysia
e-mail: alpha.agape@ieee.org, aroshas@ieee.org

A. Gaspar-Cunha et al. (Eds.): Soft Computing in Industrial Applications, AISC 96, pp. 163–172.
springerlink.com

that displays/ presents the biofeedback information to end-users, and (3) processing system that performs computation for decision making and input/ output control of the system. Biofeedback protocol provides feedback to end-users in the form of visual, auditory or tactile signal(s) (combination of one or more) [4]. These protocols have been successfully implemented to healthy young adults [15, 11, 13], healthy older adults [15, 13], frail older adults [14], stroke patients [16, 12], Unilateral Vestibular Loss (UVL) patients [5, 17], and Bilateral Peripheral Vestibular Loss (BVL) patients [9].

The importance of biofeedback systems has been greatly recognized in clinical applications. However, visual and audio biofeedback systems may interfere with end-user's visual or acoustic dependency for different ADL tasks [1]. The application of these biofeedbacks are limited to individuals who are not deaf and blind. Tactile biofeedback therefore presents itself as a generic, realistic and appropriate alternative to provide additional sensory information for rehabilitation applications. Tactile feedback technology is based on the ability of the skin to sense and communicate this modality to the CNS. There are three variations to the tactile modalities (1) electrotactile, (2) thermal and (3) vibrotactile [2]. Vibrotactile feedback has been identified to be the safest on human skin. The feedback information is presented in a simple and realistic manner, making signals easy to decipher [13]. The signals are generated utilizing vibration actuators (vibrotactors), aimed to augment the somatosensory perception. Previous studies, report successfully utilizing vibrotactors mounted at the foot [13], waist [1, 17], and head [4, 10] resulting in significant postural improvement.

Balance training is an effective intervention method to improve static postural sway and dynamic balance in humans [7]. Balance training programs that incorporated biofeedback, studied the effects of biofeedback for balance training on stable platforms/ ground using tandem stance [1, 4], single leg balance [10] and functional reach while standing [12, 13]. The potential of incorporating vibrotactile feedback with a wobble board (perturbed surface) training has not been considered, to the authors' knowledge.

Wobble board balance training routines have been shown in previous studies to significantly improve postural control by strengthening ankle proprioception [3, 8]. This work proposes a method for rehabilitation, conditioning, and strengthening ankle proprioception using a wobble board routine with real-time vibrotactile biofeedback stimulated through soft-computing methods. A negative feedback closed-loop control system was designed. This system utilized tactile biofeedback to generate warnings, based on the outputs of the Fuzzy Inference System (FIS). The FIS monitored the rotational angles experienced by the ankle and trunk along the Anterior-Posterior (AP) plane. The purpose of this study was to determine if the implementation of biofeedback improved the postural control of subjects on the wobble board. An improved postural control (stable) on the wobble board is a predictor of an effective intervention for foot proprioception training [8, 7].

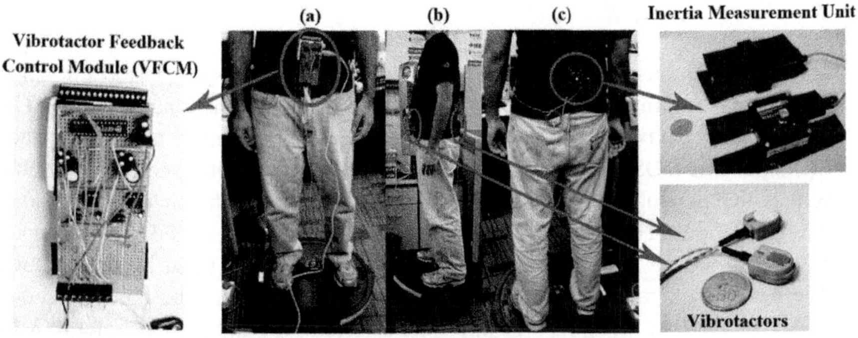

Fig. 1 (a) Front View of subject shows the VFCM attached to the elastic waist strap (b) Side View shows the vibrotactors tucked between the elastic belt and subjects stomach and lower back (c) Posterior View shows the positioning of the IMU on the trunk for trunk angle measurement

2 System Hardware

The devices integrated into this system involves two Inertial Measurement Units (IMU), a wobble board (BOSU® Balance Trainer), Vibrotactor Feedback Control Module (VFCM) and a personal computer for real-time communication and data recording. Fig. 1 depicts the placement of the hardware on the subject while standing on the balance trainer.

2.1 Inertial Measurement Unit (IMU)

The wireless IMU device used in this study was MicroStrain®'s Inertia-Link®. The sensing unit measures 41 mm x 63 mm x 24 mm and weighs 39 g, making it suitable for wearable applications without impairing natural movement. The IMU combined triaxial accelerometers, triaxial gyro, and an on-board processor with sensor fusion algorithms. The sensor has a resolution of 0.1° and supports a full 360° measurements of orientation ranges over all axes. This work utilizes two IMUs. The IMUs were set to stream Euler angles in real-time at a sampling rate of 100 Hz. The IMUs provided accurate roll and pitch measurements, with yaw measurements drifting over time. The yaw measurements were not used in this study. All communication with the IMUs were handled by the host computer via USB base stations. Readings were logged and saved on-board the computer. One IMU was mounted on the surface of the wobble board while the other IMU was mounted to the trunk of the subject. The IMUs were attached by means of hook-and-loop fasteners. The IMUs were fastened to monitor the platform and trunk movement of subjects, with and without biofeedback, in Eyes Open (EO) and Eyes Closed (EC) conditions.

2.2 *Wobble Board*

The wobble board used was the BOSU® Balance Trainer, it is a proprioception and core stability training device [6, 7]. This device has two functional surfaces. The base configuration used in this work, a convex base, provided an unstable surface to stand on. The BOSU® surface measured 635 mm in diameter and had a variable height, which depended on the amount of air in its inflatable chamber. The direction and degree of perturbation experienced was solely dependent on the subjects' body sway (perturbations were self-inflicted). The balance trainer tilted in the direction of the nett force acting on the surface of the balance trainer. The balance trainer allowed for $\pm 40^\circ$ of tilt along the Anterior-Posterior (AP) and Medial-Lateral (ML) planes. The BOSU® balance trainer enabled investigation of (1) immediate defensive postural reaction and (2) the adaptation of postural control mechanisms with and without forewarning (biofeedback) in EO and EC.

2.3 *Vibrotactor Feedback Control Module (VFCM)*

Vibrotactors are small light-weighted motors that produce vibrations when powered. The vibrotactors used in this work was an inertia transducer, Tactaid VBW32 from Audiological Engineering Corporation. The device has a resonant frequency of 250 Hz, an ideal frequency for the human somatosensory system. The VBW32 also has a very quick ring-up and ring-down period enabling it to provide rapid responsiveness, which is required for fine control in this work. The vibrotactile control module consisted of Cypress's CY8C27443 8-bit Microprocessor, a 9 V power supply, and audio amplifiers (LM3836N) that functioned as motor drivers for the vibrotactors used. The microprocessor monitored the input from the host computer to determine the magnitude of tactile feedback to be provided. Signals to the microprocessor was sent from the FIS running in the host computer via the digital lines of a data acquisition device, National Instrument's NI USB-6009. The VFCM (with a 9 V battery) weighs approximately 190 g.

3 Experimental Method

The subjects for this study consisted of six healthy subjects (3 Males and 3 Females), volunteers from the community, aged between 20 and 30 years. All participants were healthy and had no known neurological, muscular, or postural disorder. The subject group had the following average reading and ± S.D. readings, 23.69 ± 2.39 years of age with relatively similar Body Mass Indexes, 22.17 ± 2.14 kg/m^2. All subjects had no exposure to the wobble board training prior to this work. This study was reviewed and approved by the Monash University Human Research Ethics Committee (MUHREC). Written consent was obtained from all subjects who participated in this study, after the purpose and procedures were clearly explained in accordance to the guidelines set by MUHREC. Tests were conducted in EO, with and without biofeedback, and EC, with and without biofeedback. Readings were taken in the following order (1) EO with no feedback, (2) EC with no feedback, (3) EO with feedback, and (4) EC with feedback. Subjects were required to take a 2 minute break in between each acquisition, to reduce the effect of fatigue on the dataset. Each data acquisition period was carried out for 60 seconds, subjects were required to perform 3 sets of the previously mentioned acquisitions (carried out sequentially).

A stretchable waist belt was placed around the subject's waist and was held securely using hook-and-loop fasteners, Fig. 1. The surface of this belt allowed for (1) the IMU to be attached to the subject's trunk and (2) the VFCM to be attached to the subject using hook-and-loop fasteners. The vibrotactors were placed between the belt snugly at the stomach and the lower back, referring to the AP direction. The actuators were held in place by the stretchable waist belt. The objective of the wobble board training routine was to monitor the ability of subjects to maintain postural control. Postural control was gauged in relation to the postural sway, measured by trunk displacement [10]. The trunk displacement measured in radians, was converted into angles for further analysis. Good postural control is achieved by keeping the wobble board and trunk relatively stable within a defined target threshold. Readings obtained from the IMUs within the acceptable range nullifies the signals generated by the VFCM to the vibrotactors. The sensitivity limits for increasing feedback signal strengths was determined by the FIS which monitored the two IMUs.

4 Measurements and Data Collection

Once subjects mounted the wobble board and were comfortable, the host computer broadcasted a data streaming command to the IMUs, initiating data acquisition. The received data was in 32-byte string format at 100 Hz. These packets arrive at the host computer via the wireless USB base station and was temporarily buffered. The packets were validated by performing checksum calculations. Once validated, the roll, pitch and yaw component within the 32-byte string was extracted and converted into its IEEE 754 equivalent. Measurements along the pitch plane corresponds with the AP plane.

Forewarning on the platform tilt along the AP plane was conveyed to the subject via the vibrotactors. Directional information is conveyed via the activation of these actuators in the corresponding direction. Amplitude information was conveyed by varying the duty cycle of the actuator, this was implemented within the CY8C27443 using 8-bit Pulse Width Modulation (PWM). Table 1, summarizes the varying duty cycles with respect to the output from the FIS.

Table 1 Duty cycles used for varying vibration feedback sensitivity

FIS Output (X)	Feedback Level	PWM Duty Cycle (%)	Direction
$-0.20 < X < 0.20$	No Feedback [a]	0	Off
$0.21 < X < 0.50$	Low	50	Front Tactor
$0.51 < X < 1.00$	High	85	Front Tactor
$-0.50 < X < -0.21$	Low	50	Back Tactor
$-1.00 < X < -0.51$	High	85	Back Tactor

[a] The condition when feedback is not provided is also known as 'deadzone'.

The outputs from the IMUs were fed into a FIS system, that was designed to determine the amplitude of vibration for the vibrotactors. The FIS system used in this work was a Mamdani fuzzy system. The Mamdani-type inference system provides a fuzzy set output, after the aggregation of each rule in the knowledge base. The final output from the FIS is a defuzzification process of the resultant from the aggregation process. The FIS system has two input variables and one output variable. Each input variable consisted of gaussian membership functions to represent the linguistic characteristic of 'poor', 'average', and 'good'. The trunk input angles were defined to span between 45° and 120°, while the platform input angles were defined to span between -40° and 40°. The output variable consisted of one trapezoidal membership function and two gaussian membership functions 'deadzone', 'low', and 'high', spanning between -1 and 1. The negative value of the output refers to a posterior stimulation, while a positive value refers to an anterior stimulation. Membership functions that closely represent the desired distribution was selected for the FIS. A total of 15 fuzzy set rules were defined in this work. It was found that 15 fuzzy sets were sufficient to represent the problem domain in an optimized manner. Fig. 2 summarizes the fuzzy rule set defined within the FIS.

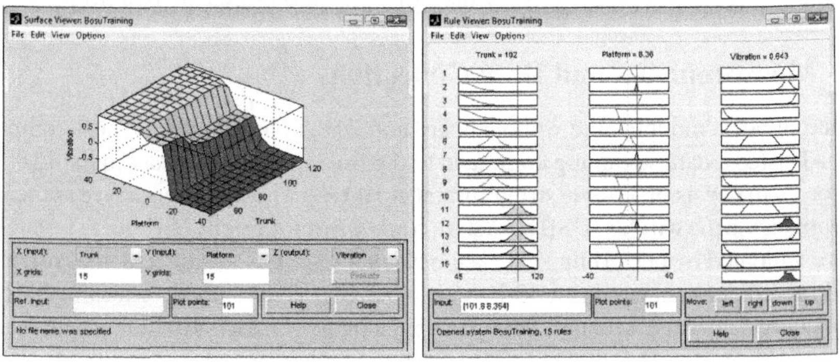

(a) *(b)*

Fig. 2 FIS system created with MATLAB 2009a Artificial Intelligence toolbox (a) A surface plot summarizing the interaction between the output (Z-axis) and the two inputs (X-axis and Y-axis) (b) A screen shot of the rule aggregation in achieving the final output

5 Results

Files stored on-board the host computer was retrieved for further analysis on the system's performance (post acquisition). The stored data acquisition files from the IMUs contained the streamed Euler angles in data columns 'ROLL' and 'PITCH'. Only the rotations along the pitch axis were extracted for analysis, representing sways along the AP (feedback was only provided along the AP plane). The mean and standard deviations of the columns were calculated for the platform and trunk angles. Comparisons were conducted according to the respective measured plane. Readings with and without biofeedback, in EO and EC, were compared against each other using an independent T-test and tabulated in Table 2. Table 2 contained the averaged measurements of the platform and trunk angles. Measurements presented has been averaged across all subjects, for each logged session. A false reject ratio $\alpha = 0.05$ was set, biofeedback was accepted as significant if p-value <0.05.

Table 2 The mean ranges of angular displacement (±S.D.) in the 'Pitch' plane measured for the platform and trunk, averaged across subjects for EO and EC balance task. Measurements presented in degrees.

Task	Platform		Trunk	
	No Feedback	Feedback[a]	No Feedback	Feedback[a]
Male EO	-2.53 ± 7.44	0.03 ± 6.86	98.56 ± 2.00	88.93 ± 0.94
Male EC	-3.55 ± 3.70	-0.75 ± 4.12	99.41 ± 1.93	87.73 ± 1.03
Female EO	-1.71 ± 6.85	0.10 ± 5.05	109.84 ± 3.83	88.72 ± 2.33
Female EC	-0.80 ± 2.45	0.41 ± 2.43	113.13 ± 2.31	89.03 ± 1.48

[a] calculated p-values<0.04, for improvement measured.

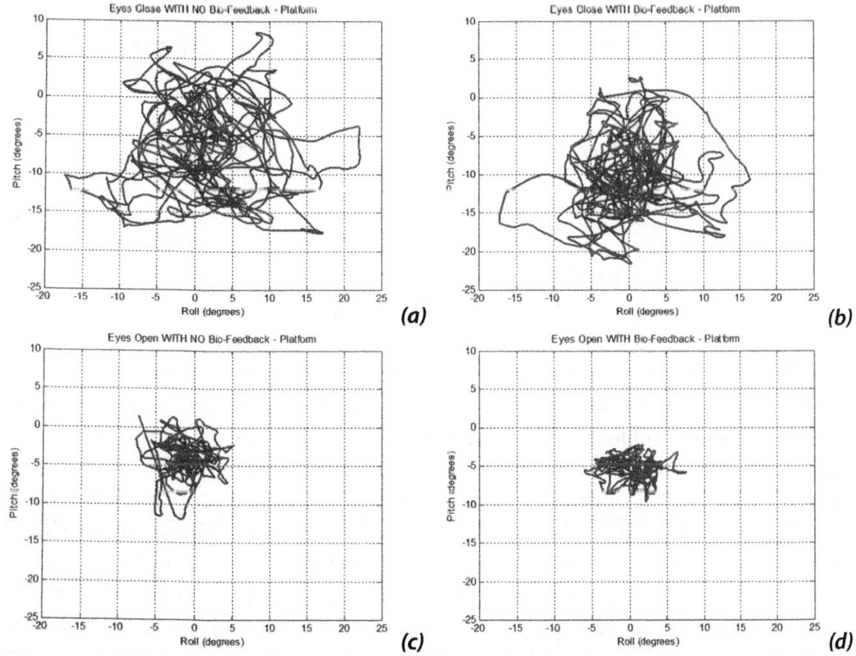

Fig. 3 Recordings of platform angular sway are depicted in EO and EC conditions, with and without biofeedback

The roll and pitch values were used to produce a X-Y plot (Roll vs. Pitch). The plot enabled qualitative assessment of the postural control exhibited by subjects while on the board, Fig. 3. The graph depicts the amount of perturbation experienced on both planes. Subject's with poor postural control exhibits a wider spread on the graph as compared to subject's with good postural control.

The FIS requires the inputs from the trunk and platform angles before determining the level of vibration to be sent to the CY8C27443. Fig. 4 illustrates a sample reading of trunk (Fig. 4(a)) and platform (Fig. 4(b)) angles which is sent into the FIS, for a single subject. The FIS generates the appropriate amplitude (Fig. 4(c)) after defuzzification of the output membership function. This output value is then passed to the VFCM to determine the magnitude of warning using the limits defined in Table 1, results shown in Fig. 4(d).

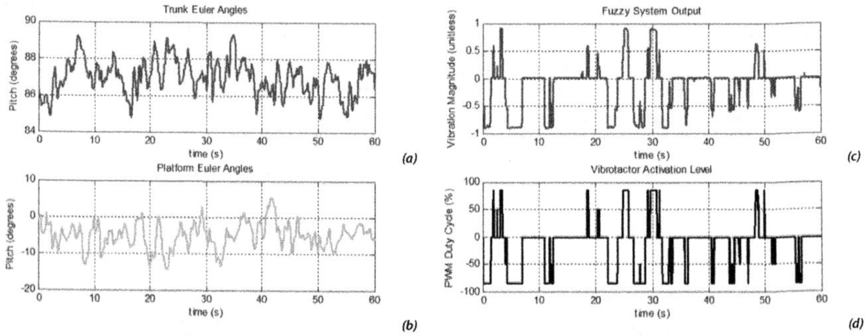

Fig. 4 Reproduced summary of inputs and output of the FIS for a subject standing on the wobble board (a) trunk Euler angles (b) platform Euler angles (c) FIS output (d) Vibrotactors' activation levels based on FIS outputs

6 Discussion and Conclusion

A fuzzy system to monitor postural control on a perturbed surface was successfully designed and tested. From the results, it is observed that the biofeedback generated by the VFCM was effective in improving postural control. The signals from this module prevented subjects from experiencing large amounts of perturbations. Subjects were observed to produced controlled and stable measurements. Subjects also demonstrated an improvement with a statistical significance of $p<0.04$. Significant testing was carried out on the data set using an independent T-test, comparing the angles recorded along the AP plane (with and without feedback - independent samples). A simple check for normality was conducted on the samples acquired, box plots were plotted for each data set to check for outliers and skewness. All data sets resulted in a normal distributed plot, with no outlier and no observable skew of the graphs.

The inclusion of the FIS system allowed for a realistic consideration of the trunk and platform angles, in determining the level and direction of feedback to be provided. The fuzzy system utilized in this work used a Mamdani fuzzy system, to allow for a continuous output function. The fuzzy system was able to provide subjects with the appropriate level of feedback, resulting in an improvement of the postural control, which was observed across all subjects.

This study has only incorporated biofeedback along the AP plane. Studies have shown that the effect of directional specific biofeedback is dependent on the nature of the stance in question [10]. The provision of AP biofeedback during a stationary stance, assists in reducing trunk sway along the AP and ML plane [11]. Biofeedback in the ML plane is more suited when performing non-stationary tasks [11]. Postural sway along the AP is influenced by the proprioception control of the human feet, while ML sway is dependent on the hip muscles strength. The wobble board training routine is suited for proprioception training [3].

This system has application potential in a rehabilitation setting. From the results observed, individuals or clinicians can use this system for postural control and proprioception training, conditioning and strengthening. The overall weight of the attachments required on the end-user is approximately 240 g. The light weight of the system ensures that the system monitors the natural postural control of the user, without introducing external factors that may influence postural control mechanism. Total weight of the system can be reduced further with the introduction of a surface mount circuit, powered by mercury battery cells. A surface mount circuit would significantly reduce the weight and size of the VFCM. Future work in this direction will involve, studying the influence of 'Gender' and 'Subject lifestyle' on the effectiveness of biofeedback systems.

Acknowledgements. This work was supported by Monash University, Sunway Campus, The Ministry of Science, Technology and Innovation (MOSTI), Malaysia, and Moves International Fitness.

References

1. Alahakone, A., Senanayake, S.: A real-time system with assitive feedback for postural control in rehabilitation. IEEE/ASME Transaction On Mechatronics 15(2), 226–233 (2010)
2. Alonso, M., Vexo, F., Thalmann, D.: Stepping into virtual reality: A practical approach, p. 151. Springer, New York (2008)
3. Clark, V.M., Burden, A.M.: A 4-week wobble board exercise programme improved muscle onset latency and perceived stability in individuals with a functionally unstable ankle. Physical Therapy in Sports 6, 181–187 (2005)
4. Davis, J., Carpenter, M., Tschanz, R., Meyes, S., Debrunner, D., Burger, J., Allum, J.: Trunk sway reductions in young and older adults using multi-modal biofeedback. Gait & Posture 31, 465–472 (2010)
5. Dozza, M., Wall, C., Peterka, R.J., Chiara, L., Horak, F.B.: Effects of practising tandem gait with and without vibrotactile biofeedback in subjects with unilateral vestibular loss. Journal of Vestibular Research: Equilibrium & Orientation 17, 195–204 (2007)
6. Ghasemzadeh, H., Jafari, R., Prabhakaran, B.: A body sensor network with electromyogram and inertial sensors: multimodal interpretation of muscular activities. IEEE Transaction on Information Technology in Biomedicine 14(2), 198–206 (2010)

7. Gopalai, A.A., Gouwanda, D., Senanayake, S.M.N.A., Chong, Y.Z.: Instrumented dynamic platform for observing human postural sway: A preliminary study. In: International Conference of Soft Computing and Pattern Recognition, pp. 621–626 (2009)
8. Gopalai, A.A., Senanayake, S.M.N.A.: The impact of fitness level on postural control when standing on a perturbed surface using an instrumented dynamic platform. In: Lim, C.T., Goh, J.C.H. (eds.) Proceedings of 6th World Congress of Biomechanics (WCB 2010). IFMBE, vol. 31, pp. 305–308. Springer, Heidelberg (2010)
9. Hegeman, J., Honegger, F., Kupper, M., Allum, J.H.: The balance control of bilateral peripheral vestibular loss subjects and its improvement with auditory prosthetic feedback. Journal of Vestibular Research: Equilibrium & Orientation 15, 109–117 (2005)
10. Huffman, J., Norton, L., Adkin, A., Allum, J.: Directional effects of biofeedback on trunk sway during stance tasks in healthy young adults. Gait & Posture 32, 62–66 (2010)
11. Janssen, L., Verhoef, L., Horlings, C., Allum, J.: Directional effects of biofeedback on trunk sway during gait task in healthy young subjects. Gait & Posture 29, 575–581 (2009)
12. Januario, F., Campos, I., Amaral, C.: Rehabilitation of postural stability in ataxic/hemiplegic patients after stroke. Disability and Rehabilitation 32(21), 1775–1779 (2010)
13. Shirogane, S., Tanaka, T., Izumi, T., Maeda, Y., Oyama, Y., Yoshida, N., Ino, S., Ifukube, T.: A feasibility study of an integrated system using a force plate and a plantar vibrotactile stimulator for fostering postural control in the elderly. Physical & Occupational Therapy in Geriatrics 28(1), 22–32 (2010)
14. Sihvonen, S., Sipila, S., Era, P.: Changes in postural balance in frail elderly women during a 4-week visual feedback training: a randomized control trial. Gerontology 50, 87–95 (2004)
15. Verhoeff, L., Horlings, C., Janssen, L., Bridenbaugh, S., Allum, J.: Effects of biofeedback on trunk sway during dual tasking in healthy young and elderly. Gait & Posture 30, 76–81 (2009)
16. Walker, C., Brouwer, B.J., Culham, E.G.: Use of visual feedback in retraining balance following acute stroke. Physical Therapy 80, 886–895 (2000)
17. Wall, C., Kentala, E.: Control of sway using vibrotactile feedback of body tilt in patients with moderate and severe postural control deficits. Journal of Vestibular Research: Equilibrium & Orientation 15, 313–325 (2005)

Part III: Bio-inspired Systems

TDMA Scheduling in Wireless Sensor Network Using Artificial Immune System

Zohreh Davarzani, Mohammah-H Yaghmaee, and Mohammad-R. Akbarzadeh-T

Abstract. Today, wireless sensor networks encompass a large volume of applications. Wireless sensor networks consisted of many nodes by low energy batteries. Therefore, they must consume power as low as possible. TDMA Protocol in these networks is designed for this goal. In this paper a multiobjective immune algorithm is proposed for finding optimal solutions to TDMA scheduling problem. The simulation results show a better performance in comparison to two algorithms using instances with different sizes.

Keywords: TDMA scheduling, wireless sensor networks, Immune System, genetic algorithm.

1 Introduction

Today, wireless sensor networks encompass a large volume, as they are increasingly used in many applications such as traffic monitoring, Earthquake and fire detection. These networks consist of a group of wireless sensor nodes. Sensors are units with sensing, processing, wireless networking capability and a battery with low energy. They can automatically collect the information from sensor nodes and report the measurements to an access point.

Since sensor nodes consisted of a battery with low energy, therefore there are some needs protocol to control sensor nodes network life time. This protocol is called MAC (Medium Access Control) and is vital in specification network lifetime. This protocol for sensor networks provides two accesses: contention based access or time division multiple access (TDMA).

Zohreh Davarzani
Cognitive Computing Lab, Center for Applied Research on Soft Computing and Intelligent Systems

Zohreh Davarzani · Mohammah-H Yaghmaee · Mohammad-R. Akbarzadeh-T
Department of Computer Engineering, Ferdowsi University of Mashhad, Mashhad, Iran
e-mail: zo.davarani@stu-mail.um.ac.ir,
yaghmaee@ieee.org,
akbarzadeh@ieee.org

A. Gaspar-Cunha et al. (Eds.): Soft Computing in Industrial Applications, AISC 96, pp. 175–185.
springerlink.com

In TDMA protocol, time is divided into equal time slots. Each time slot is allocated to nodes and is designed to conform a single packet for transmission and perception between pairs of nodes in the network [2]. A basic principle of TDMA scheduling is to assign time slots to nodes so that collisions would not happen when they are transmitting packets and total number of time slots and sensor nodes energy consumption are minimized [11]. Any nodes just can transmit its data package in time slot that allocated for it. Since sensor nodes energy is limited, this protocol can be used for saving energy of nodes.

In this area, many works have been done. Some researchers have discussed the energy saving problem [1, 2] for TDMA scheduling. Some references have studied how to minimize packet delay in aspect of time [3], how to improve fairness [7], how to maximize parallel operation [12, 13], and how to shorten the total slots to finish a set of transmission tasks [3]. The total number of time slots is minimized by using particle swarm optimization [11]. Ergen et al. [3] proposed three algorithms based on coloring method in graph theory.

This paper describes the application of an artificial immune system to a TDMA scheduling application. A new approach immune algorithm is proposed for finding optimal solutions to TDMA scheduling problems.

The reminder of this paper is organized as follows: the artificial immune system is introduced in section 2. Section 3 describes TDMA scheduling in wireless sensor networks. Section 4 describes the optimization framework, coding method and fitness function. In section 5, the computational results are given. Some concluding remarks are made in section 6.

2 Artificial Immune System

The Artificial immune system (AIS) is a complex functional system that defends the human body from foreign agents such as bacteria or viruses that are called pathogens. Patterns expressed on pathogens are called antigens. The immune system contains cells for recognizing and killing them. These cells are called antibodies. The disease procedure involves attack of an antigen and its proliferation within the human body. After the proliferation of antigen, antibodies are randomly distributed throughout the immune system. AIS has two important processes are called Cloning and affinity maturation. Combination of them is known as the Clonal Selection Principle that is shown in Fig. 1. These are used to explain how the immune system reacts to infection of antigen. This theory is one of methodologies in AIS for salving optimization problem and is a meta-heuristic which is developed based on such system. This paper aims at proposing an artificial immune algorithm to TDMA scheduling. Flowchart of AIS is shown in Fig. 2.

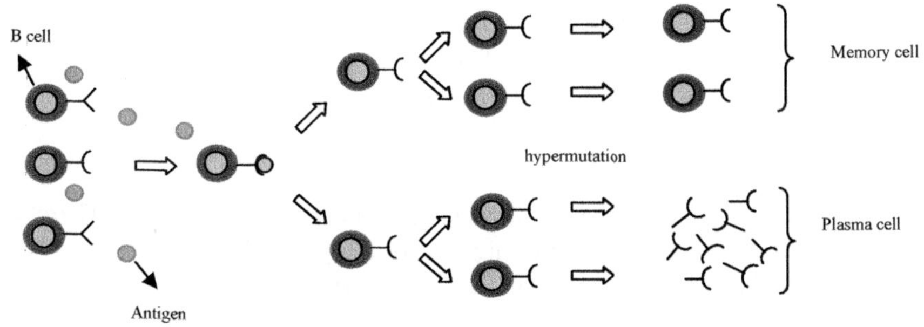

Fig. 1 Clonal Selection Principle [5]

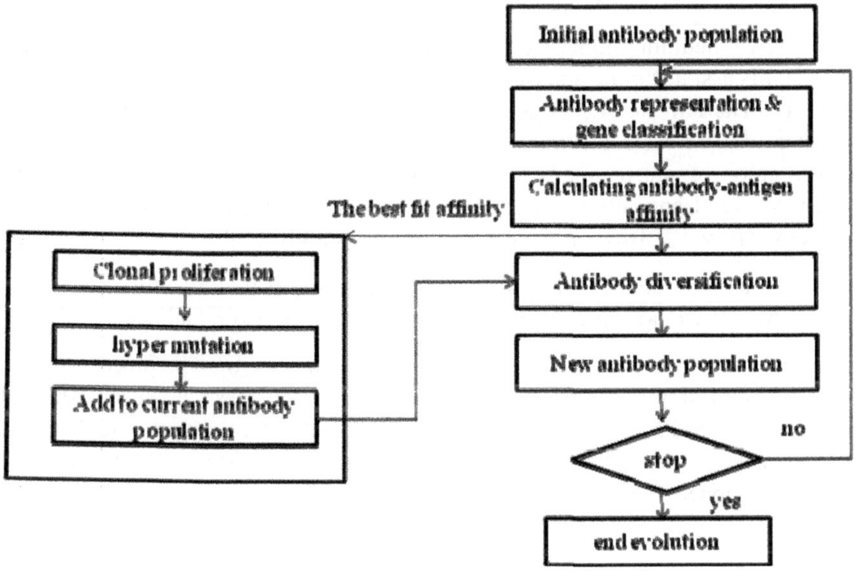

Fig. 2 Flowchart of AIS

3 Definition of TDMA Scheduling

A sensor network can be shown by graph $G=(V, E)$ where V is a set of sensor nodes and E represents the set of communication links between two neighbor nodes. One of the nodes in V is called access point *(AP)*. All traffic generated at sensors is destined for *AP*. In Graph G, distance between two nodes i and j equal to minimum number of edges between two nodes that is showed by d_{ij}. Communication links in the network can be shown by Matrix C that is $N*N$ where N is number of sensor nodes. This Matrix described as follows:

If there is an edge between node i and j

Otherwise

$$c_{ij} = \begin{cases} 1 \\ 0 \end{cases}$$

Sensor nodes may send the data packages simultaneously to AP. Therefore collision can be happened with high probability in these networks. These collisions can be described by matrix *I* which defined by:

$$I_{ij} = \begin{cases} 1 & \text{If } d_{ij}<2 \\ 0 & \text{Otherwise} \end{cases}$$

In TDMA scheduling, time is divided into equal intervals called time slots. Each time slot is assign for transmission or reception a packet between two nodes in the network. The aim of TDMA scheduling is time slots assignment such that total number of time slots for transmission and reception of data packages energy consumption of sensor nodes are minimized. Also collisions would not happen when they are transmitting packets [1, 12]. Because of the occurrence collision in the sensor network, two constraints are defined as follows [8]:

1. A node cannot have transmission and reception at the same time slot.
2. A node cannot simultaneously receive data from several adjacent nodes.

There is a set of sensors which plan to transmit packets to *AP*. The procedure that a single packet follows from its source node to *AP* is called a task. Each subtask needs one time slot for data transmission. Therefore, the goal of the problem is to determine a subsequence of the subtasks and assigning time slots to nodes such that collisions would not happen. For example of sensor network is shown in Fig. 3.

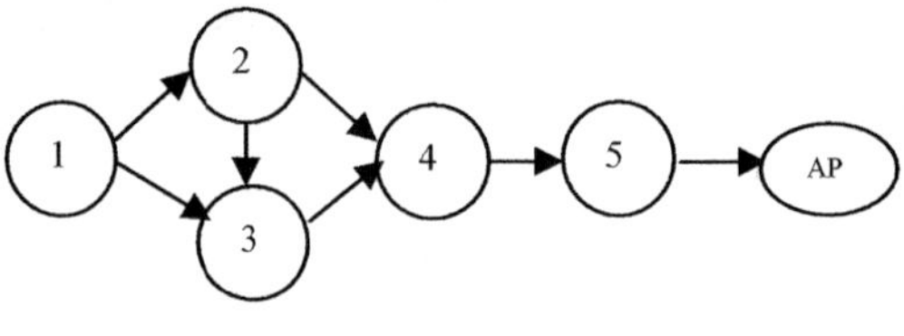

Fig. 3 An example of network

4 AIS for TDMA Scheduling

The proposed AIS for TDMA scheduling are as follows:

1. Initialization: set number of initial population (pop size) and number of clone antibody in each generation. Set initial population.

2. Objective function and affinity evaluation: for all antibodies, evaluate fitness function and affinity. Affinity of antibody is equal to Eq. (1):

$$\text{affinity}(a_i) = \frac{1}{\text{fitness}(a_i)} \quad (1)$$

Where a_i is antibody i.
3. Select *M* antibodies from population with high affinity. This subset is called *F*.
4. Cloning: select *M* copies from subset *F* by using roulette wheel rules. This subset is called *C*.
5. Select *M* antibodies from population with high affinity. This subset is called *F*.
6. Cloning: select *M* copies from subset *F* by using roulette wheel rules. This subset is called *C*.
7. Mutation and diversity operations:
   A: Mutation: select *n* antibodies from *C* and apply mutation operations that make *n* antibodies. Add *n* new antibodies to current generation.
   B: Diversity: for creation of diversity in population, we use from diversity method that is describe in below.
8. Reproducing next generation
   A: select best antibody from current generation to the next generation.
   B: select (popsize-1) antibodies from current generation by using tournament selection rules and add to next generation.
9. Repeat step 2-8 until termination criterion is obtained.

4.1 Antibody Representation

In this paper sequencing subtask is generated following the approach by Jianlin Mao [11]. In this method the length of antibody is equal to $\sum_{i=1}^{N} n_i$ that N is total number of tasks and n_i is the number of subtasks of task i. Each antibody is consists of two parts: *TaskID* and Hop-No. *TaskID* is number of task that subtask belongs to and *Hop-No* is sequence number of this subtask in all the subtasks of task.

For example of antibody representation, a random combination of subtasks is as follows:

(1, 1) (2, 1) (3, 1) (1, 2) (5, 1) (3, 2) (2, 2) (4, 1) (3, 3) (1, 3) (4, 2) (2, 3) (1, 4)

Then taking the TaskIDs out, the above sequence can be encoded as follows:

1,2,3,1,5,3,2,4,3,1,4,2,1

4.2 Mutation Operators

In this study, two mutation operators are used. One of them is one point mutation and another Precedence preserving shift mutation (PPS) operator of Lee et al. [13]. In one point mutation, two subtasks are randomly selected and their locations are changed with each other.

PPS is used to change the sequence of the operations. This operator carries out as follows:

Step 1. Selecting randomly a position i from the parent chromosome. Job which is fixed in this position is called j_r.
Step 2. Finding the leftmost position (*lmp*) and the rightmost position (*rmp*). Lmp is position that first operation of j_r is fixed in this position. Rmp is position that latest operation of j_r is fixed in this position.
Step 3. Selecting randomly a position p in the range of *lmp* to *rmp*.
Step 4. Moving the ith element of chromosome, to the position p.

Fig. 4.a shows one point mutation and Fig. 4.b shows PPS mutation.

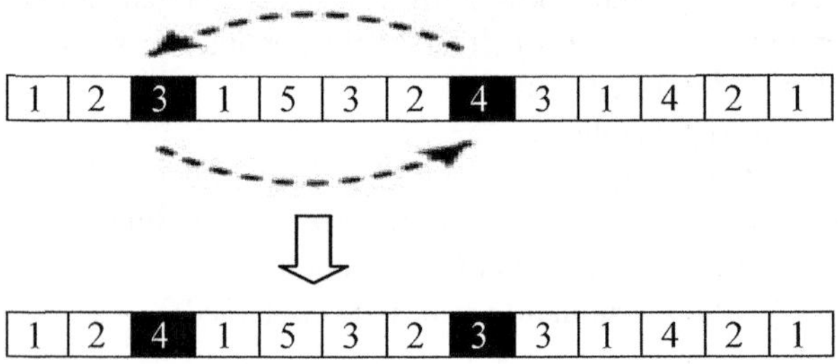

Fig. 4 a) One point mutation operator

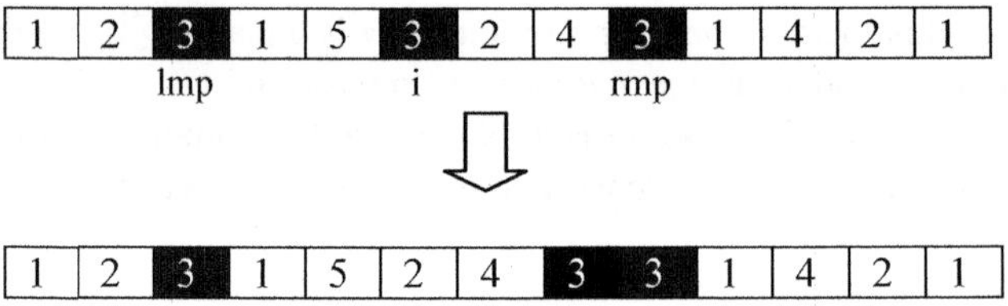

Fig. 4 b) Precedence preserving shift mutation (PPS) operator

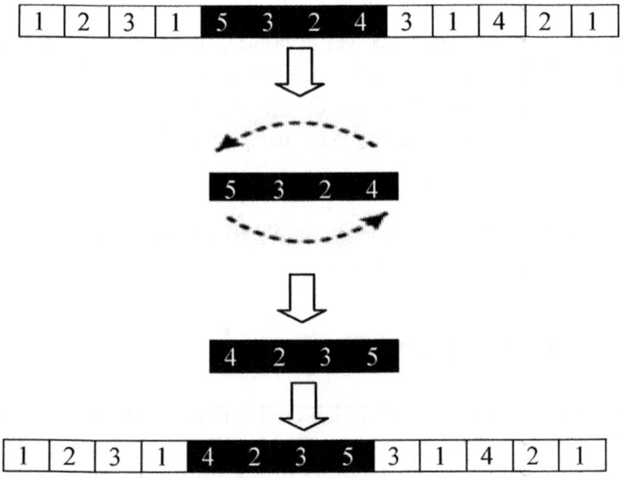

Fig. 5 Antibody inversion

4.3 Antibody Diversification

Traditional immune system often suffers from loss of diversity of the antibody that causes the search to be trapped in local optima. To avoid local optima, two diversification mechanisms of Guan et al. [5] are used in biological immune systems. All the schemes described below.

Antibody Inversion

In this mechanism, a subset of consecutive subtask is randomly chosen from antibody and inverse their location from front to rear. Inversion mechanism is shown in Fig. 5.

5 Fitness Evaluation

In wireless sensor networks, allocating time slots to sensor nodes is a NP-hard problem. In this paper, Objective function consists of two parts: first one minimizes energy consumption and another minimizes total number of time slots in the TDMA cycle. For decreasing wireless sensor nodes energy, nodes are switched off when they don't have data to transmitting and receiving [11]. In general, total energy consumption in wireless sensor network can be calculated by the following equation:

$$EC = \sum_{i=1}^{N} [p_i^{tx} * (t_i^{tx} + t_i^{s-tx}) + p_i^{rx} * (t_i^{rx} + t_i^{s-rx})] \tag{2}$$

In Eq. (2), *N* is the total number of nodes in the network. p_i^{tx} and p_i^{rx} are power consumption of transmitter and receiver at node i, respectively. t_i^{tx} and t_i^{rx} are the total work time of the transmitter and receiver at node i. p_i^{s-tx} and p_i^{s-rx} are the total transition time consumed between the sleep and active states.

Therefore, total objective function is defined as Eq. (3):

$$\min F(s) = \alpha * EC + (1 - \alpha) * \text{Totalslots} \qquad (3)$$

In this equation, *Totalslots* is the total number of required time slots and *EC* is total energy consumption in the network.

6 Experimental Setup and Simulation

We implemented the algorithm in matlab environment and run it on a PC with 2.1 GHz and 320 MB of RAM memory. Some parameters in energy aspect are similar to Ref. [3] that is as follows: the transition time between the sleep and active states is assumed to 470μsec. The power consumed in transmission and reception of a packet set to 81and 180 mW, respectively.

At first, we test proposed algorithm with network consists of 7 nodes and 1 AP. Slot allocations results are shown in Table 1. In this table tuple (i,j) expresses jth subtask of task i. Two algorithms are used for comparisons, which are hybrid GA-PSO algorithm proposed by Ziari and Davarzani [4] and hybrid PSO-GA algorithm proposed by Mao and Zhiming [11]. According to this table, proposed algorithm can get better results and has less number of times slot required for sending data packages than other methods. The reason is proposed algorithm has more time slots with 2 or 3 subtasks than other methods while Hybrid PSO-GA method has more time slot with 1 subtask. Thus it's total number of time slot is more than the other methods.

Also we test algorithm by two networks with 25 and 49 sensor nodes. These networks are proposed by [11] which are problems 1 and 2. The time and energy results of finishing a sampling round are given in Table 2. In this table α=0 means that objective is the total number of time slot performance and α=1 means the objective is the energy consumption performance. As shown in this table, by increase size of network total number of time slot and energy consumption increase. Also show that the proposed algorithm always gets better performance than other methods for different sizes and values α. This is because of the proposed algorithm has more time slots with more number of subtasks. Fig. 6.a and 6.b show average fitness function of total number of time slots and total energy consumption of these three algorithms over generation in problem1. As shown in Fig1.a hybrid GA-PSO converge quickly at the beginning, and hybrid PSO-GA converges in 220 generation while proposed algorithm improves the result after 320 generation. Fig 6.b shows that convergence speed of three algorithms is same but proposed algorithm improves the average fitness on energy aspect.

Table 1 Slot allocation of algorithms

Time slot#	1	2	3	4	5	6	7	8	9	10	11	12	13	14
Proposed algorithm	(4,1)	(3,1) (5,1)	(7,1) (2,1) (3,2)	(3,3) (2,2)	(6,1) (2,3) (1,1)	(2,4) (1,2)	(6,2) (1,3)	(4,2)	(1,4)	-	-	-	-	-
GA-PSO	(4,1)	(3,1)	(3,2) (2,1) (6,1)	(1,1) (7,1)	(5,1) (1,2)	(3,3)	(4,2)	(6,2) (2,2)	(2,3)	(1,3)	(2,4)	(1,4)	-	-
PSO-GA	(1,1)	(1,2)	(1,3)	(1,4) (2,1)	(2,2)	(3,1) (6,1)	(6,2) (3,2)	(2,3)	(4,1)	(3,3)	(4,2)	(5,1)	(7,1)	(2,4)

Table 2 Energy consumption of the algorithm

Problem number	Proposed algorithm		ga-pso		Pso-ga	
	$\alpha=0$	$\alpha=1$	$\alpha=0$	$\alpha=1$	$\alpha=0$	$\alpha=1$
Problem 1	19.05	4.4980	27.0750	5.3060	25	5.7467
Problem 2	26.01	9.11	36.04	11.09	37.124	12.87

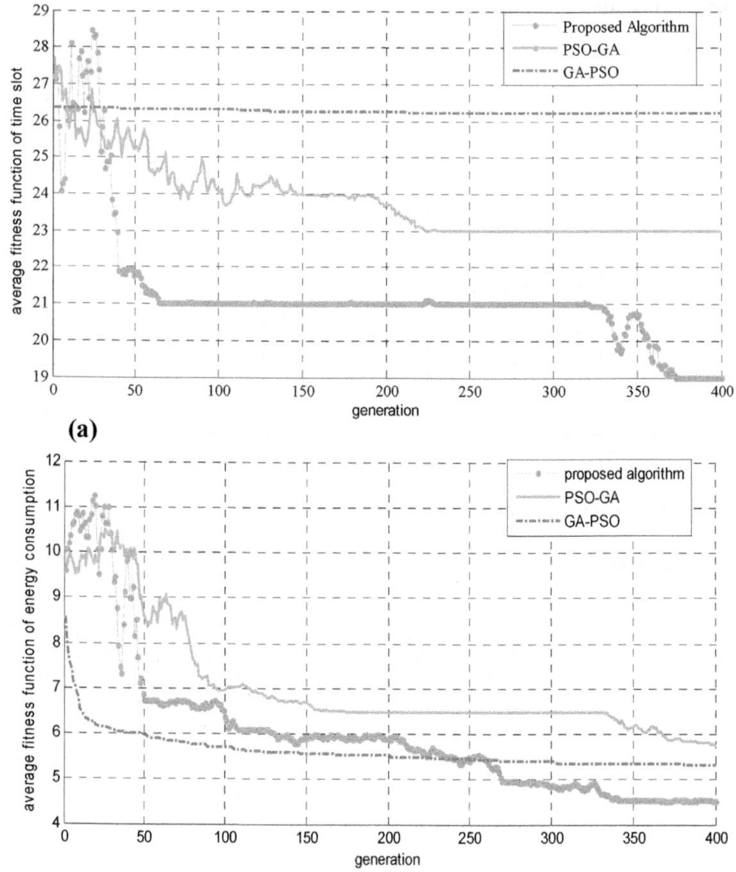

Fig. 6 (a) Average fitness function of energy consumption in problem1. (b) Average fitness function of total number of time slot in problem1

7 Conclusion

In this paper we have considered TDMA scheduling problem and discussed how to assign time slots for saving energy and time in wireless sensor network. Then we have presented a multiobjective immune algorithm for solving this problem. At the end, the proposed algorithm is compared with two algorithms. Simulation results show that proposed framework can get better results than other methods on both time and energy aspects.

References

[1] Ergen, Varaiya, P.: TDMA Scheduling Algorithms for Wireless Sensor Networks (2009)
[2] Xu, N.: A survey of sensor network applications. IEEE Communications Magazine, 102–114 (2002)
[3] Ergen, S.C., Varaiya, P.: TDMA scheduling algorithms for sensor networks. Department of Electrical Engineering and Computer Sciences, University of California (2005)
[4] Davarzani, Z., Ziari, M., Yaghmaee, M.H.: TDMA Scheduling in wireless sensor networks using hybrid of genetic Algorithm and particle swarm optimization. International network computing (2010)
[5] Luh, G., Chueh, C.: Multi-modal immune algorithm for the job-shop scheduling problem, pp. 1516–1532 (2009)
[6] Coello, C.A., Cruz Cortes, N.: An Approach to Solve Multiobjective Optimization Problems Based on an Artificial Immune System, pp. 212–221 (2002)
[7] Younis, M.: An energy-efficient scalable and collision-free MAC layer protocol for wireless sensor networks. Wireless Communications and Mobile Computing, 285–304 (2005)
[8] Shuguang, C.: Energy-delay tradeoffs for data collection in TDMA-based sensor networks. In: IEEE International Conference on Communications, pp. 16–20 (2005)
[9] Sridharan, A., Krishnamachari, B.: Max-min fair collision-free scheduling for wireless sensor networks. In: Workshop on Multihop Wireless Networks (2004)
[10] Shen, Y., Wang, M.: Broadcast scheduling in wireless sensor networks using fuzzy Hopfield neural networ. Department of Engineering Science, pp. 900–907 (2000)
[11] Mao, J., Zhiming, W., Xing, W.: A TDMA scheduling scheme for many-to-one communications. Computer Communications, 863–872 (2007)
[12] Gandham, S., Dawande, M., Prakash, R.: Link scheduling in sensor networks: Distributed edge coloring revisited, pp. 2492–2501 (2005)
[13] Veeramachaneni, K., Osadciw, L.A.: Optimal scheduling in sensor networks using swarm intelligence. In: Proceedings of 38th Annual Conference on Information Systems and Sciences, pp. 17–19 (2004)
[14] Lee, K., Yamakawa, T.: A genetic algorithm for general machine scheduling problems. International Journal of Knowledge-Based Electronic (1998)

A Memetic Algorithm for Solving the Generalized Minimum Spanning Tree Problem

Petrică Pop, Oliviu Matei, and Cosmin Sabo

Abstract. The generalized minimum spanning tree problem is a natural extension of the classical minimum spanning tree problem, looking for a tree with minimum cost, spanning exactly one node from each of a given number of predefined, mutually exclusive and exhaustive node sets. In this paper we present a memetic algorithms for solving the generalized minimum spanning tree problem that combines the population concept of genetic algorithms with a fast local improvement method. The proposed algorithm is competitive with other heuristics published to date in both solution quality and computation time. The computational results for several benchmarks problems are reported and the results point out that the memetic algorithm is an appropriate method to explore the search space of this complex problem and leads to good solutions in a reasonable amount of time.

1 Introduction

The *generalized minimum spanning tree problem* (GMSTP) was introduced by Myung *et al.* [3] and is define as follows: given $G = (V,E)$ an n-node undirected graph and $V_1,\ldots,V_m$ a partition of V into m node sets called *clusters* (i.e., $V = V_1 \cup V_2 \cup \ldots \cup V_m$ and $V_l \cap V_k = \emptyset$ for all $l,k \in \{1,\ldots,m\}$ with $l \neq k$), then the GMSTP asks for finding a minimum-cost tree T spanning a subset of nodes which

Petrică Pop
North University of Baia Mare, Victoriei 76, Baia Mare, Romania
e-mail: `petrica.pop@ubm.ro`

Oliviu Matei
North University of Baia Mare, Victoriei 76, Baia Mare, Romania
e-mail: `oliviu.matei@holisun.com`

Cosmin Sabo
North University of Baia Mare, Victoriei 76, Baia Mare, Romania
e-mail: `cosmin_sabo@prime-tech.ro`

A. Gaspar-Cunha et al. (Eds.): Soft Computing in Industrial Applications, AISC 96, pp. 187–194.
springerlink.com

includes exactly one node from each cluster V_i, $i \in \{1,...,m\}$. We will call such a tree a *generalized spanning tree*.

We assume here that edges are defined between all nodes which belong to different clusters and we denote the cost of an edge $e \in E$ by c_{ij}.

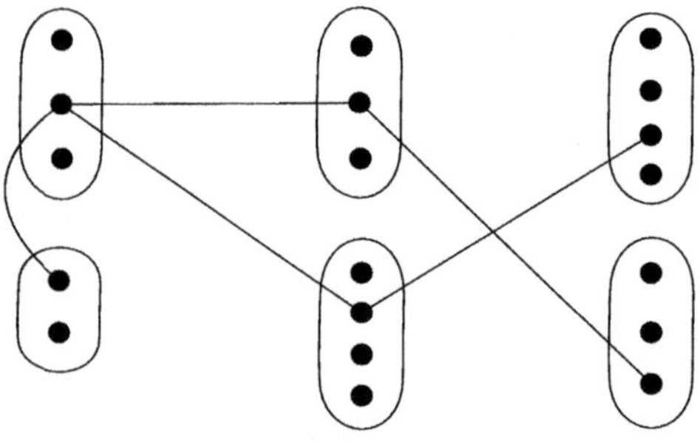

Fig. 1 Example showing a generalized spanning tree in the graph $G = (V, E)$

The GMSTP belongs to the class of generalized combinatorial optimization problems that generalize classical combinatorial optimization problems in a natural way by considering a related problem relative to a given partition of the nodes of the graph into clusters. In the literature one finds generalized problems such as: the generalized traveling salesman problem, the generalized Steiner tree problem, the generalized vehicle problem, generalized fixed-charge network design problem, generalized minimum vertex-biconnected network problem, etc.

In the literature have been considered two variants of the generalized minimum spanning tree problem:

- one in which in addition to the cost attached to the edges, we have costs attached also to the nodes, called the prize collecting generalized minimum spanning tree problem, see [9, 16] and
- the second one consists in finding a minimum cost tree spanning at least one node from each cluster, denoted by L-GMSTP and was introduced by Dror *et al.* [2]. The same authors have proven that the L-GMSTP is NP-hard.

The GMSTP has several real world applications, in what it follows we present some examples:

- design of backbones in communication networks [8];
- network design problem arising in desert environments [2];
- determining the location of the regional service centers [14].

In the present paper we confine ourselves to the problem of choosing exactly one node from each of the clusters. The MST is a special case of the GMSTP where each cluster consists of exactly one node.

By reduction from the vertex cover problem, Myung *et al.* [3] proved that the GMSTP is an NP-hard problem. Pop [14] showed a stronger result concerning the complexity of the problem, namely the GMSTP even defined on trees is NP-hard. A survey on integer linear programming formulations for the GMSTP has been recently presented by Pop [18].

Myung et al. [3] used a branch and bound procedure in order to solve the GMSTP. Their lower procedure is a heuristic method which approximates the linear programming relaxation associated with the dual of the multicommodity flow formulation of the GMSTP. They developed also a heuristic algorithm which finds a primal feasible solution for the GMSTP using the obtained dual solution and reported the exact solution of instances with up to 100 vertices. The GMSTP was solved to optimality for nodes up to 200 by Feremans et al. [4] using a branch-and-cut algorithm. More recently, Pop et al. [13] have proposed a new integer programming formulation of the GMSTP based on a distinction between local and global variables and a solution procedure that finds an optimal solution on instances with up to 240 vertices.

The difficulty of obtaining optimum solutions for the GMSTP has led to the development of several metaheuristics. The first such algorithms were the tabu search (TS) heuristic of Feremans [5] and the simulated annealing (SA) heuristic of Pop [14], an improved version of the SA was described in [17]. Two variants of a TS heuristic and four variable neighborhood search (VNS) based heuristics were later devised by Ghosh [7]. Another VNS algorithm combined with integer linear programming was proposed by Hu et al. [10]. The authors report that their VNS approach can produce solutions that are comparable to those obtained by means of the second variant of the TS heuristic of Ghosh [7]. Golden et al. [8] have devised a local search heuristic (LSH) and a genetic algorithm (GA) for the GMSTP. Both algorithms have yielded improvements on TSPLIB instances with sizes between $198 \leq n \leq 225$. On none of these instances did the LSH outperform the GA. Recently, an attribute based tabu search heuristic employing new neighborhoods was proposed by Oncan et al. [12]. The authors mention that their TS based heuristic yields the best results for all instances.

In order to solve the GMSTP we propose in this paper a memetic algorithm, that can be seen as a hybrid technique that combines the population concept of genetic algorithms with an intensification mechanism that exploits the specific knowledge of the problem.

2 The Memetic Algorithm for Solving the Generalized Minimum Spanning Tree Problem

Memetic algorithms have been introduced by Mascato [11] to denote a family of metaheuristic algorithms that emphasis on the used of a population-based approach with separate individual learning or local improvement procedures for problem search. Therefore a memetic algorithm is a genetic algorithm (GA) hybridized with a local search procedure to intensify the search space.

Genetic algorithms are not well suited for fine-tuning structures which are close to optimal solutions. Incorporating of local improvement operators into the recombination step of a GA is essential in order to obtain a competitive GA. Memetic algorithms have been recognized as a powerful algorithmic paradigm for evolutionary computing, being applied successfully to solve combinatorial optimization problems such as the VRP (Vehicle Routing Problem) and the CARP (Capacitated Arc Routing Problem) [19], etc.

We present in this section a memetic algorithm for solving the GMSTP. The proposed computational model to approach the problem is genetic algorithm combined with an intensification mechanism based on Kruskal's algorithm for finding the minimum cost tree spanning a given number of nodes.

2.1 Genetic Representation

We represent a chromosome by an array of dimension m so that the gene values correspond to the nodes selected from each of the cluster V_k, $k \in \{1,...,m\}$. Therefore, an individual is represented as a sequence of nodes $(N_{k_1}, N_{k_2}, ..., N_{k_m})$, where the node N_{k_p} is the node selected from the cluster V_{k_p}, $p \in \{1,...,m\}$.

An example of an individual in the case of a graph with 56 nodes partitioned into 7 clusters is:

(11 21 8 35 28 42 55)

and the cluster representation of the individual is as follows:

$(V_2\ V_3\ V_1\ V_5\ V_4\ V_6\ V_7)$.

2.2 Initial Population

The construction of the initial population is of great importance to the performance of genetic algorithms, since it contains most of the material the final best solution is made of. In our algorithm, we have produced 20 initial solutions generated randomly: by selecting randomly the nodes from each of the clusters (exactly one node from each cluster).

2.3 The Fitness Value

Every solution has a fitness value assigned to it, which measures its quality. In our case the, the fitness value of an individual $(N_{k_1}, N_{k_2}, ..., N_{k_m})$ is given by the minimum cost of the tree which spans the nodes: $N_{k_1}, N_{k_2}, ..., N_{k_m}$. Such a tree which is a feasible solution of the GMSTP, i.e. a generalized spanning tree, always exists because we assumed that edges are defined between all nodes which belong to different clusters. This minimum generalized spanning tree is determined using the Kruskal's algorithm.

2.4 Genetic Operators

2.4.1 Crossover

Two parents are selected from the population by the binary tournament method, i.e. the individuals are chosen from the population at random.

Offspring are produced from two parent solutions using the following crossover procedure: it creates offspring which preserve the order and position of symbols in a subsequence of one parent while preserving the relative order of the remaining symbols from the other parent. It is implemented by selecting a random cut point. The crossover operator for the set of nodes N is straightforward. We use a single cut-point. Two randomly selected parents generate two offspring as follows:

- the first offspring is made of the first part of the first parent, respectively the second part of the second parent;
- the second offspring is made of the first part of the second parent, respectively the second part of the first parent.

Next we present the application of the proposed crossover. We assume two well-structured parents chosen randomly, with the cutting point between 2 and 3:

$P_1 = (11\ 21 \mid 35\ 42\ 55)$
$P_2 = (14\ 26 \mid 31\ 44\ 53)$

The offspring are:

$O_1 = (11\ 21 \mid 31\ 44\ 53)$
$O_2 = (14\ 26 \mid 35\ 42\ 55)$

2.4.2 Mutation

We use in our algorithm a straightforward mutation operator: a random node is selected to undergo mutation. Another node belonging to the same cluster replaces the selected node and a new individual is created.

2.4.3 Selection

The selection process is deterministic. The first selection is $(\mu + \lambda)$, where μ parents produce λ offspring. The new population of $(\mu + \lambda)$ is reduced again to μ individuals by a selection based of the "survival of the fittest" principle. In other words, parents survive until they are suppressed by better offspring. It might be possible for very well adapted individuals to survive forever. This feature yields some deficiencies of the method [1]:

1. In problems with optimum moving over time, a $(\mu + \lambda)$ selection may get stuck at an outdated good location if the internal parameter setting becomes unsuitable to jump to the new field of possible improvements.
2. The same happens if the measurement of the fitness or the adjustment of the object variables are subject to noise, e.g. in experimental settings.

In order to avoid effects, Schwefel investigated the properties of (μ,λ), selection, where μ parent produce λ $(\lambda > \mu)$ and only the offspring undergo selection. In other words, the lifetime of every individual is limited to only one generation. The limited life span allows to forget the inappropriate internal parameter settings. This may lead to short periods of recession, but it avoids long stagnation phases due to unadapted strategy parameters [20]. The $(\mu+\lambda)$ and (μ,λ) selection fit into the same formal framework with the only difference being the limited life time of individuals in (μ,λ) method.

2.5 Genetic Parameters

The genetic parameters are very important for the success of the algorithm, equally important as the other aspects, such as the representation of the individuals, the initial population and the genetic operators. The most important parameters are:

- the population size μ has been set to 5 times the number of clusters. This turned out to be the best number of individuals in a generation.
- the intermediate population size λ was chosen twice the size of the population: $\lambda = 2\cdot\mu$.
- mutation probability was set at 5%.

The number of epochs used in our memetic algorithm was set to 100.

3 Computational Results

The performance of the proposed memetic algorithm for solving the GMSTP was tested on seventeen benchmark problems drawn from TSPLIB test problems containing between 229 and 724 nodes. The corresponding GTSP problems are obtained by applying the CLUSTERING procedure introduced in Fischetti *et al.* [6] and contain between 46 and 145 clusters.

The testing machine was an Intel Dual-Core 1,6 GHz and 1 GB RAM with operating system Windows XP Professional. The algorithm was developed in Java, JDK 1.6.

In the next table we report the experimental results obtained using our proposed memetic algorithm on the new TSPLIB instances described by Oncan *et al.* [12] for the GMSTP. The results are compared with the best results from the literature obtained using the Tabu Search algorithm described by Oncan *et al.*

The first column in the table represents the instances of the problem and the second column gives the number of clusters. The next columns contain the values of the objective function obtained using the tabu search algorithm proposed by Oncan *et al.* [12] and our memetic algorithm. In the last column we present the solution error obtained using our proposed memetic algorithm as a percentage of the solution provided by Oncan *et al.* using the TS algorithm.

Analyzing the computational results, it results that overall the proposed memetic algorithm performs well in comparison to the tabu search algorithm (TS) developed by Oncan *et al.* [12] in terms of solution quality: in nine out of seventeen instances we obtained the same solution and in rest the solution provided is at most 0.99 % of the solution provided by the TS algorithm.

Table 1 Computational results with TS (Oncan et al.) and our MA on TSPLIB with the center clustering procedure

Instance	No. of Clusters	*TS* Oncan et al.	*MA* Cost	Sol. error %
ali535	107	114303	114379	0.99
att532	107	12001	**12001**	0.00
d493	99	16493	16841	0.97
d657	132	19427	19811	0.98
fl417	84	7935	**7935**	0.00
gil262	53	887	910	0.97
gr229	46	59740	**59740**	0.00
gr431	87	86885	**86885**	0.00
lin318	64	18471	18561	0.99
p654	131	22209	**22209**	0.00
pcb442	89	19571	20654	0.94
pr264	53	21872	**21872**	0.00
pr299	60	20290	20662	0.98
rd400	80	5868	6069	0.96
si535	107	12791	**12791**	0.00
u574	115	15037	**15037**	0.00
u724	145	15905	**15905**	0.00

Regarding the computational times, it is difficult to make a fair comparison between algorithms, because they have been evaluated on different computers and they are implemented in different languages. However, in average our running times are comparable with those obtained using the TS algorithm.

4 Conclusions

The Generalized Minimum Spanning Tree Problem is an extension of the classical Minimum Spanning Tree Problem (MST) and consists in finding the minimum cost spanning tree containing exactly one node from a given number of predefined, mutually exclusive and exhaustive clusters.

We presented an efficient memetic algorithm for solving the GMSTP that combines the population concept of genetic algorithms with a fast local improvement method. The experimental results confirms the success of our proposed approach.

Acknowledgments. This work was cofinanced from the European Social Fund through Sectoral Operational Programme Human Resources Development 2007-2013, project number POSDRU/89/1.5/S/56287 "Postdoctoral research programs at the forefront of excellence in Information Society technologies and developing products and innovative processes", partner University of Oradea.

References

1. Back, T., Hoffmeister, F., Schwefel, H.: A survey of evolution strategies. In: Proc. of the 4th International Conference on Genetic Algorithms. Morgan Kaufmann, San Diego (1991)
2. Dror, M., Haouari, M., Chaouachi, J.: Generalized Spanning Trees. European J. Oper. Res. 120, 583–592 (2000)
3. Myung, Y.S., Lee, C.H., Tcha, D.W.: On the Generalized Minimum Spanning Tree Problem. Networks 26, 231–241 (1995)
4. Feremans, C., Labbe, M., Laporte, G.: The generalized minimum spanning tree problem: Polyhedral analysis and branch-and-cut algorithm. Networks 43(2), 71–86 (2004)
5. Feremans, C.: Generalized Spanning Trees and Extensions. PhD thesis, Universite Libre de Bruxelles, Belgium (2001)
6. Fischetti, M., Salazar, J.J., Toth, P.: A Branch-and-Cut Algorithm for the Symmetric Generalized Travelling Salesman Problem. Oper. Res. 45(3), 378–394 (1997)
7. Ghosh, D.: Solving medium to large sized Euclidean generalized minimum spanning tree problems. Technical Report NEP-CMP-2003-08-02, Indian Institute of Management, Research and Publication Department, Ahmedabad, India (2003)
8. Golden, B., Raghavan, S., Stanojevic, D.: Heuristic search for the generalized minimum spanning tree problem. INFORMS J. Comput. 17(3), 290–304 (2005)
9. Golden, B., Raghavan, S., Stanojevic, D.: The prize-collecting generalized minimum spanning tree problem. J. Heur. 14, 69–93 (2008)
10. Hu, B., Leitner, M., Raidl, G.R.: Combining variable neighborhood search with integer linear programming for the generalized minimum spanning tree problem. J. Heur. 14(5), 473–499 (2008)
11. Moscato, P.: On Evolution, Search, Optimization, Genetic Algorithms and Martial Arts: Towards Memetic Algorithms. Caltech Concurrent Computation Program, Report 826 (1989)
12. Oncan, T., Cordeau, J.-F., Laporte, G.: A Tabu Search Heuristic for the Generalized Minimum Spanning Tree Problem. European J. Oper. Res. 191, 306–319 (2008)
13. Pop, P.C., Kern, W., Still, G.: A New Relaxation Method for the Generalized Minimum Spanning Tree Problem. European J. Oper. Res. 170(3), 900–908 (2006)
14. Pop, P.C.: The Generalized Minimum Spanning Tree Problem. PhD thesis, University of Twente, The Netherlands (2002)
15. Pop, P.C., Pop Sitar, C., Zelina, I., Tascu, I.: Exact algorithms for generalized combinatorial optimization problems. In: Dress, A.W.M., Xu, Y., Zhu, B. (eds.) COCOA. LNCS, vol. 4616, pp. 154–162. Springer, Heidelberg (2007)
16. Pop, P.C.: On the Prize-Collecting Generalized Minimum Spanning Tree Problem. Ann. of Oper. Res. 150, 193–204 (2007)
17. Pop, P.C., Sabo, C., Pop Sitar, C., Craciun, M.: A Simulated Annealing Based Approach for Solving the Generalized Minimum Spanning Tree Problem. Creat. Math. Inform. 16, 42–53 (2007)
18. Pop, P.C.: A survey of different integer programming formulations of the generalized minimum spanning tree problem. Carpathian J. Math. 25(1), 104–118 (2009)
19. Prins, C., Bouchenoua, S.: A Memetic Algorithm Solving the VRP, the CARP and General Routing Problems with Nodes. Edges and Arcs. Stud. Fuzziness Soft Comput. 166, 65–85 (2004)
20. Schwefel, H.P.: Collective phenomena in evolutionary systems. In: Proc. of 31st Annual Meetting of the International Society for General System Research, pp. 1025–1033 (1987)

A Computer Algorithm to Simulate Molecular Replication

Rafael Silveira Xavier and Leandro Nunes de Castro

Abstract. Molecular replicators were introduced as a possible theory to explain the origin of life. Since their proposal they have been extensively studied from a bio- chemical perspective. This work proposes a taxonomy for the main properties of replicators that are important for building computational tools to solve complex problems as well as introduces a computer algorithm that models these entities. The simulation of this algorithm allows the observation and analysis of the behavior of replicators in light of the properties introduced. A number of experiments are performed to show that the proposed taxonomy of properties can be observed by simulating the algorithm introduced.

1 Introduction

The first studies concerning replicators were devoted to discussing molecular replication as a plausible explanation for the prebiotic evolution in chemical terms [5, 6, 1], and thus to the emergence of life. From these works began studies on the identification, characterization and classification of replicators [4, 7, 8, 11, 9, 10], which have been refined in the search for a more precise definition of these systems. Among the many concepts of replicators in the literature, the following deserve particular attention within the context of the present paper:

Rafael Silveira Xavier
Natural Computing Laboratory - LCoN, Mackenzie Presbyterian University,
São Paulo, Brazil
e-mail: rafael.xavier@mackenzista.com.br

Leandro Nunes de Castro
Natural Computing Laboratory - LCoN, Mackenzie Presbyterian University,
São Paulo, Brazil
e-mail: lnunes@mackenzie.br

A. Gaspar-Cunha et al. (Eds.): Soft Computing in Industrial Applications, AISC 96, pp. 195–206.
springerlink.com

- The replicator is a macromolecule consisting of a complex chain of various types of molecular blocks (or building blocks) which acts as a standard model, a matrix for the construction of another molecule [1].
- A replicator is an entity that passes its structure largely intact during replication [4].
- The replicator is any entity that multiplies by an autocatalytic process in which some of the products of the process are functionally equivalent to the original entity [14].

Considering the scope of this work, we deal with the replicators as polymers that act as standard models, a sort of matrix, for the construction of new molecules. In other terms, replicators are molecules that transmit or perpetuate their functionality (information) through an autocatalytic process called *molecular replication* [7].

Although the study of the properties of such systems has a great potential for the construction of new natural computing [2] applications, there are still few studies on the categorization and application of replicators' properties to computer algorithms.In this direction, one first step involves the development of a computer algorithm that, when run, allows us to observe the main properties of replicators and, then, evolve this algorithm so as to design more sophisticated computer algorithms for complex problem solving.

Based upon this line of thought, this work aims to discriminate and to investigate-qualitatively the main properties of replicators, to propose a preliminary algorithm to simulate specific processes of molecular replication and, as a result, to investigate some properties of the replicators that will be useful for the development of future natural computing tools to solve complex engineering problems.

2 Replicators: An Introduction

This section provides a basic introduction to the concept of molecular replication, which is the basis for the algorithm to be proposed here, and then introduces a number of properties for the replicators. More specifically, it is proposed that replicators can be characterized by some Structural, Conditional and Existential properties, as will be detailed in the following.

2.1 Molecular Replication

Molecular replication can be understood as a specific subset of autocatalysis [12], where the reaction product has the ability to organize the junction of reagents, thus speeding up (catalyzing) the production of a new molecule. The replication process results in copying a molecule that can be called the parent or original molecule. A simple diagram can be used to conceptualize the process of molecular replication, as shown in Fig.1. At first, the T molecule interacts with some building blocks A and B to form the ternary complex C_1. This complex brings together the building blocks A

and *B*, facilitating the reaction between them. After the precursors' reaction it is formed a binary complex C_2. The dissociation of this complex gives rise to two new molecules, copies of the original one, capable of serving as *templates* for the formation of a new cycle of replication.

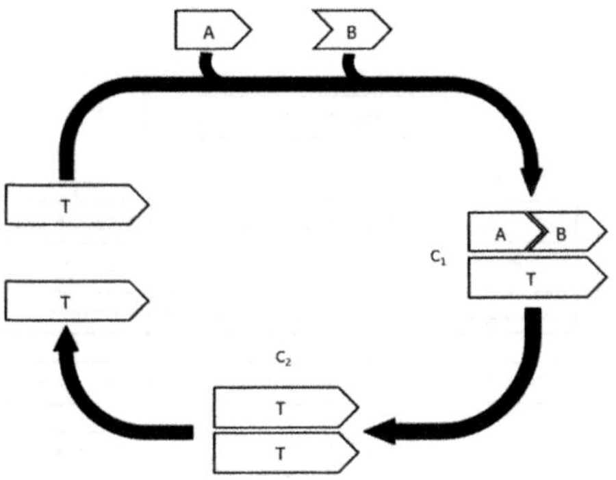

Fig. 1 Simplified molecular replication cycle

Replication as an autocatalytic process is associated with three factors: *multiplication*, *variance* and *inheritance* [14]. The multiplication of an entity or an autocatalytic cycle requires three specifications:

- *Multiplication*: there must be some material input (e.g., molecules used as building blocks) used for the construction of new entities;
- *Variance*: the original entity and the copy must be equivalent in some functional aspect within the cycle, but there may be some difference between them;
- *Inheritance*: the output of the multiplication process should contain more than one entity equivalent to the original replicator; that is, the copy must inherit some information from its parent.

Thus, multiplication can be characterized as an autocatalytic process that can generate some entities equivalent, but with slight modifications, to the original entity, thus preserving the identity of the entity's functional progenitor [14].

The notion of equivalence given above is associated with a concept of variability. Two entities can vary their structures, and still be functionally equivalent. However, the variability is also linked to the introduction of novelty and to the generation of entities that are not equivalent to a unique entity, i.e., structural changes bring new features.

Heredity involves a copying process, in which the offspring inherits the set of defining features original to the replicator. Heredity can be defined as the ability of an entity to transmit (copy) a portion or all of their structure, implying that there may be non-hereditary changes.

2.2 Replicator Properties

In order to appropriately describe the replicators so as to introduce a new computational model of replication we propose a taxonomy for the main replicator properties. These replicators' properties can be divided into three categories: *structural*, *conditional* and *existential*.

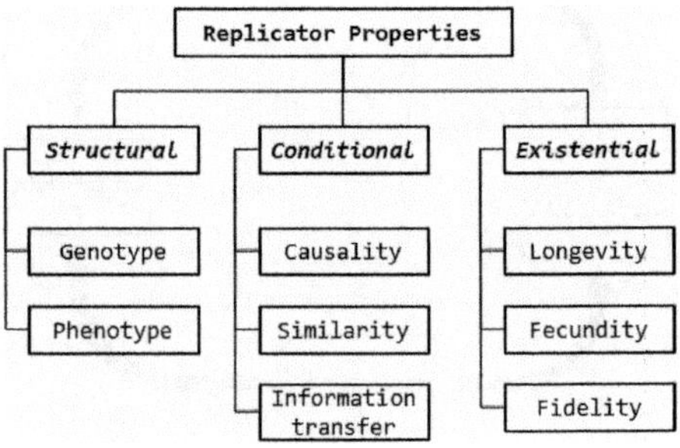

Fig. 2 Scheme of the replicator's properties.

2.2.1 Structural Properties

The structural properties of a replicator are associated with the arrangement and function of its constituent parts. Basically, a replicator has a *genotype* and a *phenotype* [14]. The genotype is the part of a replicator that encodes its form and function, i.e. its informational load. The phenotype, in turn, is the function of the replicator itself within a specific context or environment.

Based on a simplified interpretation of the Szathmary's work [14], the internal structure of a replicator can be divided into four main parts (Fig. 3):

- C: representing the complete set of features of a replicator. It corresponds to the whole replicator structure;
- V: set of features that can change state without promoting a loss of functional identity in the replicator;
- ID: subset of C that cannot be altered without promoting a loss of functionality in the replicator;
- H: the hereditary portion of a replicator's structure.

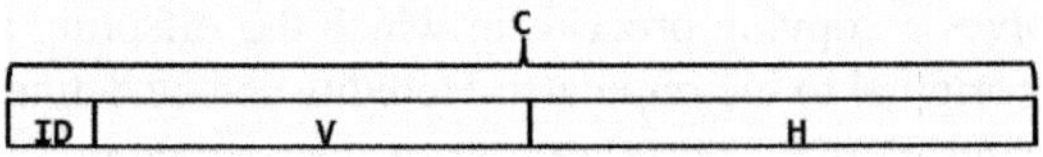

Fig. 3 The internal structure of a replicator

Some important relationships among the structural sub-sets must be highlighted. Subset H can be represented by the entire structure of the replicator, i.e., $H \subseteq C$. If $H = C$, then the replicator will have a full inheritance. If $H \subset C$ then the replicator will have a partial inheritance, and if $ID \subseteq H$ then the replicator will have a functional inheritance. Thus, inheritance is partially functional since the ID is contained within the inheritable set. If the partial inheritance is not functional, i.e., $ID \not\subseteq H$ the replicator will not be able to pass its functionality to its offspring. The classification of such replicator will be addressed in the next section. Finally, one can define the heritability of a replicator as strictly functional, i.e., $H = ID$.

2.2.2 Conditional Properties

The conditional properties characterize an entity as a replicator. They are: *causality*, *similarity* and *information transfer* [3], as detailed bellow:

- *Causality*: the original replicator is involved in the production (the cause) of the offspring;
- *Similarity*: is the equivalence relation between the original replicator and its offspring (copy). This relation divides the replicators in equivalence classes, where all the replicators in a particular class are considered functionally equivalent;
- *Information transfer*: defines that the parent replicator must pass some kind of information to its offspring. This property divides the replicators in informational and non-informational. An informational replicator can store and transmit information in a stable way, whilst a non-informational replicator cannot.

2.2.3 Existential Properties

The existential properties are associated with the existence of a replicator in a given environment [1]. These properties are described below:

- *Longevity*: corresponds to the lifetime of the replicator;
- *Fecundity*: is associated with the rate of replication of a molecule;
- *Fidelity* (or replication precision): is associated with the accuracy of the replication process.

3 An Algorithm to Simulate Replication

An algorithm was designed and implemented to simulate a number of collisions between a set of templates previously known and a set of building blocks and observe the behavior of replicators throughout the experiments. The number of copies (fecundity) of each replicator, its variations, longevity (life span) and other characteristics will be measured and analyzed in the light of the properties of the biological replicators introduced.

The reaction environment (also known as soup) is simulated using two vectors **T** and **B** representing, respectively, the collections of replicators and building blocks for a K set of collisions that simulates encounters among blocks and replicators in the soup and the following processes: *matching*, *replication* and *mutation*. The *matching* process simulates the binding between a building block and a template, which is a transition state before the replication; the *replication* process, as its name suggests, is associated with the generation of copies of replicators; and finally, the *mutation* process represents the variations that can occur in templates during replication.

In this model, the replicators are represented by binary strings with an arbitrary length L. The building blocks are formed by binary strings of varying sizes, always smaller than L, furthermore, each block and each replicator has a counter to store their copies.

Throughout K collisions for each template it is randomly selected a building block. During the collision the matching between the block and replicator is tested and it is verified if the template is able to replicate. The matching is accomplished through an exhaustive search of the binary sequence block in the entire structure of the replicator. If any region in the template is found, with a similarity greater than or equal to a threshold (μ) every bit of this region is marked to be overlooked in future comparisons and a unit is subtracted from the number of copies of the block, indicating that there was a binding between the building block and the replicator. In Fig. 4, which exemplifies this process, the dotted square in the left corner of the block and the template contains the number of copies of each one of them.

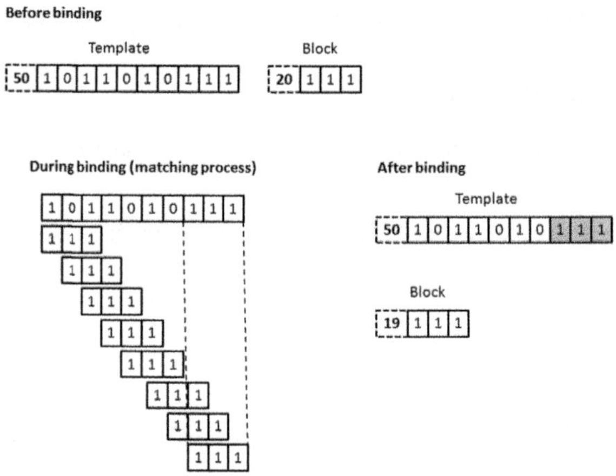

Fig. 4 Illustration of the matching process

Replication only occurs if a template has a number of marked bits greater than or equal to a replication threshold (λ). During the replication process the replicator can vary in structure given a default probability of mutation pm. After replicating all the bits that were marked in the replicator are cleared, the counter of the replicator is iterated and a new replicator is inserted into the soup. It is important to note that over collisions, the replicators and building blocks that are extinct are eliminated from the soup. The flowchart of the proposed algorithm is presented below.

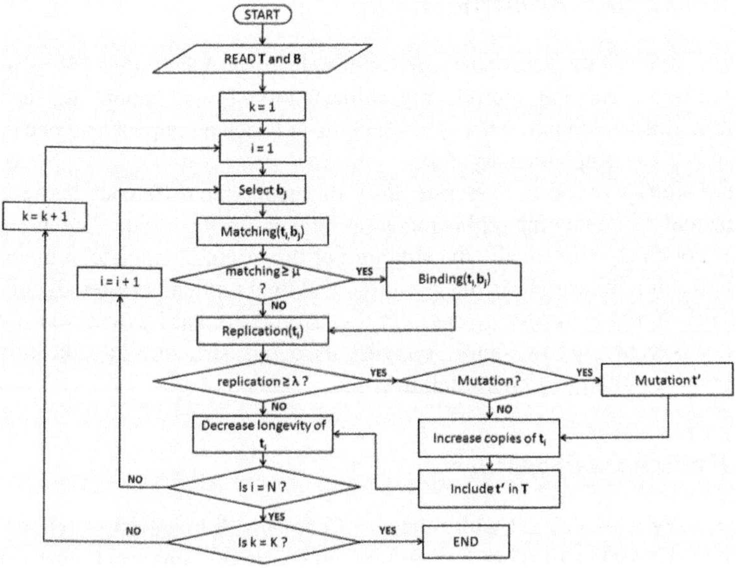

Fig. 5 Flowchart of computational replicator model

- **T**: set of templates $\mathbf{t}_i \in \mathbf{T}, i = 1, ..., N$;
- **B**: set of building blocks $\mathbf{b}_j \in \mathbf{B}, j = 1, ..., n$;
- K: total number of collisions;
- N: number of templates in **T**;
- μ: *matching* threshold between a replicator and a block;
- λ: replication threshold.

In the three sections that follow we will discuss how the proposed algorithm fits with each of the properties summarized in this paper.

3.1 Structural Properties

Despite mutations that may occur during successive replication processes,a portion of the original replicator remains without change. This portion can be seen within the proposed model, as the *ID* group, because it meets the constraint introduced in this paper that the *ID* portion of a replicator cannot be changed without promoting a loss in the replicator's functionality. The hereditary portion H of the replicators can be seen as the whole molecular structure (genotype), whilst the variable part V can be represented by the replicator region of the mutated offspring and finally the complete replicator's binary string representing C. Therefore, all structural properties are observable in the model.

3.2 Conditional Properties

The observation of the *causal property* is straightforward, because each copy of the replicator is generated by reproducing, subjected to a low mutation rate, the parent replicator. This is a consequence of a matching between a template and one or more building blocks complementary to it.

The *similarity property* is measured by the genotypic difference between each template and the remaining replicators after total collisions. As the mutation operator does not change drastically the structure of the original template, it maintains a high degree of similarity between the initial and final replicators after all collisions.

The *information transfer property* is also straightforwardly observed, for each replicator is generated by simply copying its parent structure into the offspring, subjected to a mutation with a small rate.

3.3 Existential Properties

The fecundity is associated with number of copies and longevity is related to the lifetime of the replicator. These properties can be directly observed in experiments conducted in this study. The fidelity is associated with the replication process.

4 Experimental Analysis

This section will present the experimental results and discussions that validate the computational model proposed for the replicators within the context of the various properties (*conditional*, *structural* and *existential*) introduced in this work.In a previous work [13] we introduced a different algorithm to study replication in which the replicators were incapable of creating a physical copy of themselves, but there was a replication index that allowed the counting of the number of offspring that would be generated by a replicator in case it could create a copy. The results presented in that particular case were quite different from the results to be presented here, suggesting that the physical replication is indeed important for a better understanding of how a number of replicators behave within the soup.

4.1 Materials and Methods

Three experiments were initially conducted using the same configuration parameters described below. Despite the same parameters, the stochastic processes of building blocks selection for collisions and the mutation after replication result in qualitatively and quantitatively different behaviors. Parameters:

- Total number of collisions: 1,000;
- Number of templates: 4;
- Initial number of copies of each template: 1;
- Size of templates: 10 bits;

- Initial life span of each template: 10 % of the total number of collisions;
- Number of building blocks (randomly generated): 10,000;
- Initial number of copies of each building block: random within the range [1,100];
- Size of building blocks: random within the range [2, $L/2$], where L is the templates' length.
- matching threshold (μ): 1.0;
- replication threshold (λ): 0.7;
- Mutation probability: 1%;

In the experiment the following initial templates were used:

- $t_1 = [0000000000]$;
- $t_2 = [1111111111]$;
- $t_3 = [0000011111]$;
- $t_4 = [1111100000]$.

4.2 Experimental Results

This section describes the experimental results obtained with the proposed algorithm and their interpretation in the light of the properties introduced. It will be shown the population size of building blocks and templates during collisions, the maximum, mean and minimum fecundity (number of copies) of the soup for all collisions an the maximum, mean and minimum longevity (replicator life span) of the soup during collisions (Fig. 6 to 8).

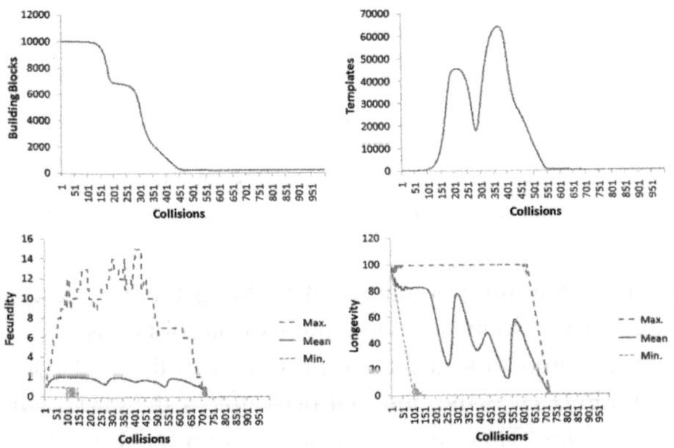

Fig. 6 Experiment 1: Population of building blocks and templates (top), fecundity and longevity during 1000 collisions (bottom).

The three experiments show similar behavior. The replicator with the largest number of copies in all experiments, is around 14 or 16 copies. The apex of the population of templates is around 70,000 templates. The use of building blocks is similar in all experiments. We note that the apex of the population of templates

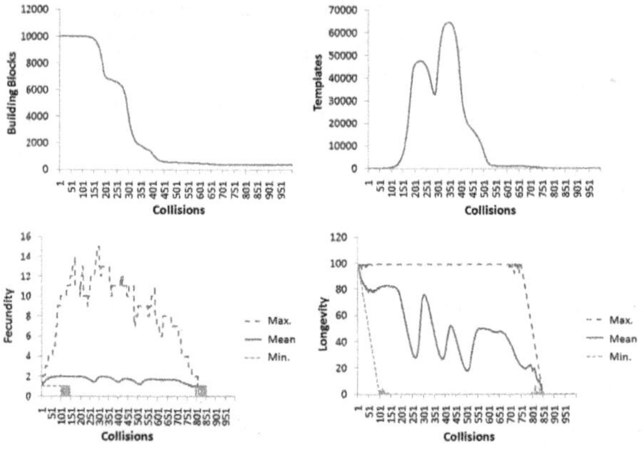

Fig. 7 Experiment 2: Population of building blocks and templates (top), fecundity and longevity during 1000 collisions (bottom).

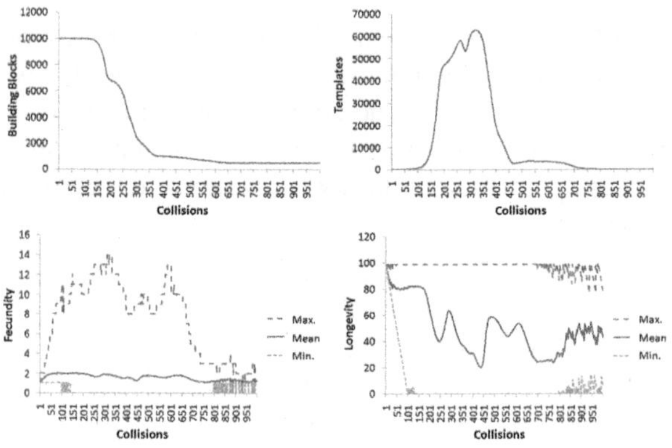

Fig. 8 Experiment 3: Population of building blocks and templates (top), fecundity and longevity during 1000 collisions (bottom).

coincided with the declining population of building blocks and found the highest fecundity, i.e., occurred around the same number of collisions.

One interesting feature, more clearly presented in the first and second experiments (Fig. 6 and Fig. 7), is the bimodal behavior of the curve which denotes the population of templates. There are two peaks or two local maxima, indicating the formation of two generations of replicators. The first generation develops (replication) to the point that this generation of replicators becomes incompatible with the existing building blocks in the soup, but the action of the mutation operator and a decrease in the population of building blocks, respectively, cause the creation of replicators more adapted to the existing blocks and increases the probability of finding favorable blocks because there are fewer blocks to be selected. Finally the scarcity of blocks associated with the reduction in the longevity of replicators caused the decline in population of templates.

Another interesting feature observed is the fact that the mean fecundity is around 2. This indicates that throughout all the collisions at least one new replicator was generated.

The results obtained from the experiments presented in this section are consistent with the architecture of the model and give us a general framework about the development of computational replicators in the proposed model.

5 Conclusion

Replicators are theoretically important structures because they serve as a plausible conceptual tool to explain the origins of life. Despite the vast literature discoursing about replicators, few models addressing specific features of replicators are available. Furthermore, no work so far has given attention to the fact that there may be a number of features that, altogether, lead to a general description of replicators with great potential for the development of novel computer algorithms for problem solving. This is exactly the line of thought proposed and followed by this paper.

Based on the molecular replicator literature we organized and introduced a taxonomy of features for the replicators and a novel, but simple, computer algorithm that can be used to simulate computational replicators. After that, some experimental scenarios were designed in order to empirically investigate the behavior of the proposed model. The results obtained allowed us to observe all replicators' properties and make important conclusions mainly about their existential properties. We could observe the longevity, fecundity and fidelity properties of the replicators in the experiments.

As future research we will perform a number of experiments to investigate the sensitivity of the algorithm to some of its tunable parameters, such as μ, λ and others. Also, we plan to adapt the model so that it accounts for replicator networks and, finally, use these networks to solve pattern recognition problems.

Acknowledgements. The authors thank CAPES, Fapesp, CNPq and Mackpesquisa for the financial support.

References

[1] Dawkins, R.: The Selfish Gene. Oxford University Press, Oxford (1976)
[2] de Castro, L.N.: Fundamentals of natural computing: An overview. Physics of Life Reviews 4(1), 1–36 (2007)
[3] Hodgson, G.M., Knudsen, T.: Information, complexity and generative replication. Biology and Philosophy 23(1), 47–65 (2008)
[4] Hull, D.L.: Individuality and selection. Annual Review of Ecology and Systematics 11, 311–332 (1980)

[5] Lewontin, R.C.: The units of selection. Annual Review of Ecology and Systematics 1, 1–18 (1970)
[6] Orgel, L.E.: The Origins of Life: Molecules and Natural Selection. John Wiley & Sons, Chichester (1973)
[7] Orgel, L.E.: Molecular replication. Nature 358, 203–209 (1992)
[8] Szathmary, E.: A classification of replicators and lambda-calculus models of biological organization. Proceedings of the Royal Society B: Biological Sciences 260(1359), 279–286 (1995)
[9] Szathmary, E.: The evolution of replicators. Philosophical Transactions of the Royal Society B: Biological Sciences 355(1403), 1669–1676 (2000)
[10] Szathmary, E.: The origin of replicators and reproducers. Philosophical Transactions of the Royal Society B: Biological Sciences 361(1474), 1761–1776 (2006)
[11] Szathmary, E., Smith, J.M.: From replicators to reproducers: the first major transitions leading to life. Journal of Theoretical Biology 187(4), 555–571 (1997), doi:10.1006/jtbi.1996.0389
[12] Wintner, E., Conn, M.M.: Studies in molecular replication. Accounts of Chemical Research 27(7), 198–203 (1994)
[13] Xavier, R.S., de Castro, L.N.: Computational replicator: A taxonomy of properties and a preliminary model. In: Second World Congress on Nature and Biologically Inspired Computing (2010)
[14] Zachar, I., Szathmary, E.: A new replicator: A theoretical framework for analysing replication. BMC Biology 8(1), 21 (2010)

Part IV: Soft Computing for Modeling, Control, and Optimization

Particle Filter with Differential Evolution for Trajectory Tracking

Leandro M. de Lima and Renato A. Krohling

Abstract. Over the last decades, Particle Filter also known as the Sampling Importance Resampling algorithm has successfully been applied to solve different problems in Engineering, e.g., trajectory tracking, non-linear estimation, and many others. Basically, the Particle Filter algorithm consists of a population of particles, which are sampled to estimate a posterior probability distribution. Unfortunately, in some cases the algorithm suffers from particle degeneracy, in which most particles converge prematurely to local minima due a loss of diversity of the population, and therefore do not contribute to estimation of the true probability distribution. In this paper, in order to tackle this drawback and to improve the performance of the standard Particle Filter we propose a modification to the algorithm by inserting a sampling mechanism inspired by Differential Evolution. Simulation results of the enhanced hybrid version are presented and compared with the standard Particle Filter algorithm and show the suitability of the proposed approach.

1 Introduction

State estimation is a very important issue not only on automatic control but also in time series prediction, tracking, robot navigation, etc. A common approach used for modeling is the state space model. If the process is linear and Gaussian there exists optimal solution in a closed form like the Kalman Filter and the Extended Kalman Filter. On the other hand, for real-world problems which present non-linear behavior and are non-Gaussian, the information available arrives sequentially and is corrupted by noise. For such kind of processes there are no closed solutions and the

Leandro M. de Lima
Department of Informatics, PPGI, Federal University of Espírito Santo,
29060-970, Vitória, Brazil
e-mail: leandromunizlima@yahoo.com.br

Renato A. Krohling
Department of Informatics, PPGI, Federal University of Espírito Santo,
29060-970, Vitória, Brazil
e-mail: krohling.renato@gmail.com

A. Gaspar-Cunha et al. (Eds.): Soft Computing in Industrial Applications, AISC 96, pp. 209–219.
springerlink.com

hidden state of the model can be estimated using Particle Filter (PF), which has its foundation on the theory of Bayesian estimation [4, 3, 1].

Particle Filter consists of an approximation of the posterior probability distribution by sampling points, called particles, which are updated as new data arrives. Particles are random samples, which follow the trajectory of the state. Particle filter uses Sequential Monte Carlo (SMC) integration method to approximate the posterior probability distribution by sampling from a known probability distribution in order to calculate the weights. A common drawback presented by sampling from a distribution is degeneracy, which means that after some iterations some of the weights tend to concentrate in a non-optimal point. To overcome this issue, we propose a modification of the Sampling Importance Resampling (SIR) algorithm incorporating Differential Evolution into it. Previous attempts to hybridize Evolutionary Algorithm (EA) with PF have already been presented in the literature, e.g., Evolution Strategy (ES)[11], Particle Swarm Optimization (PSO)[6], Evolutionary Algorithm[7], and recently Ant Colony Optimization (ACO)[12]. In this paper, we propose a hybrid approach combining PF with DE.

The rest of this paper is organized as follows: In Sect. 2, the PF is described. Section 3 presents a description of Differential Evolution. In Sect. 4, the Particle Filter combined with Differential Evolution is proposed. Section 5 presents simulation results followed by conclusions in Sect. 6.

2 Particle Filter

Consider the non-linear state space model described by

$$X_k = f(X_k, k) + V_{k-1} \tag{1}$$

$$Y_k = g(X_k, k) + U_k \tag{2}$$

where X_k is the state variable at time instant k, Y_k are the observations at time instant k, f and g are non-linear functions, V_k is the system noise and U_k is the noise measurement. One assumes that the system noise and the measurement noise are normal random variables with covariance Q_k and R_k, respectively. The problem considered here consists on obtaining the best estimate for the state variable X_k when only data of observations are available, i.e., $Y_{1:k} = \{Y_1, Y_2, \ldots, Y_k\}$. In this case, the state variable is modeled as a hidden Markov Process. The problem of the estimation of X_k can be solved calculating the posterior probability density (pdf) of X_k based on the observations Y_k. So, the estimation problem can be formulated as minimization of the Mean Squared Error (MSE) estimate or the Maximum A Posteriori (MAP) probability given by:

$$\hat{X}_k = E[X_k|Y_{1:k}] = \int X_k p(X_k|Y_{1:k})dX_k \tag{3}$$

$$\hat{X}_k = \arg\max_{X_k} p(X_k|Y_{1:k}) \tag{4}$$

where E stands for the expected value.

In order to estimate the hidden states a frequently used approach is the Sequential Monte Carlo, which has its fundamentals on the Bayesian filtering theory. The posterior pdf $p(X_k|Y_{1:k})$ of X_k can be evaluated recursively from a priory pdf of the initial state X_0 of the system according to basically a *prediction* and an *observation* step. In the prediction step, one is interested to propagate into the next time step k by means of the transition density given by

$$p(X_k|Y_{1:k-1}) = \int p(X_k|X_{k-1})p(X_{k-1}|Y_{1:k-1})dX_{k-1}. \tag{5}$$

The observation (or update) step, involves the application of the Bayes theorem when new data arrives and is calculated according to

$$p(X_k|Y_{1:k}) = \frac{p(Y_k|X_k)p(X_k|Y_{1:k-1})}{p(Y_k|Y_{1:k-1})}. \tag{6}$$

These two steps provide the optimal solution to the estimation problem, but unfortunately the solution of the multidimensional integration is difficult to be obtained analytically. An alternative approach provides the Sequential Monte Carlo method [4, 3, 1]. In this paper, the approximation of the posterior pdf is solved by means of the Sampling Importance Resampling algorithm.

The SIR algorithm consists of three steps: first, all particle are sampled. Next, the importance weight is calculated for each of them. Then, the particles are resampled in order to discard particles with low importance and explore the region where particles have high importance. Figure 1 shows the particles as dark bubbles and its weights are represented by bubbles' size, whereas bubbles with bigger size correspond to higher weights. The importance is calculated by particle's posterior density. A detailed description of the SIR is beyond the scope of this paper and the reader is referred to [4, 3, 1] for more details.

3 Differential Evolution

Differential Evolution is a method of optimization of multidimensional functions. It is an evolutionary algorithm composed of a population of possible solutions. The initial population may be started randomly if no prior knowledge is available about the solution space. Assuming a population $P_G^{(i)}$, a vector of size N_P with components $p_{j,G}^{(i)}$, where i is an index for each individual in $P_G^{(i)}$, j is the position in D-dimensional individual and G is the generation that the population belongs.

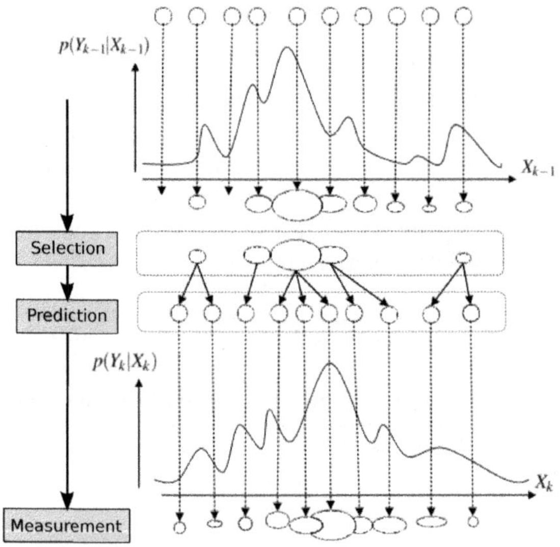

Fig. 1 Illustration of the two steps of the PF algorithm showing particles with different weights

Successive generations can be created by adding to an element, chosen at random, the weighted difference of two others, also selected randomly. For standard DE (DE/rand/1/bin) mutation, crossover and selection operators are directly defined as follows.

For each possible solution vector $P_G^{(i)}$ in generation G a mutant vector $M_G^{(i)}$ is calculated by

$$M_G^{(i)} = P_G^{(r_1)} + F(P_G^{(r_2)} - P_G^{(r_3)}) \tag{7}$$

where $i = 1,2,\dots,N_P$ and r_1, r_2 and r_3 are mutually different random integer indexes selected from $\{1,2,\dots,N_p\}$. Further, it implies that $N_p \geq 4$ is required. $F \in [0,2]$ is a real constant, which determines the amplification of the added differential variation of $(P_G^{(r_2)} - P_G^{(r_3)})$. Larger values for F result in higher diversity in the generated population and lower values cause faster convergence [8].

DE utilizes the crossover operation to generate new solutions by shuffling each vector with a mutant one and also to increase the diversity of the population.

To calculate the trial vector's elements we use

$$z_{j,G}^{(i)} = \begin{cases} m_{j,G}^{(i)}, & \text{if } rand_j(0,1) \leq C_r \text{ or } j = k \\ p_{j,G}^{(i)}, & \text{otherwise.} \end{cases} \tag{8}$$

The constant $C_r \in (0,1)$ is a user-defined crossover rate, which controls the fraction of mutant values that are used. For each j a uniform random number is generated within the interval $[0,1]$, called here $rand_j(0,1)$. The index $k \in \{1,2,\ldots,D\}$ is a random parameter index, chosen once for each i to make sure that at least one parameter is always selected from the mutated vector $M_G^{(i)}$. Most used values for C_r are within the interval $[0.4,1]$ according to [2].

In selection, $Z_G^{(i)}$ or $P_G^{(i)}$ vector is selected to be a member of the next generation $G+1$ if it has the higher objective function value than the other one, for a maximization problem.

$$P_{G+1}^{(i)} = \begin{cases} Z_G^{(i)}, \text{ if } f(Z_G^{(i)}) \geq f(P_G^{(i)}) \\ P_G^{(i)}, \text{ otherwise.} \end{cases} \tag{9}$$

There are other variants based on different mutation and crossover strategies [10]. The Differential Evolution algorithm is presented as a pseudocode in Algorithm 1.

Algorithm 1. Differential Evolution

Input: Population size N_P

```
repeat
    for each individual in DE i = 1,...,N_P do
        Select P^(r1), P^(r2) and P^(r3), where r1 ≠ r2 ≠ r3 ≠ i
        // D is the dimension of a particle
        j_rand = floor(D * rand(0,1))
        for each individual component j = 1,...,D do
            if rand(0,1) ≤ C_r or j = j_rand then
                z_j^(i) = p_j^(r1) + F * (p_j^(r2) − p_j^(r3))
            else
                z_j^(i) = p_j^(i)
            end if
        end for
    end for
    // select next generation
    for each individual i = 1,...,N_P do
        if f(Z^(i)) ≥ f(P^(i)) then
            P^(i) = Z^(i)
        end if
    end for
until termination condition met
```

4 Particle Filter Using Differential Evolution

The degeneracy issue in Particle Filter happens when there are a small amount of particles with very high importance and the other ones have low relevance. As we want to sample from the posterior density, the proposal density need to be close to the posterior one, that means importance weights variance be close to 0 to well estimate it.

An attempt to tackle the problem of degeneracy of PF is to create a new group of particles that generate higher weights. These new particles replace those that currently have less relevance to the original set, i.e., with lower weights. In this paper, this new group of particles has the fittest individuals, those that will have high weights, found by Differential Evolution. In this context, the individuals in DE are those particles estimated in the actual step in PF, and the fitness function in DE is the function that calculates the weight of a particle.

In lines 6–13 of Algorithm 2 is shown the required change in the Particle Filter algorithm to be integrated into the DE. For the DE be able to take advantage of the computational effort already spent, the current particles found by the Particle Filter (X_{orig}) are used as initial population, instead of initializing its population at random. As output the DE algorithm (Algorithm 1) returns a DE modified set (X_{de}). Then, those two sets are sorted in a way that merging the $(M-N)$ last particles in X_{orig} with the N first ones in X_{de} will replace the N particles with lower weights by N fittest ones found in DE. This new set is used as the particles set in this time step.

Algorithm 2. Particle Filter with Differential Evolution

Input: Population size M

  for each particle $i = 1,\ldots,M$ **do**

    // generation of particles (samples)

    $X_k^{(i)} \sim p(X_k|X_{k-1})$

  end for

5: **repeat**

    $Xorig = \left\{X_k^{(i)}, W_k^{(i)}\right\}_{i=1}^{M}$

    // creation of better particles through DE

    $Xde = DifferentialEvolution(Xorig)$ // Call Algorithm 1

    Sort $Xorig$ in ascending order

10:   Sort Xde in descending order

    // replaces the N lower weighted particles in $Xorig$

    // with the N higher weighted ones in Xde

    $X = \{Xde\}_{i=1}^{N} \cup \{Xorig\}_{i=N+1}^{M}$

    for each particle $i = 1,\ldots,M$ **do**

15:     // computation of the weights

      $W_k^{(i)} = W_k^{(i)} p(Y_k|X_k^{(i)})$

      // normalization of the weights

      $W_k^{(i)} = \dfrac{W_k^{(i)}}{\sum_{i=1}^{M} W_k^{(i)}}$

      // resampling

20:     $\left\{X_k^{(i)}, \frac{1}{M}\right\}_{i=1}^{M} = \left\{X_k^{(i)}, W_k^{(i)}\right\}_{i=1}^{M}$

      Replication of particles in proportion to their weights

    end for

  until termination condition met

---

5 Results

The algorithms were tested on a truck-trailer problem [9, 5, 7] with multiple trailers. As shown in Fig. 2, it is the case where we have one truck connected with 3 trailers. Its dynamics model is defined by the following equations:

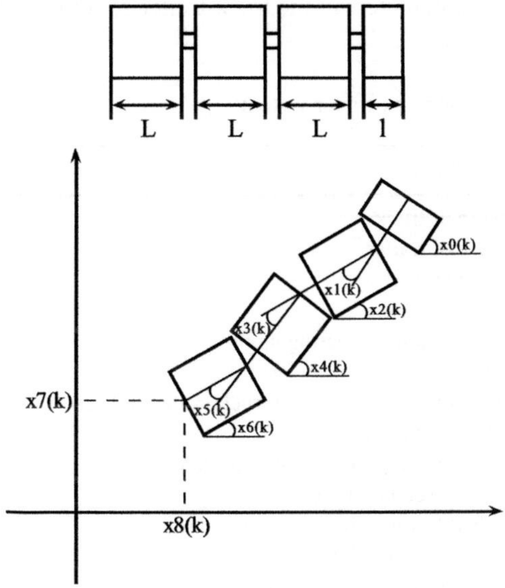

Fig. 2 Truck-trailer system with three trailers

$$x_0(k+1) = x_0(k) + \frac{vT}{l}\tan(u(k)) + w_0(k) \tag{10}$$

$$x_1(k) = x_0(k) - x_2(k) \tag{11}$$

$$x_2(k+1) = x_2(k) + \frac{vT}{L}\sin(x_1(k)) + w_2(k) \tag{12}$$

$$x_1(k) = x_2(k) - x_4(k) \tag{13}$$

$$x_4(k+1) = x_4(k) + \frac{vT}{L}\sin(x_3(k)) + w_4(k) \tag{14}$$

$$x_5(k) = x_4(k) - x_6(k) \tag{15}$$

$$x_6(k+1) = x_6(k) + \frac{vT}{L}\sin(x_5(k)) + w_6(k) \tag{16}$$

$$x_7(k+1) = x_7(k) + vT\cos(x_5(k))\sin\left(\frac{x_6(k+1)+x_6(k)}{2}\right) + w_7(k) \quad (17)$$

$$x_8(k+1) = x_8(k) + vT\cos(x_5(k))\cos\left(\frac{x_6(k+1)+x_6(k)}{2}\right) + w_8(k) \quad (18)$$

The variables and parameters of the truck-trailer system [9] are given in Table 1.

Table 1 Parameters of the truck-trailer system

Name	Description
l	Length of truck
L	Length of trailer
T	Sampling time
v	Speed of truck
$u(k)$	Control input to steering angle
$x_0(k)$	Angle of truck
$x_1(k)$	Angle difference between truck trailer and first trailer
$x_2(k)$	Angle of first trailer
$x_3(k)$	Angle difference between first trailer and second trailer
$x_4(k)$	Angle of second trailer
$x_5(k)$	Angle difference between second trailer and third trailer
$x_6(k)$	Angle of third trailer
$x_7(k)$	Vertical position of third trailer
$x_8(k)$	Horizontal position of third trailer

We define $w(k) = [w_0(k)\, w_2(k)\, w_4(k)\, w_6(k)\, w_7(k)\, w_8(k)]^T \sim N(0,Q)$ and $v(k) \sim N(0,R)$ as being mutually independent Gaussians. The covariance matrices of control noise and measurement noise are $Q = diag(1^2,\ldots,1^2)$ and $R = diag(1^2,\ldots,1^2)$, respectively, with appropriate dimensions. We assume that the measurement signal is missing for a while and the target truck trailer moves fast during the measurement missing. In this simulation the sampling time is 0.5 seconds and the number of particles used in the experiments was set to 100, 500 and 1000. Moreover, the trailers have $5.0m$ each, the truck has $2.8m$ and its speed is $2.0m/s$. The crossover rate C_r, the mutation factor F and number of generations in DE methods are set to 0.8, 0.75 and 200, respectively. During the simulation, white Gaussian noise was the control input to steering angle, $u(k)$. The start value of vector $\mathbf{x}(0) = [x_0(0)\, x_1(0)\ldots x_7(0)\, x_8(0)]^T$ is $[0\,0\,0\,0\,0\,0\,0\,10\,10]^T$.

It is shown in Fig. 3 the simulation result of x_8 for a specific set of state values. Since this system has nine states, only x_8 is depicted for simplicity[7]. The performance for other states are summarized in Table 2. The values in the Table 2 are the average of the MSE over 10 runs.

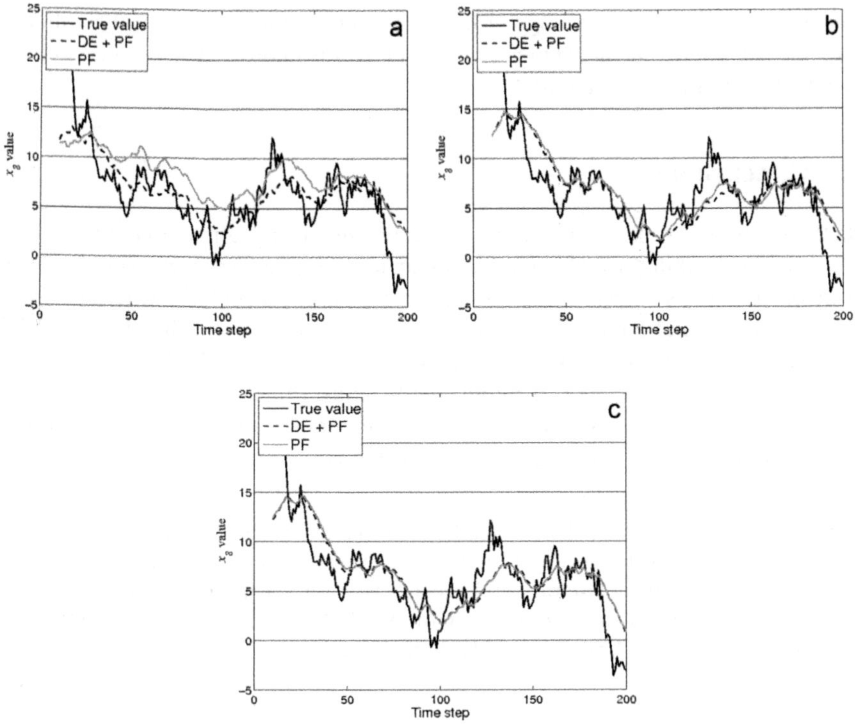

Fig. 3 Estimation results for the state x_8. **a** 100 particles. **b** 500 particles. **c** 1000 particles

Table 2 Average MSE for PF and PF+DE with 100, 500 and 1000 particles

Particles	100		500		1000	
Algorithm	PF	PF+DE	PF	PF+DE	PF	PF+DE
$x_0(k)$	3.549	8.493	1.803	1.285	1.130	0.615
$x_1(k)$	7.350	14.258	4.535	3.390	2.799	2.991
$x_2(k)$	2.384	3.191	1.086	2.301	1.028	0.603
$x_3(k)$	3.993	11.057	1.970	2.709	2.450	1.014
$x_4(k)$	3.837	8.507	4.293	4.382	6.177	4.793
$x_5(k)$	133.369	26.936	1.260	0.570	0.872	0.255
$x_6(k)$	122.791	17.780	0.548	0.180	0.357	0.235
$x_7(k)$	18.260	7.954	5.717	6.750	5.642	7.004
$x_8(k)$	200.212	39.072	27.698	19.238	16.597	14.701
sum	495.745	**137.247**	48.911	**40.806**	37.053	**32.209**

As can be seen in Fig. 3 and in the last row of Table 2, that represents the summation of all dimensions average MSE, the hybrid approach combining PF with DE obtained an improvement over the canonical PF. This improvement is noticeable for each number of particles tested.

6 Conclusions

In this work, we proposed an improved Particle Filter using Differential Evolution to decrease the tracking error. In simulations, an improvement was seen in the modified versions of PF, as shown in the simulation results for a problem where the state change rapidly and abruptly, avoiding premature convergence. The hybridized PF with DE presents better tracking performance over the standard PF.

For future work, methods to automatically adjust (tune) some parameters of the algorithm are under investigation. Also, there is room to reduce the computational time, thus allowing its use in real-time applications.

Acknowledgements. The authors would like to thank Dr. Nando de Freitas (University of British Columbia, CA) for using part of his public Matlab code for simulating PF. We also would like to thank Rainer Storn, Ken Price and Jim Van Zandt for their Matlab code for simulating DE. This work was supported in part by FAPES under grant no. 056/2009 (L. M. de Lima) and grant no. 37286374/2007 (R. A. Krohling).

References

[1] Arulampalam, M.S., Maskell, S., Gordon, N., Clapp, T.: A tutorial on particle filters for online nonlinear/non-gaussian bayesian tracking. IEEE Transactions on Signal Processing 50, 174–188 (2002)

[2] Das, S., Konar, A., Chakraborty, U.: Improved differential evolution algorithms for handling noisy optimization problems. IEEE Congress on Evolutionary Computation, 1691–1698 (2005)

[3] Doucet, A., de Freitas, N., Gordon, N.: Sequential Monte Carlo Methods in Practice. Springer, Berlin (2001)

[4] Gordon, N.J., Salmond, D.J., Smith, A.F.M.: Novel approach to nonlinear/non-gaussian bayesian state estimation. IEE Proc. F In Radar and Signal Processing 140, 107–113 (1993)

[5] Kiyuna, A., Nakazono, K., Kinjo, H., Yamamoto, T.: Simulation study of multitrailer control system using neurocontrollers evolved by a modified genetic algorithm. In: IEEE International Symposium on Computational Intelligence in Robotics and Automation, pp.1364–1368 (2003)

[6] Krohling, R.A.: Gaussian particle swarm and particle filter for nonlinear state estimation. In: Proc. of the International Conference on Artificial Intelligence and Soft Computing, Spain, pp. 399–404 (2004)

[7] Park, S., Hwang, J.P., Kim, E., Kang, H.J.: A new evolutionary particle filter for the prevention of sample impoverishment. IEEE Transactions on Evolutionary Computation, 801–809 (2009)

[8] Rahnamayan, S., Tizhoosh, H.R., Salama, M.M.A.: Opposition-based differential evolution. IEEE Transactions on Evolutionary Computation, 64–79 (2008)
[9] Simon, D.: Kalman filtering for fuzzy discrete time dynamic systems. Appl. Soft Computing 3(3), 191–207 (2003)
[10] Storn, R., Price, K.: Differential evolution – a simple and efficient heuristic for glob. optimization over continuous spaces. J. of Global Optimization 11(4), 341–359 (1997)
[11] Uosaki, K., Kimura, Y., Hatanaka, T.: Nonlinear state estimation by evolution strategies based particle filters. IEEE Congress on Evolutionary Computation, 2102–2109 (2003)
[12] Xu, B., Zhu, J., Xu, H.: An ant stochastic decision based particle filter and its convergence. Signal Processing 90(8), 2731–2748 (2010)

A Novel Normal Parameter Reduction Algorithm of Soft Sets

Xiuqin Ma, Norrozila Sulaiman, Hongwu Qin, and Tutut Herawan

Abstract. In this paper, we propose a novel normal parameter reduction algorithm of soft sets based on the oriented-parameter sum, which can be carried out without parameter important degree and decision partition. We present some new related definitions and proved theorems of normal parameter reduction. The comparison result on a Boolean-valued dataset shows that, the proposed algorithm involves relatively less computation and is easier to implement and understand as compared with the soft set-based algorithm of normal parameter reduction.

Keywords: Oriented-parameter sum, soft sets, reduction, normal parameter reduction, Boolean-valued dataset.

1 Introduction

In recent years, there has been a rapid growth in interest in soft set theory and its applications. Soft set theory was firstly proposed by a Russian Mathematician Molodtsov [1] in 1999. It is a new mathematical tool for dealing with uncertainties, while a wide variety of theories such as probability theory, fuzzy sets [2], and rough sets [3] so on are applicable to modeling vagueness, each of which has its inherent difficulties given in [4]. In contrast to all these theories, soft set theory is free from the above limitations and has no problem of setting the membership function, which makes it very convenient and easy to apply in practice. Therefore, many applications based on soft set theory have already been demonstrated by Molodtsov [1], such as the smoothness of functions, game theory, operations research, Perron integration, probability theory, and measurement theory.

Presently, great progresses of study on soft set theory have been made [5,6,7,8,9,10,11,12]. And it is worthwhile to mention that some effort has been done to such issues concerning reduction of soft sets. Maji et al. [13] employed soft sets to solve the decision-making problem. Later, Chen et al. [14] pointed out that the conclusion of soft set reduction offered in [13] was incorrect, and then

Xiuqin Ma · Norrozila Sulaiman · Hongwu Qin · Tutut Herawan
Faculty of Computer Systems and Software Engineering, Universiti Malaysia Pahang, Lebuh Raya Tun Razak, Gambang 26300, Kuantan, Malaysia
e-mail: qhwump@gmail.com

A. Gaspar-Cunha et al. (Eds.): Soft Computing in Industrial Applications, AISC 96, pp. 221–229.
springerlink.com

presented a new notion of parameterization reduction in soft sets in comparison with the definition to the related concept of attributes reduction in rough set theory. The concept of normal parameter reduction was introduced in [15], which overcome the problem of suboptimal choice and added parameter set of soft sets. An algorithm for normal parameter reduction was also presented in [15]. However, the algorithm is hard to understand and involves a great amount of computation. In order to make reduction of soft sets easy to implement and reduce computation, in this paper, we propose a simpler and more easily understandable algorithm which is referred to as a novel normal parameter reduction algorithm of soft sets.

The rest of this paper is organized as follows. Section 2 reviews the basic notions of soft set theory and analyses the normal parameter reduction of soft set put forward in [15]. Section 3 gives some theory and then proposes a novel normal parameter reduction algorithm of soft sets based on our theory. Section 4 shows the comparison results between the former algorithm in [15] and our algorithm. Finally Section 5 presents the conclusion from our study.

2 Analysis of the Normal Parameter Reduction of Soft Sets

In this section, we briefly review the definition of soft set and discuss the normal parameter reduction of soft sets which was presented by Kong et al. [15].

2.1 Soft Set Theory

Let U be a non-empty initial universe of objects, E be a set of parameters in relation to objects in U, $P(U)$ be the power set of U, and $A \subset E$. The definition of soft set is given as follows.

Definition 2.1 (See [4]). *A pair (F, A) is called a soft set over U, where F is a mapping given by F: A→P(U).*

That is, a soft set over U is a parameterized family of subsets of the universe U.

2.2 The Normal Parameter Reduction of Soft Sets

Suppose $U = \{h_1, h_2, \cdots, h_n\}$, $E = \{e_1, e_2, \cdots, e_m\}$, (F, E) is a soft set with tabular representation. Define $f_E(h_i) = \sum_j h_{ij}$, where h_{ij} are the entries in the table of (F, E).

Definition 2.2 (See [15]). *With every subset of parameters $B \subseteq A$, an indiscernibility relation IND(B) is defined by*

$$IND(B) = \{(h_i, h_j) \in U \times U : f_B(h_i) = f_B(h_j)\}$$

For soft sets (F, E), $U = \{h_1, h_2, \cdots, h_n\}$, the decision partition is referred to as

$$C_E = \{\{h_1, h_2, ..., h_i\}_{f_1}, \{h_{i+1}, ..., h_j\}_{f_2}, ..., \{h_k, ..., h_n\}_{f_s}\}$$

Where for subclass $\{h_v, h_{v+1}, \ldots, h_{v+w}\}_{f_i}$, $f_E(h_v) = f_E(h_{v+1}) = \ldots = f_E(h_{v+w}) = f_i$, and $f_1 \geq f_2 \geq \ldots \geq f_s$, s is the number of subclasses. In other words, objects in U are classified and ranked according to value of $f_E(.)$ based on the indiscernibility relation.

Definition 2.3 (See [15]). *Denote $A = \{e'_1, e'_2, \cdots, e'_p\} \subset E$ as a subset, if there exist a subset A satisfying $f_A(h_1) = f_A(h_2) = \cdots = f_A(h_n)$, then A is dispensable, otherwise, A is indispensable. $B \subset E$ is defined as a normal parameter reduction of E, if the two conditions as follows are satisfied*

(1) B is indispensable

(2) $f_{E-B}(h_1) = f_{E-B}(h_2) = \ldots = f_{E-B}(h_n)$

2.3 Algorithm of Normal Parameter Reduction

Definition 2.4 (See [15]). *Decision partition above mentioned $C_E = \{E_{f_1}, E_{f_2}, \cdots, E_{f_s}\}$, similarly, decision partition deleted e_i is figured as*

$$C_{E-e_i} = \{\overline{E-e_i}_{f'_1}, \overline{E-e_i}_{f'_2}, \cdots, \overline{E-e_i}_{f'_s}\}$$

The importance degree of e_i for the decision partition is defined by

$$r_{e_i} = \frac{1}{|U|}\left(\alpha_{1,e_i} + \alpha_{2,e_i} + \cdots + \alpha_{s,e_i}\right)$$

where $|.|$ denotes the cardinality of set and

$$\alpha_{k,e_i} = \begin{cases} \left| E_{f_k} - \overline{E-e_i}_{f'_z} \right|, & \text{if there exist } z' \text{ such that } f_k = f_{z'}, 1 \leq z' \leq s', 1 \leq k \leq s. \\ \left| E_{f_k} \right| & , \text{otherwise.} \end{cases}$$

Based on the parameter importance degree, Kong et al. [15] presented the algorithm of normal parameter reduction as follows:

1. Input the soft set (F, E) and the parameter set E;
2. Compute parameter importance degree r_{e_i}, $1 \leq i \leq m$;
3. Find maximum subset $A = \{e'_1, e'_2, \cdots, e'_p\} \subset E$ in which that sum of $r_{e'_i}$, for $1 \leq i \leq p$ is nonnegative integer, then put the A into a feasible parameter reduction set;
4. Filter in the feasible parameter reduction set, if $f_A(h_1) = f_A(h_2) = \ldots = f_A(h_n)$, then $E - A$ is the normal parameter reduction, otherwise A is deleted.
5. Get the maximum cardinality of A in feasible parameter reduction set.
6. Compute $E - A$ as the optimal normal parameter reduction.

3 A Novel Normal Parameter Reduction Algorithm

3.1 The Proposed Technique

Given a soft set *(F, E)* with a tabular presentation, $U=\{h_1,h_2,\cdots,h_n\}$, $E=\{e_1,e_2,\cdots,e_m\}$, and h_{ij} are the entries in the table of *(F, E)*.

Definition 3.1. *For soft set (F, E),* $U=\{h_1,h_2,\cdots,h_n\}$, $E=\{e_1,e_2,\cdots,e_m\}$ *and* h_{ij} *are the entries in the table of (F, E). We denote* $f_E(h_i)=\sum_j h_{ij}$ *as oriented-object sum.*

Definition 3.2. *For soft set (F, E),* $U=\{h_1,h_2,\cdots,h_n\}$, $E=\{e_1,e_2,\cdots,e_m\}$ *and* h_{ij} *are the entries in the table of (F, E). We denote* $S(e_j)=\sum_i h_{ij}$ *as oriented-parameter sum.*

Definition 3.3. *For soft set (F, E),* $U=\{h_1,h_2,\cdots,h_n\}$, $E=\{e_1,e_2,\cdots,e_m\}$ *and* h_{ij} *are the entries in the table of (F, E). We denote* $S_A=\sum_j S(e_j)$, *for* $A\subseteq E$ *as overall sum of A.*

Theorem 3.1. *For soft set (F, E),* $U=\{h_1,h_2,\cdots,h_n\}$, $E=\{e_1,e_2,\cdots,e_m\}$, *if there exists a subset* $A=\{e_1',e_2',\cdots,e_P'\}\subset E$, *such that* $E-A$ *is the normal parameter reduction of E, then we have* $S_A=qn$, *for q=0,1,2…,m, where n is the number of the universe U.*

Proof. Suppose $A=\{e_1',e_2',\cdots,e_P'\}\subset E$. According to Definition 2.3, if $B\subset E$ is defined as a normal parameter reduction of *E*, then $f_{E-B}(h_1)=f_{E-B}(h_2)=\cdots=f_{E-B}(h_n)$. In other words, if *A*=*E* −*B* can be reduced, then $f_A(h_1)=f_A(h_2)=\cdots=f_A(h_n)$. Therefore the following equation must be satisfied.

$$\begin{gathered}h_{11}'+h_{12}'+\cdots+h_{1P}'=q\\ h_{21}'+h_{22}'+\cdots+h_{2P}'=q\\ \vdots\\ h_{n1}'+h_{n2}'+\cdots+h_{nP}'=q\,.\end{gathered}$$

We can easily get

$$\begin{aligned}S_A&=S(e_1')+S(e_2')+...+S(e_P')\\ &=(h_{11}'+h_{21}'+...+h_{n1}')+(h_{12}'+h_{22}'+...+h_{n2}')+...+(h_{1P}'+h_{2P}'+...+h_{nP}')\\ &=(h_{11}'+h_{12}'+...+h_{1P}')+(h_{21}'+h_{22}'+...+h_{2P}')+...+(h_{n1}'+h_{n2}'+...+h_{nP}')\\ &=n\cdot q\end{aligned}$$

Namely, S_A is a multiple of *n*. This completes the proof. □

Definition 3.4. *For soft set (F, E),* $U=\{h_1,h_2,\cdots,h_n\}$, $E=\{e_1,e_2,\cdots,e_m\}$. *Define* h_{ij} *is an entry in the table of* (F,E). *For* $e_j \in E$, *if* $h_{1j}=h_{2j}=\cdots=h_{nj}=1$, *we denote* e_j *as* e_j^1.

Definition 3.5. *For soft set (F, E),* $U=\{h_1,h_2,\cdots,h_n\}$, $E=\{e_1,e_2,\cdots,e_m\}$. *Define* h_{ij} *is an entry in the table of* (F,E). *For* $e_j \in E$, *if* $h_{1j}=h_{2j}=\cdots=h_{nj}=0$, *we denote* e_j *as* e_j^0.

3.2 The Proposed Algorithm

Based on above theorems and definitions, we give our algorithm as follows:

1. Input the soft set (*F*, *E*) and the parameter set *E*;
2. If there exists e_j^1 and e_j^0, they will be put into the reduced parameter set denoted by *C* and a new soft set (F,E') will be established without e_j^1 and e_j^0, where $U=\{h_1,h_2,\cdots,h_n\}$, $E'=\{e_{1'},e_{2'},\cdots,e_{t'}\}$;
3. For the soft set (F,E'), calculate $S(e_{j'})$ of $e_{j'}$ (that is, oriented-parameter sum), for $j'=1',2',\cdots,t'$;
4. Find the subset $A \subset E'$ in which S_A is a multiple of $|U|$, then put the *A* into a candidate parameter reduction set;
5. Check every *A* in the candidate parameter reduction set if $f_A(h_1)=f_A(h_2)=\cdots=f_A(h_n)$,it will be kept; otherwise it will be omitted;
6. Find the maximum cardinality of *A* in the candidate parameter reduction set, then $E-A-C$ as the optimal normal parameter reduction.

4 The Comparison Result

In this section, a comparison for capturing the normal parameterization reduction is elaborated through a Boolean data set as in Table 1. Both algorithms are implemented in C++ program. They are executed sequentially on a processor Intel Core 2 Duo CPUs. The total main memory is 1 gigabyte and the operating system is Windows XP Professional SP3.

Example 4.1. Let (*F*, *E*) be a soft set with the tabular representation displayed in Table 1. Suppose that $U=\{h_1,h_2,h_3,h_4,h_5,h_6\}$, and $E=\{e_1,e_2,e_3,e_4,e_5,e_6,e_7,e_8\}$.

Table 1 A soft set (F, E)

h	e_1	e_2	e_3	e_4	e_5	e_6	e_7	e_8	$f(.)$
h_1	1	0	0	0	0	1	0	1	3
h_2	0	1	0	0	1	1	0	1	4
h_3	0	0	1	1	1	0	0	1	4
h_4	1	0	0	1	0	0	0	1	3
h_5	0	1	0	1	1	0	0	1	4
h_6	0	0	1	1	0	1	0	1	4
$S(e_j)$	2	2	2	4	3	3	0	6	S_E=22

a. The results from the algorithm in [15]

Step 1: Figuring out the oriented-object sum and then getting decision partition.

$$C_E = \{\{h_1, h_4\}_3, \{h_2, h_3, h_5, h_6\}_4\},$$

Step 2: Figuring out the oriented-object sum deleted e_i and the decision partition deleted e_i.

$$C_{E-e_1} = \{\{h_1, h_4\}_2, \{h_2, h_3, h_5, h_6\}_4\}$$
$$C_{E-e_2} = \{\{h_1, h_2, h_4, h_5\}_3, \{h_3, h_6\}_4\}$$
$$C_{E-e_3} = \{\{h_1, h_3, h_4, h_6\}_3, \{h_2, h_5\}_4\}$$
$$C_{E-e_4} = \{\{h_4\}_2, \{h_1, h_3, h_5, h_6\}_3, \{h_2\}_4\}$$
$$C_{E-e_5} = \{\{h_1, h_2, h_3, h_4, h_5\}_3, \{h_6\}_4\}$$
$$C_{E-e_6} = \{\{h_1\}_2, \{h_2, h_4, h_6\}_3, \{h_3, h_5\}_4\}$$
$$C_{E-e_7} = \{\{h_1, h_4\}_2, \{h_2, h_3, h_5, h_6\}_4\}$$
$$C_{E-e_8} = \{\{h_1, h_4\}_2, \{h_2, h_3, h_5, h_6\}_3\}$$

Step 3: Getting the importance degree of e_i. Thus, $r_{e_1} = \frac{2}{6}$, $r_{e_2} = \frac{2}{6}$, $r_{e_3} = \frac{2}{6}$, $r_{e_4} = \frac{4}{6}$, $r_{e_5} = \frac{3}{6}$, $r_{e_6} = \frac{3}{6}$, $r_{e_7} = 0$, $r_{e_8} = 1$,

Step 4: Finding maximum subset $A = \{e'_1, e'_2, \cdots, e'_p\} \subset E$ in which that sum of $r_{e'_i}$, for $1 \le i \le p$ is nonnegative integer. As a result, we get 39 subsets which are put into a feasible parameter reduction set, such as $\{e_1,e_4\}$,$\{e_1,e_2,e_3\}$, $\{e_3,e_4,e_5,e_6\}$ and so on.

Step 5: Filtering in the feasible parameter reduction set. We get the 11 subsets such as $\{e_8\}$, $\{e_1,e_2,e_3\}$ and $\{e_1,e_4,e_5,e_6\}$ so on, satisfying $f_A(h_1)=f_A(h_2)=\cdots=f_A(h_n)$ and the remainders are deleted.

Step 6: Getting the maximum cardinality of A in feasible parameter reduction set be $\{e_1,e_4,e_5,e_6,e_7,e_8\}$. So, the set $\{e_2,e_3\}$ is the optimal normal parameter reduction.

b. The results from the proposed algorithm

Step 1: Because there exists e_7^0 and e_8^1, they are put into the reduced parameter set denoted by C and a new soft set (F,E') is established without e_j^1 and e_j^0, where and $E'=\{e_1,e_2,e_3,e_4,e_5,e_6\}$.

Step 2: calculating the oriented-parameter sum $S(e_{j'})$ of $e_{j'}$ showed in Table 1.

Step 3: Finding the subset $A\subset E'$ in which S_A is a multiple of $|U|=6$. As a result we then put 9 subsets such as $\{e_1, e_4\}$,$\{e_1, e_2, e_3\}$, $\{e_3, e_4, e_5, e_6\}$ and so on into a candidate parameter reduction set.

Step 4: Filtering in the candidate parameter reduction set. We get $\{e_1, e_2, e_3\}$ and $\{e_1, e_4, e_5, e_6\}$ satisfying $f_A(h_1)=f_A(h_2)=\cdots=f_A(h_n)$ and the remainders are deleted.

Step 5: Finding the maximum cardinality of A in the candidate parameter reduction set, then $E-A-C=\{e_2,e_3\}$ as the optimal normal parameter reduction.

We can draw conclusions from the above example:

1. In order to obtain all the decision partitions, the data in the Table 1 are accessed 9 times for the oriented-object sums and partitions by means of algorithm in [15]. However, the data in Table 1 are accessed only 1 time for the oriented-parameter sums in the proposed algorithm. Consequently our algorithm involves relatively much less computation compared with the former algorithm.
2. Due to considering e_j^1 and e_j^0, the number of subsets in the candidate parameter reduction set of the proposed algorithm is much less than that of subsets in the feasible parameter reduction set of algorithm in [15]. Hence computation is reduced.
3. It is necessary to calculate the oriented-object sums, classify objects according to the oriented-object sums and then compute the importance degree in the algorithm in [15], whereas our algorithm only needs to calculate the oriented-parameter sums. As a result our algorithm is easier to implement and understand compared with the former algorithm.

Some details on the comparison result for this example are clearly depicted in Table 2.

Table 2 The comparison result

Comparison	The algorithm [15]	The proposed algorithm	Improvement %
Optimal normal parameter reduction	$\{e_2, e_3\}$	$\{e_2, e_3\}$	The same
The number of entry access	534	48	91.01%
The number of candidate parameter reduction set	39	9	76.92%
The involved operation	Addition, set operation, classification for parameter importance degree	Only Addition for oriented-parameter sum	-

Besides the soft set constructed in this section, we also experimented on some other soft sets with larger amount of data and drew the same conclusion: (1) two algorithms can obtain the same result; (2)due to much decrease on the number of entry access and the number of candidate parameter reduction set, our algorithm involves relatively much less computation compared with the algorithm[15] and then save the time; (3) because of reduce on the involved operation, our algorithm is easier to implement and understand compared with the algorithm[15]. So the conclusion can be generalized.

5 Conclusions

The related definitions and algorithm on the normal parameter reduction of soft sets have been proposed. In this paper, some new theorems are presented and proved. Based on the theorems, we propose a novel normal parameter reduction algorithm of soft sets, which can be carried out without parameter important degree and decision partition. As a result, it can involve relatively less computation and is simpler and easier to understand, compared with the algorithm of normal parameter reduction [15].

References

[1] Molodtsov, D.: Soft set theory_First results. Comput. Math. Appl. 37, 19–31 (1999)
[2] Zadeh, L.A.: Fuzzy sets. Information and Control 8, 338–353 (1965)
[3] Pawlak, Z.: Rough sets. International Journal Information Computer Science 11, 341–356 (1982)
[4] Molodtsov, D.: The Theory of Soft Sets. URSS Publishers, Moscow (2004) (in Russian)
[5] Pei, D., Miao, D.: From soft sets to information systems. In: The proceeding of IEEE International Conference on Granular Computing, IEEE GrC 2005, pp. 617–621. IEEE Press (2005)

[6] Herawan, T., Mat, D.M.: A direct proof of every rough set is a soft set. In: Proceeding of the Third Asia International Conference on Modeling and Simulation, pp. 119–124. AMS, Bali
[7] Feng, F., Li, C., Davvaz, B., Ali, M.I.: Soft sets combined with fuzzy sets and rough sets: A tentative approach. In: Soft Computing - A Fusion of Foundations, Methodologies and Applications, pp. 899–911. Springer, Heidelberg (2009)
[8] Feng, F.: Generalized rough fuzzy sets based on soft sets. In: the Proceeding of International Workshop on Intelligent Systems and Applications, Wuhan, China, ISA, pp. 1–4 (2009)
[9] Ali, M.I., Feng, F., Liu, X., Min, W.K., Shabira, M.: On some new operations in soft set theory. Comput. Math. Appl. 57, 1547–1553 (2009)
[10] Zou, Y., Xiao, Z.: Data analysis approaches of soft sets under incomplete information. Knowl.-Based Syst. 21, 941–945 (2008)
[11] Xiao, Z., Gong, K., Xia, S., Zou, Y.: Exclusive disjunctive soft sets. Comput. Math. Appl. 59, 2128–2137 (2010)
[12] Herawan, T., Mat, D.M.: A Soft Set Approach for Association Rules Mining. Knowl.-Based Syst. (2010), doi:10.1016/j.knosys.2010.08.005
[13] Maji, P.K., Roy, A.R.: An application of soft sets in a decision making problem. Comput. Math. Appl. 44, 1077–1083 (2002)
[14] Chen, D., Tsang, E.C.C., Yeung, D.S., Wang, X.: The parameterization reduction of soft sets and its applications. Comput. Math. Appl. 49, 757–763 (2005)
[15] Kong, Z., Gao, L., Wang, L., Li, S.: The normal parameter reduction of soft sets and its algorithm. Comput. Math. Appl. 56, 3029–3037 (2008)

Integrating Cognitive Pairwise Comparison to Data Envelopment Analysis

Kevin Kam Fung Yuen

Abstract. Data Envelopment Analysis (DEA) is one of the popular approaches of decision analysis. Parametric Settings for DEA is one of the essential steps for the decision making. This research proposes the method to apply Cognitive Pairwise Comparison (CPC) to the determination of the parametric settings in DEA. The usability and applicability of the enhanced DEA are demonstrated in a resource allocation problem on the basis of quality-cost balance.

1 Introduction

Data Envelopment Analysis (DEA) is an optimization technique of solving a variety of practical decision problems that arise in different fields. Data Envelopment Analysis (DEA) was initially proposed by Charnes, Cooper, and Rhodes (CCR) [3] in 1978. The CCR was extended by Banker, Charnes, and Cooper, (BCC) [2] in 1984. CCR and BCC are the two fundamental models of DEA.

Alder et al.[1] reviewed DEA context in six areas: cross-efficiency, super-efficiency, benchmarking, statistical techniques, ranking inefficient units, and multicriteria decision making methodologies. The research of DEA has also been extended to the fuzzy soft computing [4-6,8].

Whilst a number of studies discuss the improvement of the optimization model, a few studies address the assessment for utility estimation for the input of DEA model. Regarding the utility estimation for the parameters, the pairwise reciprocal matrix of the Analytic Hierarchy Process (AHP) [7] is one of the methods. Yuen [9] has indicated two major queries on this method: cognitive misrepresentation of the pairwise reciprocal matrix using the ratio scale, and the uncertainty of the prioritization methods. Yuen [9] proposed cognitive pairwise comparison (CPC) as the alternative. CPC can produce either crisp or fuzzy data. For the simplicity, this study only considers crisp output from the CPC. The applications of CPC can also be found in [9-11].

Kevin Kam Fung Yuen
School of Science and Technology, The Open University of Hong Kong
e-mail: kevinkf.yuen@gmail.com

A. Gaspar-Cunha et al. (Eds.): Soft Computing in Industrial Applications, AISC 96, pp. 231–238.
springerlink.com

This paper proposed the integrated DEA using CPC to solve the real world problem. The organization is structured as follows. Section 2 illustrates the notion of Cognitive Pairwise Comparison. Section 3 presents the model of the enhanced DEA using CPC. To demonstrate the usability and applicability, a resource allocation problem on the basis of quality-cost balance is demonstrated in section 4. Conclusion is drawn in section 5.

2 Cognitive Pairwise Comparisons

The Cognitive Pairwise Matrix (Pairwise Opposite Matrix or Cognitive Comparison Matrix) B of the objective O with respective to the criteria $\{c_i\}$, i.e. $Clst(O,\{c_i\})$, is of the form.

$$B_O = \varphi\left(Clst\left(O,\{c_i\}\right)\right) = \begin{array}{c} \\ c_1 \\ c_2 \\ \vdots \\ c_n \end{array} \begin{array}{c} \begin{array}{cccc} c_1 & c_2 & \cdots & c_n \end{array} \\ \begin{bmatrix} b_{11} & b_{12} & \cdots & b_{1n} \\ b_{21} & b_{22} & \cdots & b_{2n} \\ \vdots & \vdots & \ddots & \vdots \\ b_{n1} & b_{n2} & \cdots & b_{nn} \end{bmatrix} \end{array}, \tag{1}$$

$b_{ij} = v_i - v_j$, $\forall i, j \in (1,\ldots,n)$. b_{ij} is the comparison score from the cognitive rating scale $\aleph$ (Table 1). $\aleph$ comprises linguistic form such as equally, weakly, moderately, …, extremely, and the numerical representation form such as from 0 to κ, and the opposite form is from - κ to 0. The opposite form means that "if A dominates B, then B dominates A." $Clst$ is a cluster. φ is the Cogntive Assessment Function (CAF) performed by expert.

A Cognitive Pairwise Matrix B is validated by the Accordant Index of the form:

$$AI = \frac{1}{n^2}\sqrt{\sum_{i=1}^{n}\sum_{j=1}^{n} d_{ij}},$$

$$d_{ij} = \sqrt{Mean\left(\left(\frac{1}{\kappa}\left(B_i + B_j^T - b_{ij}\right)\right)^2\right)}, \ \forall i, j \in (1,\ldots,n), \tag{2}$$

where $AI \geq 0$, and κ is the normal utility such that $\left(b_{ik} + b_{kj}\right) \in [-\kappa, \kappa]$, $\forall i, j, k \in (1,\ldots,n)$.

The maximal value of the numerical rating scale is the default setting of κ if v_i is a non-negative value. Otherwise, κ is increased such that $v_i \geq 0$.

If $AI = 0$, then B is perfectly accordant; If $0 < AI \leq 0.1$, then B is satisfactory, then. If $AI > 0.1$, then B is unsatisfactory .

Table 1 Scale schemas: pairwise reciprocal comparison and pairwise opposite comparison

i	Verbal scales ℵ	Semantic Comparison Scales
0	Equally	0
1	Weakly	$\kappa/8$
2	Moderately	$\kappa/4$
3	Moderately plus	$3\kappa/8$
4	Strongly	$\kappa/2$
5	Strong Plus	$5\kappa/8$
6	Very Strongly	$3\kappa/4$
7	Very, very strongly	$7\kappa/8$
8	Extremely	κ
{-i}	opposites of Above	(from $-\kappa$ to 0)

Yuen [9] has proposed several cognitive prioritization operators. Two operators are recommended: Least Penalty Squares (LPS) and the Row Average plus the normal Utility (RAU). The comprehensive numerical analyses conclude that, in most cases, if $\text{AI} \le 0.1$, the results of RAU and LPS are the same or very close [9].

If $\text{AI} \le 0.1$, the vector of individual utilities can be derived by the Primitive Least Squares optimization program which is of the form:

$$\text{PLS}\left(B^{+},\kappa\right)=$$

$$\text{Min} \quad \bar{\Delta}=\sum_{i=1}^{n}\sum_{j=i+1}^{n}\left(b_{ij}-v_i+v_j\right)^2 \tag{3}$$

$$\text{s.t.} \quad \sum_{i=1}^{n} v_i = n\kappa,\ n=\left|\{v_i\}\right|, \text{ and } \kappa \text{ is the normal utility.}$$

The solution of the closed form can be solved manually and is *RAU*, given by:

$$v_i=\left(\frac{1}{n}\sum_{j=1}^{n} b_{ij}\right)+\kappa,\ \forall i\in\{1,\ldots,n\} \tag{4}$$

If $\exists v_i \in V$ is less than 0, κ can be increased such that $\{v_i \in V : v_i \ge 0\}$, or Least Penalty Squares (LPS) operator is applied as an alternative, as follows:

$\mathrm{LPS}\left(B^{+},\kappa\right)=$

$$\text{Min}\quad \hat{\Delta}=\sum_{i=1}^{n}\sum_{j=i+1}^{n}\beta_{ij}\cdot\left(b_{ij}-v_i+v_j\right)^2\ \text{,where}$$

$$\beta_{ij}=\begin{cases}\beta_1, & \mathrm{v}_i>v_j\ \&\ \mathrm{b}_{ij}>0\\ & \text{or } \mathrm{v}_i<v_j\ \&\ \mathrm{b}_{ij}<0\\ \beta_2, & \mathrm{v}_i=v_j\ \&\ \mathrm{b}_{ij}\neq 0\\ & \text{or } v_i\neq v_j\ \&\ \mathrm{b}_{ij}=0\\ \beta_3, & \textit{otherwise}\end{cases},\ 1=\beta_1\leq\beta_2\leq\beta_3 \tag{5}$$

$$\text{s.t.}\quad \sum_{i=1}^{n}v_i=n\kappa;\ \mathrm{v}_i\geq 0, i=1,2,\ldots,n.$$

For the most decision problems, if $\sum_{i=1}^{n}w_i=1$, W is said to be a normalized priority vector (or a priority vector in short) and has the following form.

$$W=\left\{w_i:w_i=\frac{v_i}{n\kappa},\forall i\in\{1,\ldots,n\}\right\},\ \sum_{i\in\{1,\ldots,n\}}v_i=n\kappa \tag{6}$$

W is the special case of V such that $n\kappa=1$.

3 Data Envelopment Analysis

Data Envelopment Analysis (DEA) is an optimization programming to determine the weights and efficiency for a decision making unit o ($\mathrm{DMU_o}$) by maximizing the ratio:

$$\frac{\textit{virtual output}}{\textit{virtual input}}=\frac{\sum_{r=1}^{s}u_{r0}y_{r0}}{\sum_{i=1}^{m}v_{i0}x_{i0}} \tag{7}$$

The input matrix and the output matrix can respectively be arranged as follows:

$$Input=\left[V\mid X\right]=\begin{matrix} & T_1 & \cdots & T_j & \cdots & T_n & T_1 & \cdots & T_j & \cdots & T_n\\ a_1 & v_{11} & \cdots & v_{1j} & \cdots & v_{1n} & x_{11} & \cdots & x_{1j} & \cdots & x_{1n}\\ \vdots & \vdots & \ddots & \vdots & \cdot^{\cdot^{\cdot}} & \vdots & \vdots & \ddots & \vdots & \cdot^{\cdot^{\cdot}} & \vdots\\ a_i & v_{i1} & \cdots & v_{ij} & \cdots & v_{in} & x_{i1} & \cdots & x_{ij} & \cdots & x_{in}\\ \vdots & \vdots & \cdot^{\cdot^{\cdot}} & \vdots & \ddots & \vdots & \vdots & \cdot^{\cdot^{\cdot}} & \vdots & \ddots & \vdots\\ a_m & v_{m1} & \cdots & v_{mj} & \cdots & v_{mn} & x_{m1} & \cdots & x_{mj} & \cdots & x_{mn}\end{matrix} \tag{8}$$

$$Output = [U \mid Y] = \begin{matrix} & T_1 & \cdots & T_j & \cdots & T_n \\ b_1 & u_{11} & \cdots & u_{1j} & \cdots & u_{1n} \\ \vdots & \vdots & \ddots & \vdots & \unicode{x22F0} & \vdots \\ b_r & u_{r1} & \cdots & u_{rj} & \cdots & u_{rn} \\ \vdots & \vdots & \unicode{x22F0} & \vdots & \ddots & \vdots \\ b_s & u_{s1} & \cdots & u_{sj} & \cdots & u_{sn} \end{matrix} \begin{matrix} T_1 & \cdots & T_j & \cdots & T_n \\ y_{11} & \cdots & y_{1j} & \cdots & y_{1n} \\ \vdots & \ddots & \vdots & \unicode{x22F0} & \vdots \\ y_{r1} & \cdots & y_{rj} & \cdots & y_{rn} \\ \vdots & \unicode{x22F0} & \vdots & \ddots & \vdots \\ y_{s1} & \cdots & y_{sj} & \cdots & y_{sn} \end{matrix} \tag{9}$$

X and Y represent the input data matrix and output data matrix respectively for a set of Decision Making Units (DMUs) denoted by $T = \{T_j\}$. X and Y can be achieved by Cognitive Pairwise Comparisons in section 2.

V and U are the weight matrices corresponding to X and Y respectively and usually are solved by the optimization model, together with the efficiency output. Although there are various types of optimization models for DEA, this research only considers the original one, CCR [3], as the motivation. CCR of one DMU_o or T_o (with minor modification) is of the form:

$$\begin{aligned} & CCR(\{x_{ij}\},\{y_{rj}\},o) = \\ & max \quad h_o = \frac{\sum_{r=1}^{s} u_{ro} y_{ro}}{\sum_{i=1}^{m} v_{io} x_{io}} \\ & s.t. \quad \frac{\sum_{r=1}^{s} u_{ro} y_{rj}}{\sum_{i=1}^{m} v_{io} x_{ij}} \le 1, \quad j = 1,2,\cdots,n \\ & \quad u_{ro} \ge \varepsilon > 0, \quad r = 1,2,\cdots,s \\ & \quad v_{io} \ge \varepsilon > 0, \quad i = 1,2,\cdots,m \end{aligned} \tag{10}$$

If h_o=1, u_{ro}>0, $\forall r$ and v_{io}>0, $\forall i$, DMU_o is efficient. Otherwise, DMU_o is inefficient. The above fractional program is equivalent to the linear program [4] as follows:

$$\begin{aligned} & LPCCR(\{x_{ij}\},\{y_{rj}\},o) = \\ & max \quad h'_o = \sum_{r=1}^{s} u_{ro} y_{r0} \\ & s.t. \quad \sum_{i=1}^{m} v_{io} x_{io} = 1 \\ & \quad \sum_{r=1}^{s} u_{ro} y_{rj} \le \sum_{i=1}^{m} v_{io} x_{ij}, \quad j = 1,2,\cdots,n \\ & \quad u_{ro} \ge \varepsilon > 0, \quad r = 1,2,\cdots,s \\ & \quad v_{io} \ge \varepsilon > 0, \quad i = 1,2,\cdots,m \end{aligned} \tag{11}$$

4 Numerical Example

This example illustrates the decision problem about recourse allocation considering the balance between cost and quality. In real life, if the service quality needs to be increased, the cost on the quality improvement is needed. Extra cost means that the company probably earns less. Without extra cost to improve the service quality, the company may suffer future loss, or lose competition capability of its service products. Finally lower quality also means that company earns less again. The proposed enhanced DEA using CPC can evaluate this problem.

Consider a problem that the efficiency is measured by the ratio of Quality Y and Cost X. Cost X is measured by three dimensions: quality system implementation cost (X_1), promotion cost (X_2), and customer service cost (X_3). Quality Y is measured by perception of service quality Y_1. Five proposals, $T=\{T_1, T_2, T_3, T_4, T_5\}$, are considered for the corporate strategy. The company would like to identify the inefficient and efficient proposals, and choose the best one in the latter action.

The company firstly performs cognitive pairwise comparisons for all proposals with respective to each dimension. Four pairwise opposite matrices, B_1,B_2,B_3,B_4, are respectively given in table 2 according to X_1, X_2, X_3,and Y_1. The utility values for the pairwise opposite matrices, which are shown in table 3, are derived by the cognitive prioritization operator, RAU since RAU produces similar result to the LPS, which requires much higher computational effort than RAU. κ=8 is set for the rating scales.

Table 2 Pairwise Opposite matrices $B_j, j=1,\dots,4$

	T_1	T_2	T_3	T_4	T_5	T_1	T_2	T_3	T_4	T_5
		B_1, AI=0.072					B_2, AI=0.059			
T_1	0	3	4	6	5	0	-3	2	3	-1
T_2	-3	0	1	2	2	3	0	4	5	2
T_3	-4	-1	0	1	1	-2	-4	0	0	-3
T_4	-6	-2	-1	0	-2	-3	-5	0	0	-3
T_5	-5	-2	-1	2	0	1	-2	3	3	0
		B_3, AI= 0.061					B_4, AI= 0			
T_1	0	5	3	4	2	0	3	2	3	1
T_2	-5	0	-3	-1	-3	-3	0	-1	0	-2
T_3	-3	3	0	0	-1	-2	1	0	1	-1
T_4	-4	1	0	0	-2	-3	0	-1	0	-2
T_5	-2	3	1	2	0	-1	2	1	2	0

The weights and efficiencies of the proposals are shown in table 4. In this example, only T_4 achieves the ideal efficiency, i.e. h_4=1, u_{r4}>0, $\forall r$ and v_{i4}>0, $\forall i$. The other four proposals are inefficient and should be revised or rejected.

It is possible to have more than one alternatives achieving ideal efficiency. The proposed CPC-DEA method is especially useful for screening and removing inefficient proposals in this scenario.

Table 3 Utility values for Decision making units

	X_1	X_2	X_3	Y_1
T_1	0.29	0.205	0.27	0.245
T_2	0.21	0.27	0.14	0.17
T_3	0.185	0.155	0.195	0.195
T_4	0.145	0.145	0.175	0.17
T_5	0.17	0.225	0.22	0.22

Table 4 Utility values for Decision making units

	v_1	v_2	v_3	u_1	h_o
T_1	3.877	0.000	4.878	0.000	*0.95**
T_2	5.882	*0.000**	1.112	4.999	1
T_3	5.128	*0.000**	0.969	4.358	1
***T_4**	5.882	1.785	1.378	3.092	1
T_5	4.545	3.619	1.710	*0.000**	1

5 Conclusion

This research proposes the enhanced DEA using CPC. For the validity and applicability, this research demonstrates the application of the proposed method to address the allocation problem considering the balance between cost and quality.

There are several motivation directions for the future study. More DEA optimization models can further be evaluated with extension of CPC. Applications based on DEA can also be investigated with extension of CPC. A case study considering both the objective evaluations and operational data can be extended and discussed. Extension of CPC as fuzzy CPC is applied to fuzzy DEA.

Regarding contributions, the proposed method can be applied to other application areas by using the cognitive pairwise comparisons to determine the parameter settings for a decision matrix, and the DEA to filter the inefficiencies.

References

1. Adler, N., Friedman, L., Sinuany-Stern, Z.: Review of ranking methods in the data envelopment analysis context. European Journal of Operational Research 140, 249–265 (2002)
2. Banker, R.D., Charnes, A., Cooper, W.W.: Some Models for Estimating Technical and Scale Inefficiencies in Data Envelopment Analysis. Management Science 30, 1078–1092 (1984)

3. Charnes, A., Cooper, W.W., Rhodes, E.: Measuring the efficiency of decision-making units. European Journal of Operational Research 2, 429–444 (1978)
4. Guo, P., Tanaka, H.: DEA: a perceptual evaluation method. Fuzzy Sets and Systems 119, 149–160 (2001)
5. Kao, C., Liu, S.-T.: Fuzzy efficiency measures in data envelopment analysis. Fuzzy Sets and Systems 113, 427–437 (2000)
6. Lertworasirikul, S., Fang, S.-C., Joines, J.A., Nuttle, H.L.W.: Fuzzy data envelopment analysis (DEA): a possibility approach. Fuzzy Sets and Systems 139, 379–394 (2003)
7. Saaty, T.L.: Analytic Hierarchy Process: Planning, Priority, Setting, Resource Allocation. McGraw-Hill, New York (1980)
8. Wu, D.D., Yang, Z., Liang, L.: Efficiency analysis of cross-region bank branches using fuzzy data envelopment analysis. Applied Mathematics and Computation 181, 271–281 (2006)
9. Yuen, K.K.F.: Cognitive network process with fuzzy soft computing technique for collective decision aiding. The Hong Kong Polytechnic University, Phd thesis (2009)
10. Yuen, K.K.F.: Enhancement of TOPSIS using the Compound Linguistic Ordinal Scale and Cognitive Pairwise Comparison. In: Proceedings of IEEE International Conference on Fuzzy Systems, pp. 649–654 (2009)
11. Yuen, K.K.F.: Enhancement of ELECTRE I using Compound Linguistic Ordinal Scale and the Cognitive Pairwise Comparison. In: Proceedings of IEEE International Conference on Systems, Man, and Cybernetics, pp. 5009–5014 (2009)

On the Multi-mode, Multi-skill Resource Constrained Project Scheduling Problem – A Software Application

Mónica A. Santos and Anabela P. Tereso

Abstract. We consider an extension of the Resource-Constrained Project Scheduling Problem (RCPSP) to multi-level (or multi-mode) activities. Each activity must be allocated exactly one unit of each required resource and the resource unit may be used at any of its specified levels. The processing time of an activity is given by the maximum of the durations that would result from a specific allocation of resources. The objective is to find the optimal solution that minimizes the overall project cost which includes a penalty for tardiness beyond the specified delivery date as well as a bonus for early delivery. We give some of the most important solution details and we report on the preliminary results obtained. The implementation was designed using the C# language.

1 Introduction

This paper is concerned with an extension of the Resource-Constrained Project Scheduling Problem (RCPSP) which belongs to the NP-hard class of problems. In the several resource constrained scheduling problem models found in the literature, there are two important aspects present in any model: the objective and the constraints. The objective may be based on time, such as minimize the project duration, or on economic aspects, such as minimize the project cost. However, success relative to time does not imply success in economic terms. Often, time-based objectives are in conflict with cost-based objectives. A recurrent situation encountered in practice is the need to complete a project by its due date and maximize profit. Ozdmar and Ulusoy [1] reported in

Mónica A. Santos
University of Minho, 4710-057 Braga, Portugal
e-mail: pg13713@alunos.uminho.pt

Anabela P. Tereso
University of Minho, 4710-057 Braga, Portugal
e-mail: anabelat@dps.uminho.pt

A. Gaspar-Cunha et al. (Eds.): Soft Computing in Industrial Applications, AISC 96, pp. 239–248.
springerlink.com

their survey of the literature, studies where the NPV is maximized while the due date is a 'hard' constraint (Patterson et al. [2][3]). There are several other multi-objective studies in the literature where efficient solutions regarding time and cost targets are generated. Guldemond et al. [4] presented a study related to the problem of scheduling projects with hard deadline jobs, defined as a Time-Constrained Project Scheduling Problem (TCPSP). They used a non-regular objective function.

Researchers agree that a project cannot be insulated from its costs, or executed without the scheduling of activities. As the costs depend on the activities in progress and scheduling is related to other constraints than monetary, the researchers explicitly included cash-flows-resources-constraints in their formulations. Elmaghraby and Herroelen [5] lay down the following property of an optimal solution that maximizes the NPV: the activities with positive cash flows should be scheduled as soon as possible and those with negative cash flow as late as possible. They concluded that the faster conclusion of the project is not necessarily the optimal solution with regard to maximizing the NPV. In Mika et al. [6] study, a positive flow is associated to each activity. The objective is to maximize the NPV of all cash flows of the project. They use two meta-heuristics that are widely used in research: Simulated Annealing (SA) and Tabu Search (TS).

Tereso et al.'s research ([7][8][9]) is included in the minimum-cost class problems. A recent metaheuristic, the Electromagnetism-Like Mechanism (EM), developed by Birbil and Fang [10], was implemented in Tereso et al. [7] in Matlab for multimodal activities projects, with stochastic work. Improved results in terms of computing performance were presented later in Tereso et al. [8] with an enhanced implementation using the JAVA programming language and in Tereso et al. [9], where a dynamic programming model was developed on a distributed platform.

Constraints complicate the efficient optimization of problems, and the more accurately they describe the real problem, the more difficult it is to handle it. Recent models include most of the requirements described by Willis [11] for modeling realistic resources. These requirements include the variable need of resources according to the duration of the activities, variable availability of resources over the project duration and different operational modes for the activities.

A discrete time/resource function implies the representation of an activity in different modes of operation. Each mode of operation has its own duration and amount of renewable and non-renewable resources requirement. Boctor [12] presented a heuristic procedure for the scheduling of non-preemptive resource-limited projects, although renewable from period to period. Each activity had a set of possible durations and resource requirements. The objective was to minimize the project duration. A general framework to solve large-scale problems was suggested. The heuristic rules that can be used in this framework were evaluated, and a strategy to solve these problems efficiently was designed.

Heilmann [13] also worked with the multi-mode case in order to minimize the duration of the project. In his work, besides the different modes of execution of each activity, there is specified a maximum and minimum delay between activities. He presented a priority rule-based heuristic. Basnet [14] presented a "filtered beam" search technique to generate makespan minimizing schedules, for multi-mode single resource constrained projects, where there is a single renewable resource to consider and the multi-mode consists essentially of how many people can be employed to finish an activity.

In a previous paper [15] we provided a formal model to the multi-mode, multi-skill resource constrained project scheduling (MRCPSP-MS) problem and a breadth-first procedure description, for an optimal allocation of resources in a project, with multi-mode activities, minimizing its total cost, while respecting all the restrictions. We implemented a procedure using the object oriented paradigm language, JAVA and achieved the optimal solution for a simple 3 activities project network, by obtaining all possible solutions and search the best between them. The plan was to complete an adaptation of a "filtered beam" search algorithm to this problem in the future; this report addresses this issue.

1.1 Problem Description

Consider a project network in the activity-on-arc (AoA) mode of representation: $G = (N, A)$, with $/N/ = n$ (representing the events) and $/A/ = m$ (representing the activities). Each activity may require the simultaneous use of several resources with different resource consumption according to the selected execution mode - each resource may be deployed at a different level. It is desired to determine the optimal resources allocation to the activities that minimizes the total cost of the project (resources + penalty for tardiness + bonus for earliness). We follow the dictum that an activity should be initiated as soon as it is sequence-feasible.

There are $/R/= \rho$ resources. A resource has a capacity of several units (say w workers or m/c's) and may be used at different levels, such as a 'resource' of electricians of different skill levels, or a 'resource' of milling machines but of different capacities and ages. A *level* may also be the amount of hours used by a resource; for example, half-time, normal time or extra-time. An activity normally requires the simultaneous utilization of more than one resource for its execution.

The problem presented here belongs to the class of the optimization scheduling problems with multi-level activities. This means that the activities can be scheduled at different modes, each mode using a different resource level, implying different costs and durations. Each activity must be allocated exactly one unit of each required resource and the resource unit may be used at any of its specified levels. The processing time of an activity is given by the maximum of the durations that would result from a specific allocation of the resources required by the activity. The objective is to find the optimal solution that minimizes the overall project cost, while respecting a delivery date. Briefly, the constraints of this problem are:

- Respect the precedence among the activities.
- A unit of the resource is allocated to at most one activity at any time at a particular level (the unit of the resource may be idle during an interval).
- Respect the capacity of the resource availability: The total units allocated at any time should not exceed the capacity of the resource to which these units belong.
- An activity can be started only when it is sequence-feasible and all the requisite resources are available, each perhaps at its own level, and must continue at the same levels of all the resources without interruption or preemption.

Figure 1 presents the mathematical model for the problem. For more information on this model refer to our previous paper [15].

Let:

- $G(N, A)$: Project network in AoA representation, with a set of N nodes, representing the events and A activities.
- n: number of nodes; $n = |N|$.
- m: number of arcs or number of activities; $m = |A|$.
- a : activity, which may also be represented by arc (i, j).
- r: resource $r \in |R|$
- C^k: the kth uniformly directed cutset *(udc)* of the project network that is traversed by the project progression; $k = 1, \ldots, K$.
- l: level at which a resource is applied to an activity.
- $x_{(a,r,l)}$: a binary variable, of value 1 if resource r is allocated to activity a at level l, and 0 otherwise.
- $p(a, r, l)$: the processing time of activity a when resource r is allocated at level l.
- $p(a)$: processing time of the activity a (considering all resources).
- $c(a, r, l)$: resource cost of activity a when resource r is allocated at level l.
- $c_R(a)$: resource cost of the activity a (considering all resources).
- η_a : the count of resources required by activity a.
- ρ : number of resources, $\rho = |R|$.
- b_r : capacity of resource r.
- $\gamma(r, l)$: : marginal cost of resource r at level l.
- γ_E : marginal gain from early completion of the project.
- γ_L : marginal loss (penalty) from late completion of the project.
- t_i : time of realization of node i (AoA representation), where node 1 is the "start node" of the project and node n its "end node".
- T_s : target completion time of the project.
- c_E: earliness cost.
- c_T: tardiness cost.
- c_{ET}: earliness-tardiness cost.
- c_R : total resource cost for all project activities.
- TC: total cost of the project.

Minimize TC

Subject to:

$$p(a) \geq p(a, r, l) \text{ for all a, r and l}$$

$$t_j - t_i \geq p(a), \forall\, a \in A$$

$$\sum_{a \in C^k} x_{(a,r,l)} \leq b_r, \forall\, r \in R$$

$$\sum_{for all\, l} x(a, r, l) = 1, \forall\, a, \forall\, r \in R$$

$$\eta_a - \sum_{r \in R} \sum_{for all\, l} x(a, r, l) = 0, \forall\, a \in C^K$$

Where:

$$TC = C_R + C_{ET}$$

$$c_R = \sum_{a \in A} c_R(a)$$

$$c_{ET} = c_E + c_T = \gamma_E \cdot e + \gamma_L \cdot d$$

$$c_R(a) = \sum_{r \in R} c(a, r, l)$$

$$c(a, r, l) = \gamma(r, l) * p(a, r, l)$$

$$e \geq T_s - t_n$$

$$d \geq t_n - T_s$$

$$e, d \geq 0$$

Fig. 1 Mathematical Model

2 Solution Details

The initial procedure we adopted, applied to a small project, was based in a breadth first search (BFS) algorithm. All the nodes (partial solutions) in the search tree were evaluated at each stage before going any deeper, subsequently implementing an *exhaustive search* that visits all nodes of the search tree. This strategy can be applied for small projects but becomes infeasible for larger ones.

The branch and bound (BaB) search technique allows reducing the number of nodes being explored. It can be seen as a *polished* breadth first search, since it applies some criteria in order to reduce the BFS complexity. Usually it consists of keeping track of the best solution found so far and checking if the solution given by that node is greater than the best known solution. So if that node cannot offer a better solution than the solution obtained so far, the node is discarded. The BaB process consists of two procedures: subset generation and subset elimination. The former (the subset generation) is accomplished by *branching*, where a set of descendent nodes, form a tree-like structure. The latter (subset elimination) is realized through *bounding*, where upper and lower bounds are calculated for the "value" of each node. The bounding function can be strong, which is usually harder to calculate but faster in finding the solution, or weak, which is easier to calculate but slower in finding the solution. The BaB approach is more efficient if the bounds can be made very tight. In our case, the objective of our problem is to minimize the total cost of the project, that gets a bonus or a penalty cost while respecting or exceeding the specified due date; respectively. As a result, finding a strong bounding function would depend on the three project parameters cited: the penalty cost, bonus cost and due date. The feat of the bounding function is simply in reducing the search while not discarding potentially desirable branches. A "filtered beam" search is a heuristic BaB procedure that uses breadth first search but only the top "best" nodes are kept. At each stage of the tree, it generates all successors for the selected nodes at the current stage, but only stores a predetermined number of descendent nodes at each stage, called the *beam width*. This paper is concerned with the study of the adaptation of the initial algorithm, presented below, to a "filtered beam" search procedure.

2.1 Procedure Description

The procedure to be executed can be based either on the BFS algorithm or on the Beam Search Algorithm. If the latter is the one adopted a beam width value must be defined. We consider that activities can be in one of four states: "to begin", "pending", "active" and "finished". To get the first activities with which to initiate the process, we search all activities that do not have any predecessors. These activities are set to state "to begin". All others are set to the state "pending".

Activities in the state "to begin" are analyzed in order to check resources availability. If we have enough resources, all activities in the state "to begin" modify the state to "in progress", otherwise we apply, in sequence, the following rules, until resources conflict are resolved:

1. Give priority to activities precedents of a larger number of "pending activities".
2. Give priority to activities that use fewer resources.
3. Give priority to activities in sequence of arrival to the state "to begin".

An "*event*" represents the starting time of one or more activities and the project begins at event 0. Each activity must be allocated exactly one unit of each resource. For each *active* activity, we calculate all the possible combinations of resources levels. Then we join all activities combinations, getting the initial combinations of allocation modes for all *active* activities. These initial combinations form branches through which we will get possible solutions for the project. All combinations have a copy of resources availability information, and activities' current state.

If the algorithm set to find best solution is the *Beam Search Algorithm* then:

1. If the number of combinations is less than the beam width value, all combinations are kept.

2. Otherwise, the set of combinations must be reduced to the beam width value. In this case some combinations need to be discarded using the possible rules to evaluate the ones in the top best:
   Select top best combinations that have: -Minimum Duration.
   -Minimum Cost.
   -Minimum Cost/Duration.

In either case, we continue applying the following procedure to each combination:

3. To all activities in progress, we find the ones that will be finished first, and set that time as the next *event*.
4. We update activities found in step 1 to state "finished", and release all the resources being used by them.
5. For all activities in the state "to begin", we check if they can begin, the same way we did when initiating the project. Activities in the state "to begin" are analyzed in order to check resources availability. If no resource conflicts exists, all activities in the state "to begin" are set to state "active" and resources are set as being used, otherwise we apply in sequence, the rules described above.
6. For all activities in the state "pending", we check for precedence relationships. For all activities that are precedence-feasible their state is updated to state "to begin". These activities aren't combined to the previous set of "to begin" activities to give priority to activities that entered first in this state.
7. If there are resources available and any pending activities were set "to begin" we apply step 5 again.
8. For all new activities "in progress" we set their start time to the next *event* found in step 3, and determine all the possible combinations of its resources levels. Then we join all found combinations for these activities, getting new combinations to join to the actual combination being analyzed. This forms new branches to process in order to get the project solution.
9. We continue by applying step 1 (or 3) to each new combination until all activities are set to state "finished".
10. Once all activities in a combination are set to state "finished", we have a valid project solution.

When the project final solutions are found, we evaluate, for each one, the finishing time of the project and the total project cost, choosing the best one.

The BaB and the Beam Search procedures are typical methods applied to the RCPSP. The differentiating aspects of our approach are, on one hand the definition of the set of states followed by the activities, combined with the priority rules used to solve resource conflicts, and on the other hand the alternative evaluation rules used to discard undesirable "branches".

2.2 Application Development

The software was developed in C# language using Visual Studio 2010. To construct the project network (in AoN), we use Graph#, an open source library for

.Net/WPF applications that is based on a previous library QuickGraph. These libraries support GraphML that is an XML-based file format for graphs, although we didn't make use of this format. The graph is automatically generated for each project loaded in the application. To save/load existing projects we define an *xml* file that embodies all project characteristics for this problem.

Three main classes were defined for the application. The base class is *NetProject* that keeps all project required information: name, activities, resources, due date, bonus and penalty cost. Then we have the *Resource* class that keeps the resource identification availability and levels. Each resource level has a unitary cost. The *Activity* class has activity identification, resources requirement and its precedents. The referred classes are the most relevant to represent the project, additional classes are used to support the evaluation of the project solution.

2.2.1 Functionalities

The application provides the functionalities described next.

- Load a Project - The project must be saved as an xml file, using a structure that represents the project components (activities, resources, etc.).
- Create a Project - There are two main steps to create a new project:
  - First the project "skeleton" is built through a wizard that initiates asking the project name and the number of resources and activities. Next the resource data is introduced i.e. the availability of each resource and the number of associated levels. Finally the activities information is introduced, that is the identification and precedents of each activity.
  - Secondly it generates the project graph and a project grid where the remaining project information can be introduced.
- Edit/Save a Project.
- Determinate best solution - This can be achieved using either BFS based Algorithm or Beam Search Algorithm.
- Save solution to a txt file.

We present next the application look, using some prints.

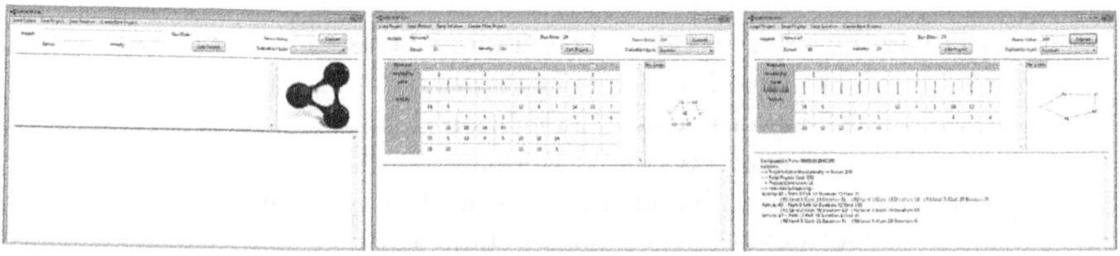

Fig. 2 Application prints

3 Preliminary Results

The next computational tests were performed on an Intel® Pentium® M @1.20GHz 1.25GB RAM. Consider the following data for a 3 activities network.

Table 1 Resource Requirements, Processing Times and Resource Costs of the Project.

RESOURCE →	**1**			**2**			**3**			**4**			
AVAILABILITY	2			1			3			2			
↓ Activity \ Levels →	**1**	**2**		**1**	**2**	**3**	**1**	**2**	**3**	**1**	**2**	**3**	
Unitary costs	2	4		3	5	7	1	4	5	1	3	5	η_j
A1(Processing time)	14	6	-	-	-	-	12	8	5	18	12	7	3
A1(Resource cost)	28	24	-	-	-	-	12	32	25	18	36	35	
A2(Processing time)	-	-	-	7	5	3	-	-	-	8	5	4	2
A2(Resource cost)	-	-	-	21	25	21	-	-	-	8	15	20	
A3(Processing time)	20	12	-	22	16	10	-	-	-	-	-	-	2
A3(Resource cost)	40	48	-	66	80	70	-	-	-	-	-	-	

Assume the following rates for earliness and lateness costs: $\gamma_E = -10$, $\gamma_L = 20$ and the due date $T_S = 24$.

Using the BFS Algorithm the project obtained solution is presented in table 2.

Table 2 Solution totals, obtained using BFS Algorithm.

t_n	C_E	C_T	C_R	TC	Runtime (ms)
16,0	80,0	0,0	230	150,0	66

Since the due date was 24, a bonus is applied. Activities execution modes are:

Activity A1 - Start: 0 End: 12 Duration: 12 Cost: 71
| R1 Level 2 (Cost: 24 Duration: 6)
| R3 Level 1 (Cost: 12 Duration: 12)
| R4 Level 3 (Cost: 35 Duration: 7)
Activity A3 - Start: 0 End: 12 Duration: 12 Cost: 118
| R1 Level 2 (Cost: 48 Duration: 12)
| R2 Level 3 (Cost: 70 Duration: 10)
Activity A2 - Start: 12 End: 16 Duration: 4 Cost: 41
| R2 Level 3 (Cost: 21 Duration: 3)
| R4 Level 3 (Cost: 20 Duration: 4)

Table 3 Solution totals, obtained using Beam Search Algorithm.

Beam Width	Cost						Duration						Cost/Duration					
	t_n	C_E	C_T	C_R	TC	Runtime (ms)	t_n	C_E	C_T	C_R	TC	Runtime (ms)	t_n	C_E	C_T	C_R	TC	Runtime (ms)
150	26	0	40	201	241	33	16	80	0	230	150	52	26	0	40	201	241	42
200	26	0	40	201	241	48	16	80	0	230	150	68	26	0	40	201	241	69
700	20	40	0	240	200	253	16	80	0	230	150	319	20	40	0	240	200	372
900	16	80	0	231	151	419	16	80	0	230	150	482	16	80	0	231	151	607

The BFS Algorithm generates 972 combinations for the three activity network. We used a beam width between 150 and 900. As we can see by the results exhibited in table 3, the duration evaluation type was the best for this network, achieving the same result as the BFS Algorithm, even with the lowest beam width. The other evaluation types gave both the same result.

4 Conclusions

We developed a practical tool, useful to represent multi-mode projects, and to find a solution for the problem on hand – select the best mode for each resource in each activity in order to minimize the total cost, considering the resource cost, a penalty for tardiness and a bonus for early completion. We must continue testing the tool, in order to evaluate the quality of the solution obtained, since the heuristic used doesn't guarantee the optimum. Further experiments will also allow specifying the limits of its applicability in terms of the number of activities, the number of resources, and the number of alternative levels of resource application. Another useful effort is to compare as well the solutions obtained with both algorithms, trying to define a recommended *beam width* and evaluation type.

References

1. Ozdamar, L., Ulusoy, G.: A Survey on the Resource-Constrained Project Scheduling Problem. IIE Transactions 27, 574–586 (1995)
2. Patterson, J.H., Slowinski, R., Talbot, F.B., Weglarz, J.: An algorithm for a general class of precedence and resource constrained scheduling problems. In: Advances in Project Scheduling, Amsterdam, pp. 3–28 (1989)

3. Patterson, J.H., Talbot, F.B., et al.: Computational experience with a backtracking algorithm for solving a general class of precedence and resource constrained scheduling problems. Eur. J. Oper. Res. 49, 68–79 (1990)
4. Guldemond, T., Hurink, J., Paulus, J., Schutten, J.: Time-constrained project scheduling. J. Sched. 11(2), 137–148 (2008)
5. Elmaghraby, S.E., Herroelen, W.S.: The scheduling of activities to maximize the net present value of projects. Eur. J. Oper. Res. 49, 35–40 (1990)
6. Mika, M., Waligora, G., Weglarz, G.: Simulated annealing and tabu search for multi-mode resource-constrained project scheduling with positive discounted cash flows and different payment models. Eur. J. Oper. Res. 164(3), 639–668 (2005)
7. Tereso, A.P., Araújo, M.M., Elmaghraby, S.E.: The Optimal Resource Allocation in Stochastic Activity Networks via The Electromagnetism Approach. In: Ninth International Workshop on PMS 2004, Nancy-France, pp. 26–28 (2004)
8. Tereso, A.P., Araújo, M.M., Elmaghraby, S.E.: Optimal resource allocation in stochastic activity networks via the electromagnetic approach: a platform implementation in Java. Control Cybern. 38, 745–782 (2009)
9. Tereso, A.P., Mota, J.R., Lameiro, R.J.: Adaptive Resource Allocation Technique to Stochastic Multimodal Projects: a distributed platform implementation in JAVA. Control Cybern. 35, 661–686 (2006)
10. Birbil, S.I., Fang, S.C.: An Electromagnetism like Mechanism for Global Optimization. J. Global Optim. 25, 263–282 (2003)
11. Willis, R.J.: Critical path analysis and resource constrained project scheduling theory and practice. Eur. J. Oper. Res. 21, 149–155 (1985)
12. Boctor, F.F.: Heuristics for scheduling projects with resource restrictions and several resource-duration modes. Int. J. Prod. Res. 31, 2547–2558 (1993)
13. Heilmann, R.: Resource–constrained project scheduling: a heuristic for the multi–mode case. OR Spectrum 23(3), 335–357 (2001)
14. Basnet, C., Tang, G., Yamaguchi, T.: A Beam Search Heuristic for Multi-Mode Single Resource Constrained Project Scheduling. In: Proceedings of the 36th Annual ORSNZ Conference Christchurch NZ, November– December 1-8 (2001)
15. Santos, M.A., Tereso, A.P.: On the Multi-Mode, Multi-Skill Resource Constraint Project Scheduling Problem (MRCPSP-MS). Eng. Opt., Lisbon Portugal September 6-9 (2010)

Strict Authentication of Multimodal Biometric Images Using Near Sets

Lamiaa M. El Bakrawy, Neveen I. Ghali, Aboul Ella Hassanien, and James F. Peters

Abstract. In this paper, a strict authentication watermarking scheme based on multi-modal biometric images and near sets was designed and introduced. The proposed scheme has a number of stages including feature enrolment for extracting the human facial features. Three human facial features which are nose length, nose width and distance between eyes balls are extracted. The near sets approach is adapted to choose the best feature among the considered features. The watermark is generated from hashing the extracted facial features that then encrypted using Advanced Encryption Standard (AES) technique and embedding the encrypted value into the human fingerprint image in order to confirm the integrity of respective biometric data. The experimental result shows that the proposed scheme guarantees the security assurance.

Keywords: Biometrics, encryption, facial feature, image authentication, near sets, security assurance.

1 Introduction

Biometrics refers to the automated identification of a person depending on his/her physiological or behavioral characteristics. Biometric characteristics

Lamiaa M. El Bakrawy · Neveen I. Ghali
Faculty of Science, Al-Azhar University, Cairo-Egypt
e-mail: lamiaabak@yahoo.com,nev_ghali@yahoo.com

Aboul Ella Hassanien
Faculty of Computers and Information, Cairo University
e-mail: aboitcairo@gmail.com

James F. Peters
Computational Intelligence Laboratory, Department of Electrical & Computer Engineering, Univ. of Manitoba, E1-526, 75A Chancellor's Circle, Winnipeg, MB R3T 5V6
e-mail: jfpeters@ee.umanitoba.ca

A. Gaspar-Cunha et al. (Eds.): Soft Computing in Industrial Applications, AISC 96, pp. 249–258.
springerlink.com

include fingerprint, face, DNA, ear, facial thermogram, hand thermogram, iris, retina, voice, etc. [6,13]. Biometrics have been gradually increased with the current wide employment since the conventional knowledge- based technology which can be forgotten, like a password, or lost or stolen, like an personality card has a defect [12,16]. Multi-modal biometrics refers to using of multiple biometric indicators for identifying persons, it has been shown to raise accuracy [11] and population treatment, while decreasing vulnerability to spoofing [25,26].

Strict image authentication methods do not allow any modifying in the image data. These methods can be classified into two extensive categories depending on the techniques that are used: methods based on conventional cryptography and methods that use fragile watermarking [7].

In this paper, near set is used for feature partitioning and selection. Hash function is applied to compact selected features to a fixed size. Finally, conventional cryptography and watermarking technique are employed to propose strict authentication technique for multimodal biometric image.

The remainder of this paper is ordered as follows. Section (2) reviews the related works. Brief introduction of near sets and nearness approximation spaces, hash function and Advanced Encryption Standard are respectively introduced in Section (3). The details of the proposed method is presented in Section (4). Section (5) shows the experimental results. Conclusions are discussed in Section (6).

2 Related Work

With the broad increase employment of biometric identification systems, the authenticity of biometric data itself has appeared as an essential research topic. Jain and Uludag in [14] introduced two applications of an amplitude modulation-based watermarking method, in which they secrete a user's biometric data in a diversity of images. This method has the capability to enhance the security of both the hidden biometric data (e.g., eigen-face coefficients) and host images (e.g., fingerprints). Vatsa et al. in [26] proposed a new biometric image watermarking algorithm which synergistically combines the DWT and LSB based algorithms for enhanced robustness and resiliency when subjected to both geometric and frequency attacks. Kim et al. in [16] proposed multimodal biometric image watermarking scheme through a two-stage integrity verification method using the hidden thumbnail feature vectors for safe authentication of multimodal biometrics data, face and fingerprint, respectively. It is based on the robust image watermarking scheme.

3 An Overview

3.1 Near Sets and Nearness Approximation Spaces

Near set approach is used to compare object descriptions. Sets of objects X,X' are called near each other if the sets include objects with at least partial matching descriptions [19, 8, 20, 21].

Let $\sim B$ denote $\{(x, x')|f(x) = f(x')\forall f \in B\}$ (called the indiscernibility relation).

Let X, $X' \subseteq O, B \subseteq F$. Set X is near X' if, and only if there exists $x \in X, x' \in X', \emptyset_i \in B$ such that $x \sim \emptyset_i x'$

The original generalized approximation space (GAS) model [24] has been extended as a result of current work on nearness of objects (see, e.g.,[21,20]). A nearness approximation space (NAS) is a tuple $NAS = (O, F, \sim B_r, N_r, \nu_{Nr})$, defined using set of perceived objects O, set of probe functions F indicating object features, indiscernibility relation $\sim B_r$ defined relative to $B_r \subseteq B \subset F$, family of neighbourhoods N_r, and neighbourhood overlap function ν_{Nr} . The relation $\sim B_r$ is the common indiscernibility relation from rough set theory limited to a subset $B_r \subseteq B$. The subscript r denotes the cardinality of the limited subset B_r, where we consider $(|B|, r)$ i.e., $|B|$ functions $\emptyset_i \in F$ taken r at a time to define the relation $\sim B_r$. This relation defines a partition of O into nonempty, pairwise disjoint subsets that are equivalence classes denoted by $[x]_{B_r}$, Where

$$O[x]_{B_r} = \{x' \in O|x \sim_{B_r} x'\} \tag{1}$$

These classes form a new set called the quotient set $O/ \sim B_r$, Where

$$O/ \sim B_r = \{[x]_{B_r}|x \in O\} \tag{2}$$

In effect, each choice of probe functions B_r defines a partition ξ_{B_r} on a set of objects O, namely,

$$\xi_{B_r} = O/ \sim B_r \tag{3}$$

Every choice of the set B_r leads to a new partition of O. The overlap function ν_{Nr} is defined by

$$\nu_{Nr} : P(O) * P(O) \rightarrow [0, 1] \tag{4}$$

where $P(O)$ is the power set of O. The overlap function ν_{Nr} maps a pair of sets to a number in [0,1] indicating the degree of overlap between sets of objects with features defined by probe functions $B_r \subseteq B$. For each subset $B_r \subseteq B$ of probe functions, define the binary relation $\sim_{B_r} = \{(x,\ x') \in \mathcal{O} \times \mathcal{O} : \forall \phi_i \in B_r, \phi_i(x) = \phi_i(x')\}$. Since each $\sim_{B_r}$ is, in fact, the usual indiscernibility relation, for $B_r \subseteq B$ and $x \in \mathcal{O}$, let $[x]_B$ denote the equivalence class containing x, *i.e.*,

$$[x]_{B_r} = \{x' \in O \mid \forall f \in B_r, f(x') = f(x)\} \tag{5}$$

If $(x, x') \in\sim B_r$ (also written $x \sim B_r x'$), then x and x' are said to be B-indiscernible with respect to all feature probe functions in B_r. Then term a group of partitions $N_r(B)$ (families of neighborhoods), where

$$N_r(B) = \{\xi_{B_r} \mid B_r \in B\} \tag{6}$$

Families of neighborhoods are constructed for each combination of probe functions in B using $(|B|, r)$, i.e., $|B|$ probe functions taken r at a time. The family of neighborhoods $N_r(B)$ contains a set of precepts.

$N_r(B)$ contains a set of precepts. A percept is a by product of perception, i.e., something that has been observed. For example, a class in $N_r(B)$ indicates what has been perceived about objects belonging to a neighbourhood, i.e., observed objects with matching probe function values.

3.2 Hash Functions

Hash functions are used to compact inputs of arbitrary length to outputs of a fixed size [1,23], and it is frequently used in data integrity, message authentication, and digital signature. It was first introduced in early 1950s. Since then different Hash algorithms were designed, and widely studied, in which the mainly broadly used were MD5 and $SHA-1$, invented by Rivest in 1992 and the NIST (National Institute of Standards and Technology) in 1995 respectively[9,10,17]. The SHA-1 algorithm is based on principles like to MD4 message digest algorithm. It operates on message blocks of 512 bits for which a 160-bit digest is created. SHA-1 is called secure because it is computationally infeasible to discover a message which corresponds to a known message digest, or to discover two diverse messages which create the same message digest. Any modify to a message in transit will, with very high probability, leads to a diverse message digest, and the signature will fail to confirm [3, 22, 27].

3.3 Advanced Encryption Standard

The Advanced Encryption Standard is lately approved by NIST (National Institute of Standards and Technology). It will swap the earlier standards because of its good characteristics according to security, cost, and efficient implementations so it replaced Data Encryption Standard (DES) as a worldwide standard for symmetric key encryption [5,15,18]. AES is an iterated block cipher which consists of sequences of 128 bits (digits with values of 0 or 1). These sequences will occasionally be referred to as blocks and the number of bits which will be referred to as their length. The Cipher Key for the AES algorithm is a sequence of $128, 192 or 256$ bits [2,4].

4 Proposed Scheme

In the proposed scheme the facial features of a face image are extracted then near set approach is used to choose the best feature among the extracted features, which considerably improves image authentication. After that a hash function is applied on the selected feature to create hash value, which is encrypted and embedded in fingerprint image in order to confirm the integrity of respective biometric data, the overall explanation of the proposed system are given in Figure 1.

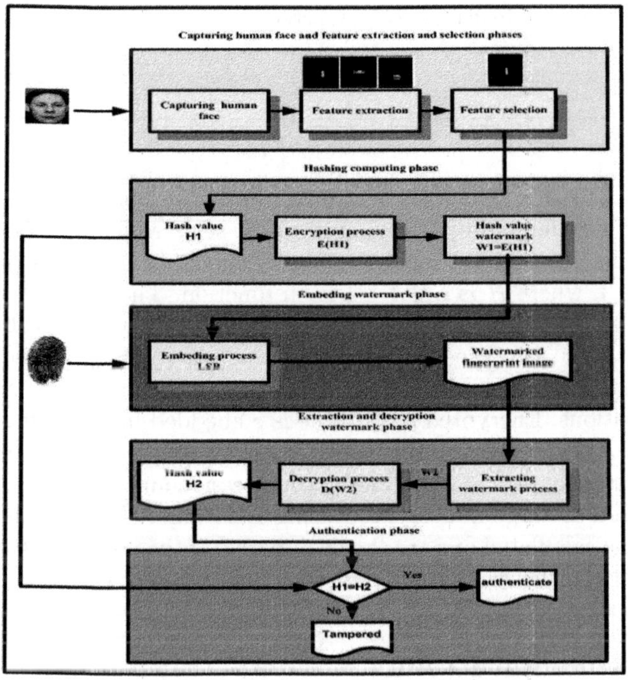

Fig. 1 The overview of the proposed algorithm

Facial features of a face image such as Nose Length, Nose Width and distance between eyes balls are extracted using Algorithm 1. Near set-based algorithms to partition every combination of 3 features and to carry out feature selection are given in [20].

Algorithm 1. Features Extraction

Input: Face images I
Output:Facial features: Nose Length (NL), Nose Width (NW) and distance between eyes balls(ED)
While True do

    For (i = 0; i $\leq$ I) do
        Initialize NL(i) =0, NW(i) =0 and ED(i) =0;
        Repeat (for number of features)
         Retrieve image i
         Extract Region Of Interest (ROI) for each feature;
         NL(i)=NL(i)$_b$−NL(i)$_a$
         NW(i)=NW(i)$_b$−NW(i)$_a$
         ED(i)=ED(i)$_b$−ED(i)$_a$
        Until all features are extracted.
    End
End

Since $SHA-1$ is significantly stronger hash function against attacks, the watermark is generated by applying it on the selected feature (Nose Length) of the face image to give a fixed size hash value h, called the message digest, $h = H\ (NL)$, where H is $SHA-1$ hash function. The size of h is regularly much smaller than the size of NL. For more security the hash value was encrypted using Advanced Encryption Standard (AES). We used AES because of its excellent characteristics according to security, cost, and efficient implementations. Encrypted hash value is embedded into the fingerprint image using Least Significant Bit (LSB). LSB is used because it is robust to geometric attacks such as JPEG lossy compression and Gaussian noise. Then the recovered message is decrypted using AES algorithm and the extracted hash and the calculated one are then compared. If the two values are matched then image is authenticated otherwise it is not authenticated.

5 Experimental Results

In our authentication system, we used a face image with size 92×112, and a fingerprint image with size 225×325. The calculated hash value of the selected feature is defined as:

030DD6CB3F13BC578337410B45AC7573A7F513AE,

and the extracted decrypted one was identical with the calculated, which means the image is authenticated.

After applying JPEG lossy compression (quality level = 0) on the watermarked image, the decrypted value is:

776F8FCF829163F37D8B6945662B30CE749B775E,

so the decrypted extracted hash value is distinct from the hash value of the selected feature, that means there is no authenticate.

Also after adding Gaussian noise ($\sigma = 20$) to the watermarked image, and the decrypted value is

3EC2AF54BF06778FF00563F4978E923D1576C208,

so the decrypted value is distinct from the hash value of the selected feature i.e. It is not authenticate.

For integrity verification, biometric image watermarking scheme is first employed without any tampered as shown in Fig. 2. Fig. 2(a) and 2(b) are original biometric data. Fig. 2(c). watermarked fingerprint image. Fig. 3(a) and 3(b) show the watermarked image after applying JPEG lossy compression and adding Gaussian noise respectively.

(a) (b) (c)

Fig. 2 Biometric image watermarking:(a) face image, (b) fingerprint image, (c) watermarked fingerprint image.

(a) (b)

Fig. 3 Attacked watermarked image using (a) JPEG lossy compression, (b) adding Gaussian noise

The proposed method is applied for another face image and fingerprint image to prove that the proposed method can be simply applied in diverse kinds of images. Fig. 4(a) and 4(b)are original biometric data. Fig. 4(c).

watermarked fingerprint image. It is noticed that the hash value of the selected feature is

030DD6CB3F13BC578337410B45AC7573A7F513AE

After encryption and embedding the encrypted value in fingerprint image. The recovered message is decrypted and the result is

030DD6CB3F13BC578337410B45AC7573A7F513AE

The extracted hash and the calculated one are matched so image is authenticate. After applying JPEG lossy compression (quality level = 0) on the watermarked image, The decrypted value is defined as:

776F8FCF829163F37D8B6945662B30CE435A6555

So the decrypted value is distinct from the hash value of the selected feature i.e, It is not authenticate.

Also after adding Gaussian noise ($\sigma = 20$) to the watermarked image, the decrypted value is defined as:

A048D8DC281239FC4A8ECC2E0CBC32FDFF489E82

So the decrypted value is distinct from the hash value of the selected feature (i.e. it is not authenticate). Fig. 5(a) and 5(b) show the watermarked image after applying JPEG lossy compression and adding Gaussian noise respectively.

(a) (b) (c)

Fig. 4 Biometric image watermarking:(a) face image, (b) fingerprint image, (c) watermarked fingerprint image.

(a) (b)

Fig. 5 Attacked watermarked image using (a) JPEG lossy compression, (b) adding Gaussian noise

6 Conclusions

In this paper, we propose Strict image authentication algorithm for multimodal biometric data, face and fingerprint, respectively. First, facial features of a face image (Nose Length,Nose Width and distance between eyes balls) are extracted then near set approach is used to choose the best feature among the considered one which considerably improves image authentication. Second, a hash function is applied on the selected feature to create hash value. Hash value is encrypted and embedded in fingerprint image in order to confirm the integrity of respective biometric data.

Also JPEG lossy compression (quality level = 0) and Gaussian noise (σ = 20) are applied on the watermarked image, the decrypted value is distinct from the hash value of the selected feature i.e, It is not authenticate. Experimental results showed that the proposed method guarantees the security assurance.

References

1. Andreeva, E., Preneel, B.: A Three-Property-Secure Hash Function. In: Avanzi, R.M., Keliher, L., Sica, F. (eds.) SAC 2008. LNCS, vol. 5381, pp. 228–244. Springer, Heidelberg (2009)
2. Dunkelman, O., Keller, N.: The effects of the omission of last rounds MixColumns on AES. Information Processing Letters 110, 304–308 (2010)
3. Eastlake, D., Jones, P.: US secure hash algorithm-1 (SHA1). Network Working Group. Category: Informational, pp. 1–22 (2001)
4. Farhan, S.M., Khan, S.A., Jamal, H.: An 8-bit systolic AES architecture for moderate data rate applications. Microprocessors and Microsystems 33, 221–231 (2009)
5. Federal Information Processing Standards Publication 197 Advanced encryption standard (AES). NIST (2001), `http://csrc.nist.gov/publications/fips/fips197/fips-197.pdf`
6. Feng, H., Wah, C.C.: Private key generation from on-line handwritten signatures. Information Management & Computer Security, 159–164 (2002)
7. Haouzia, A., Noumeir, R.: Methods for image authentication: a survey. Multimed Tools Appl. 39, 1–46 (2008)
8. Hassanien, A., Abraham, A., Kacprzyk, J., Peters, J.F.: Computational Intelligence in Multimedia Processing: Foundation and Trends. In: Computational Intelligence in Multimedia Processing: Foundation and Trends. SCI, vol. 96, pp. 3–49 (2008)
9. He, W., Peng, X., Qin, W., Meng, X.: The keyed optical Hash function based on cascaded phase-truncated Fourier transforms. Optics Communications 283, 2328–2332 (2010)
10. HouZhen, W., HuanGuol, Z., QianHong, W., Yu, Z., ChunLei, L., XinYu, Z.: Design theory and method of multivariate hash function. Information Sciences 53, 1977–1987 (2010)
11. Jain, A.K., Bolle, R., Pankanti, S.: Biometrics: personal identification in networked society. Kluwer Academic Publishers, Dordrecht (1999)

12. Jain, A.K., Pankanti, S.: Fingerprint classification and matching. In: Bovic, A. (ed.) Image and Video Processing Handbook. Academic Press, NewYork (2000)
13. Jain, A.K., Ross, A., Prabakhar, S.: An introduction to biometric recognition. IEEE Transactions on Circuits and Systems for Video Technology. Special Issue on Image and Video Based Biometrics 14(1) (January 2004)
14. Jain, A.K., Uludag, U.: Hiding Biometric Data. IEEE Transactions on Pattern Analysis and Machine Intelligence 25, 1494–1498 (2003)
15. Kermani, M.M., Masoleh, A.R.: Fault detection structures of the S-boxes and the inverse S-boxes for the advanced encryption standard. J. Electron Test 25, 225–245 (2009)
16. Kim, W.G., Lee, H.: Multimodal biometric image watermarking using two-stage integrity verification. Signal Processing 89, 2385–2399 (2009)
17. Lian, S., Sun, J., Wang, Z.: Secure hash function based on neural network. Neurocomputing 69, 2346–2350 (2006)
18. Mukhopadhyay, D.: An improved fault based attack of the advanced encryption standard. In: Preneel, B. (ed.) AFRICACRYPT 2009. LNCS, vol. 5580, pp. 421–434. Springer, Heidelberg (2009)
19. Peters, J.F.: Fuzzy sets, near sets, and rough sets for your computational intelligence toolbox. In: Hassanien, A.-E., Abraham, A., Herrera, F. (eds.) Foundations of Computational Intelligence Volume 2. Studies in Computational Intelligence, vol. 202, pp. 3–25. Springer, Heidelberg (2009)
20. Peters, J.F.: Near Sets. General Theory About Nearness of Objects. Applied Mathematical Sciences 1, 2609–2629 (2007)
21. Peters, J.F., Ramanna, S.: Feature Selection: Near Set Approach. In: Raś, Z.W., Tsumoto, S., Zighed, D.A. (eds.) MCD 2007. LNCS (LNAI), vol. 4944, pp. 57–71. Springer, Heidelberg (2008)
22. Pongyupinpanich, S., Choomchuay, S.: An Architecture for a SHA-1 Applied for DSA. In: Proceeding of 3rd Asian International Mobile Computing Conference (AMOC), Thailand, May 26-28, 2004, pp. 8–12 (2004)
23. Seoa, M.S., Haitsmab, J., Kalkerb, T., Yooa, C.D.: A robust image fingerprinting system using the Radon transform. Signal Processing: Image Communication 19, 325–339 (2004)
24. Skowron, A., Stepaniuk, J.: Generalized approximation spaces. In: Lin, T.Y., Wildberger, A.M. (eds.) Soft Computing. Simulation Councils, San Diego, pp. 18–21 (1995)
25. Snelick1, R., Uludag, U., Mink1, A., Indovina1, M., Jain, A.: Large scale evaluation of multimodal biometric authentication using state-of-the-art systems. IEEE Transactions on Pattern Analysis and Machine Intelligence 27, 450–455 (2005)
26. Vatsa, M., Singh, R., Noore, A., Houck, M.M., Morris, K.: Robust biometric image watermarking for fingerprint and face template protection. IEICE Electronics Express 2, 23–28 (2006)
27. Wen, C.Y., Yang, K.T.: Image authentication for digital image evidence. Forensic Science Journal 5, 1–11 (2006)

Part V: Soft Computing for Data Mining

Document Management with Ant Colony Optimization Metaheuristic: A Fuzzy Text Clustering Approach Using Pheromone Trails

Angel Cobo and Rocio Rocha

Abstract. This paper proposes an ant colony optimization (ACO) algorithm to deal with fuzzy document clustering problems. A specialized glossary and a thesaurus are used in order to extract features of the documents and to obtain a language-independent vector representation that can be used to measure similarities between documents written in different languages. The pheromone trails obtained in the ACO process are used to determine membership values in a fuzzy clustering. To illustrate the behavior of the algorithm, it was applied to a corpus of bilingual documents in different areas of economic and management.

1 Introduction

In the last few decades, the textual revolution has seen a tremendous change in the availability of online information, and production of textual documents in digital form has increased exponentially. Despite the widespread use of digital texts, handling them is a difficult task because the large amount of data necessary to represent them. Otherwise, the ever increasing amount of text documents written in different languages and the ever increasing dependence of organisations on such information require effective multilingual document management mechanisms. Text mining is concerned with the task of extracting relevant information from natural language text and to search for interesting relationships between the extracted entities. Text categorization and clustering are two basic techniques in this area. In the first case, we focus on a classification problem of deciding whether a document belongs to a

Angel Cobo
Department Applied Mathematics and Computational Sciences,
University of Cantabria, Spain
e-mail: acobo@unican.es

Rocio Rocha
Department Business Administration, University of Cantabria, Spain
e-mail: rochar@unican.es

A. Gaspar-Cunha et al. (Eds.): Soft Computing in Industrial Applications, AISC 96, pp. 261–270.
springerlink.com

set of pre-specified classes of documents. If we have a collection of documents with no known structure, these semantic categories are unknown and the classification process has to discover them; this is the basic principle of clustering. Text clustering has become one of the key methods for organizing information, but it is one of the more difficult text-mining problems, since it deals with very high-dimensional data sets and the subjectivity of classification complicate matters. Sometimes a document can be hard to classify even for a human expert. Categories are rarely well separated, and hence, the class partition is best described by fuzzy memberships. In fuzzy clustering, documents can belong to more than one cluster, and associated with each element is a set of membership levels that indicate the strength of the association between that document and a particular cluster. Fuzzy document clustering can be seen as the task of determining a membership value to each entry of the decision matrix $M = (u_{ij})_{1\leq i\leq n, 1\leq j\leq k}$, where k is the number of clusters, n the number of documents and $u_{ij} \in [0,1]$ is the degree of membership of document d_i in cluster $\mathscr{C}_j$. Surveys of fuzzy set theory applied in cluster analysis can be found in the specialized bibliography [8]. [6] presents a short overview of methods for fuzzy clustering and states desired properties for an optimal fuzzy document clustering algorithm. Integration of fuzzy logic in data mining has become a powerful tool in handling natural data and fuzzy clustering techniques have also become the major techniques in cluster analysis.

In this paper we use an Ant Colony Optimization (ACO) approach to calculate membership levels using the pheromone trails of an ACO process. ACO is amongst the most well-known Swarm Intelligence models [3,4]. Swarm Intelligence (SI) [1] refers to techniques based on the idea that groups of extremely simple agents with little or no organisation can exhibit complex and intelligent behaviour by using simple local rules and communication mechanisms. ACO mimics the way real ants find shortest routes between a food source and their nest. In ACO principles of communicative behaviour occurring in real ant colonies are used, and several generations of artificial ants search for good solutions. Every ant of a generation builds up a solution using information provided by the previous ants (pheromone trails) and heuristic information that represents a priori information about the problem.

Within the discovery phase of text mining, other models have been applied in recent years that take their inspiration from behaviour observed in groups of living beings. Ant-based sorting and clustering algorithms [2,7] were among the first techniques to be inspired by the behaviour of ants. In the ant clustering algorithms the clustering operation happens on a toroidal bidimensional grid, where the objects (documents) are placed randomly and a set of artificial ants explore the grid picking and dropping them. The probabilities of picking and dropping are based on the disparity between that document and other documents in its neighbourhood. In the specific case of applying SI techniques to text mining problems, even if they have not been the subject of such an intense application as that produced in other fields, it is indeed possible to find interesting references that make use of these kinds of strategies to extract knowledge on documental collections. Examples of the application of SI to various text mining problems are [5, 11, 12]. ACO algorithms have also been used in data clustering problems. By example, [9, 10] present

ACO algorithms for optimally clustering objects into clusters and compare the performance with other popular methods. However, both works focus on clustering of low-dimensional datasets with low sample size. In this work, we apply ACO techniques to a high-dimensional problem and we use the pheromone trails as measure of the membership of the documents to the groups. The designed algorithm has also important differences; it uses similarities between documents with different types of features as heuristic information. Pheromone trails have been used to accelerate the convergence of clustering algorithms in text mining, by example in [13] an ant-based fast text clustering approach is presented. This approach aims at solving the problem about ants random moving using pheromone in an ant clustering algorithm, this approach leads to a hard clustering, that is, each document is assigned to a single cluster. Our approach uses a different ant-based metaheuristic (ACO) and uses the pheromone trails to deal with fuzzy text clustering problems.

2 ACO Clustering Approach

This section describes the proposed ant algorithm to solve a document clustering problem (see Algorithm 1). Given a predefined number of clusters k, we can see the clustering problem as a combinatorial optimization problem where we have to decide the document distribution between the clusters. The aim of the optimization process is to obtain an assignment such that the mean similarity between each document and the centroid of the belonging cluster is maximized. Each assignment can be represented as a assignment vector $\mathscr{A} = (a_1, a_2, \ldots, a_n)$ where $a_i \in \{0, 1, \ldots, k\}$ indicates that document d_i is assigned to cluster $\mathscr{C}_{a_i}$. The value $a_i = 0$ corresponds to the case of an empty assignment. The quality of assignment $\mathscr{A}$ is measured in terms of the value of the following objective function:

$$Q(\mathscr{A}) = \frac{1}{n} \sum_{i=1}^{n} Sim(d_i, centroid(\mathscr{C}_{a_i})) \tag{1}$$

The algorithm considers N_a ants to build assignments. Each iteration of the algorithm corresponds to one generation with an assignment built by each ant. An ant starts with an empty assignment vector $\mathscr{A} = (0, 0, \ldots, 0)$ and builds a clustering solution step by step going through several decisions. Each decision implies a random selection of a non-classified document d_i such that $a_i = 0$, the decision of the cluster $\mathscr{C}_j$ where the document will be placed and the assignment $a_i = j$. To construct a solution, the ant uses the heuristic information composed by the similarity matrix $\mathbf{S}$, and pheromone trail information based on the quality of solutions previously found. The similarity matrix is a symmetrical matrix with $s_{ij} = Sim(d_i, d_j)$. The pheromone trail value τ_{ij} represents the pheromone concentration of document d_i associated to cluster $\mathscr{C}_j$ and the pheromone matrix $\tau = (\tau_{ij})_{1 \le i \le n; 1 \le j \le k}$ allows to associate to each document k pheromone concentrations. All entries of the pheromone matrix are initialized to some small value $\tau_0 \in (0, 1)$ randomly selected, and with the progress of iterations, its values are updated depending upon the quality of the assignments produced. Guided by both matrices, the ants build improved solutions within each iteration/generation. When an ant is building its solution $\mathscr{A}$, the probability that the ant places document d_i in cluster $\mathscr{C}_j$ is defined by

$$p_{ij} = \frac{\tau_{ij}^{\alpha}\, ms(i,j)^{\beta}}{\sum_{r=1}^{r=k} \tau_{ir}^{\alpha}\, ms(i,r)^{\beta}} \tag{2}$$

where α and β are positive real parameters whose values determine the relative importance of the pheromone and the heuristic information in the decision, and $ms(i,j)$ is a function that computes the mean similarity of document d_i with the documents that are assigned to cluster $\mathscr{C}_j$ in solution $\mathscr{A}$. When zero documents have been assigned to cluster $\mathscr{C}_j$ the value of $ms(i,j)$ is 1. Once an ant has constructed its solution $\mathscr{A}$, the pheromone matrix is updated according to the following rule:

$$\tau_{ij} = \begin{cases} (1-\rho)\tau_{ij} + Q(\mathscr{A})^2 & \text{if } a_i = j \\ (1-\rho)\tau_{ij} & \text{if } a_i \neq j \end{cases} \tag{3}$$

where ρ is an evaporation rate that avoids old pheromone from having a too strong influence on future decisions. Additionally, we use an elitist strategy, at the end of every generation, the pheromone trails laid on the clustering solution with best quality found by the ants are reinforced to facilitate the search around this solution.

The clustering built by each ant is a hard clustering, in which each document is assigned to exactly one cluster, but the pheromone matrix can be used to generate the fuzzy degree of membership u_{ij} of document d_i in cluster $\mathscr{C}_j$ according to:

$$u_{ij} = \frac{\tau_{ij}}{\sum_{r=1}^{r=k} \tau_{ir}} \tag{4}$$

Algorithm 1. ACO clustering algorithm

Initialization of pheromone matrix τ: $\tau_{ij} = \tau_0 \in (0,1) \quad \forall i,j$
repeat
  for all ant in $Colony$ **do**
    Initialize $\mathscr{A} = (0,0,\ldots,0)$
    repeat
      Random selection of document index i such that $a_i = 0$
      Selection of cluster index j_s for document d_i according to (2) for $j = 1,\ldots,k$
      $a_i = j_s$
    until $a_r \neq 0 \;\forall r$
    Evaluate objective function (1)
    Pheromone update using (3)
  end for
  The current iteration best ant in the colony performs an additional pheromone update (3)
until termination criterion attained
Select the best solution
Compute membership degrees using (4)
Return solution

2.1 Vector Representation of Documents and Similarity Measure

In order to apply the described algorithm we need a vector representation of documents and a function $Sim(\cdot,\cdot)$ that computes similarities between documents. Traditionally, every document is represented by a vector of weighted terms (features). Using word based features is the most popular and, despite its simplicity, a very effective feature construction method. This representation involves an indexing and a pre-process to extract the terms. In the indexing phase the text is tokenized and stopwords are removed to keep only potentially interesting words. The remaining terms can be lemmatized or stemmed. Given a set of index terms, not all of which are equally useful for describing the contents of a particular document, numerical weights $w_{ij} \geq 0$ are assigned to each index term or keyword k_i of a document d_j. These weights can be computed using a traditional *tf-idf* schema, that computes the weight of a term in a document as the product of two factors; the first one measures the raw frequency of the term inside the document, and the second one is motivated by the fact that a term which appears in many documents is not very useful for distinguishing documents. The *tf-idf* weighting is defined by

$$w_{ij} = f_{ij} \times idf_i = \frac{freq_{ij}}{\max_p freq_{pj}} \log \frac{n}{n_i} \tag{5}$$

where $freq_{ij}$ is the number of times that k_i appears in document d_j, n the total number of documents in the collection and n_i the number of documents is which the keyword k_i appears. The vector that represents a document can be normalized in order to facilitate the computation of similarities as we will show below.

Using word based features is the most popular, however, with multilingual collections alternative approaches are needed in order to implement language independent text mining systems. The first alternative is the use of standard machine translation techniques to create a monolingual corpus or the use of bilingual dictionaries. In spite of the inaccuracies introduced by the machine translation, the translated documents can be classified correctly using appropriated text mining techniques. Others alternatives are the extraction of language-independent features (names of places, people and organizations, dates, numerical expressions, cognates), the representation over conceptual spaces or the use of descriptor terms of multilingual thesaurus.

In order to obtain a language-independent representation of the documents, in our approach we use several linguistic resources and feature extraction strategies. Firstly, we use the Eurovoc thesaurus[1], a thesaurus covering the fields in which the European Communities are active. It provides a means of indexing the documents in the documentation systems of the European Institutions and of their users. Eurovoc exists in 22 official languages of the European Union and it is a structured list of more than 6600 descriptors and 127 microthesauri. Another linguistic resource used in this work is a multilingual economic glossary. This glossary contains over 11500 records of terms that include words, phrases, and institutional titles commonly

[1] http://eurovoc.europa.eu

encountered in documents of the International Monetary Fund (IMF)[2] in areas such as money and banking, public finance, balance of payments, and economic growth. The multilingual economic glossary of the IMF has been extended with terms of a business glossary, the total number of terms in the constructed glossary is more than 18000. Using these resources we employ a modified vector-space model to represent a document. We extract four kinds of features from every document:

- Verbs, adjectives and nouns in the native language identified using TreeTagger[3].
- Proper names identified by TreeTagger.
- Terms in the economic multilingual glossary.
- Microthesauri of Eurovoc. We look for Eurovoc descriptors inside the document text and associate to the document the most appropriate microthesaurus.

In this way a document is represented as $d = (\mathbf{v}_{nl}(d); \mathbf{v}_{pn}(d); \mathbf{v}_{gl}(d); \mathbf{v}_{th}(d))$, each subvector is weighted using expression (5).

The similarity between two documents can be computed by the cosine of the angle between their vectors, also known as angular separation. In our case, as we have four vectors of features, a convex linear combination of similarities is used:

$$\begin{aligned} Sim(d_1,d_2) = & \lambda_1 as(\mathbf{v}_{nl}(d_1),\mathbf{v}_{nl}(d_2)) + \lambda_2 as(\mathbf{v}_{pn}(d_1),\mathbf{v}_{pn}(d_2)) + \\ & \lambda_3 as(\mathbf{v}_{gl}(d_1),\mathbf{v}_{gl}(d_2)) + \lambda_4 as(\mathbf{v}_{th}(d_1),\mathbf{v}_{th}(d_2)) \end{aligned} \tag{6}$$

with $\lambda_i \in [0,1]$ and $\sum_{i=1}^{4} \lambda_i = 1$. The function $as(\cdot,\cdot)$ computes the angular separation between two vectors, that is $as(\mathbf{x},\mathbf{y}) = \frac{\mathbf{x} \circ \mathbf{y}}{\|\mathbf{x}\|\|\mathbf{y}\|}$. If the vectors are normalized, this score is computed as the inner product of the vectors. Since the weights are non-negative, $Sim(d_1,d_2)$ varies from 0 to 1. When the similarity is 0 the two vectors are totally dissimilar and when the similarity is 1 the two vectors are totally equal. This metric is relatively simple to compute and experimental results have indicated that it tends to lead to better results than Euclidean distance in text mining problems. In our approach we consider a convex linear combination of similarities, however it would be possible to use a multi-objective optimization approach.

3 Evaluation and Implementation Issues

In order to obtain an evaluation of the proposed approach, we have implemented the same as part of a document management system developed using open source technologies. To illustrate the fuzzy clustering approach a non-parallel bilingual corpus of selected economic research papers was constructed. The corpus includes 250 categorized documents extracted from databases of scientific journals in management and economics. The papers are published in international journals of the involved areas. We have selected documents in Spanish and English of five different functional

[2] http://www.imf.org/external/np/term/index.asp

[3] http://www.ims.uni-stuttgart.de/projekte/corplex/TreeTagger/

areas: marketing (MKT), accounting and finance (ACC), information systems (IS), economic theory (ECO) and human resource management (HRM). We have 25 articles in each language and functional area. All documents were preprocessed with TreeTagger, weighted using (5), and the similarity matrices for each type of feature were constructed. The dictionary size of this corpus is determined by 26601 native language words (verbs, adjectives and nouns), 16179 proper names, 127 Eurovoc microthesauri and 18724 economic glossary terms.

The scenario that has been considered for the simulation includes the following parameter settings: number of clusters to create $k = 5$, number of artificial ants 10, parameter for volatility of pheromone trail $\rho = 0.01$, and parameters that determine the relative importance of pheromone value and heuristic information $\alpha = 2.5$ and $\beta = 5.0$. One of the main difficulties of applying heuristic methods to a given problem is to decide on an appropriate set of parameter values. These parameter settings were found to be good when applying the algorithm to similar problems using other document corpora. As termination criterion we consider a maximum number of iterations, this value is set to 30. Finally, the definition of the similarity function (6) uses four coefficients λ_i that have to be fixed. In our case, we use the values $\lambda_1 = \lambda_2 = 0.05$ and $\lambda_3 = \lambda_4 = 0.45$, placing greater emphasis on language-independent features (microthesauri of Eurovoc and economic glossary terms).

Figure 1 shows the evolution of the pheromone matrix at 0, 5, 10, 15, 20 and 25 iterations in a run of the algorithm. Each graph corresponds to a density plot of the matrix, dark cells represent assignments of documents to clusters with high pheromone values. In the density plots the cluster structure of the collection can be observed; it should be noted that the documents were arranged according to the subject category of the journal of which they were extracted. The final pheromone values can generate membership levels u_{ij} using expression (4). The density plot of these values can be shown in the last graph of Figure 1.

The clustering solution constructed by the ACO algorithm is summarized in Table 1. As can be observed, each cluster has a dominant category, according to the a priori classification based on the source journal of each document. In order to analyze the quality of solutions, three external quality measures were used: purity, F-measure and entropy. The purity measures how much a cluster is specialized in a category; and is defined as the ratio of the number of documents in the dominant category to the total number of documents in the cluster. The F-measure uses the ideas of precision and recall from information retrieval, defined as $Precision(i,j) = n_{ij}/n_j$ and $Recall(i,j) = n_{ij}/n_i$, where n_{ij} is the number of documents of category i in cluster j, n_j is the number of documents in cluster j and n_i the total number of documents in category i in the corpus. Using these definitions the F-measure is

$$F = \sum_i \frac{n_i}{n} \max_j \{F(i,j)\} \quad \text{where} \quad F(i,j) = 2\frac{Precision(i,j) \times Recall(i,j)}{Precision(i,j) + Recall(i,j)} \qquad (7)$$

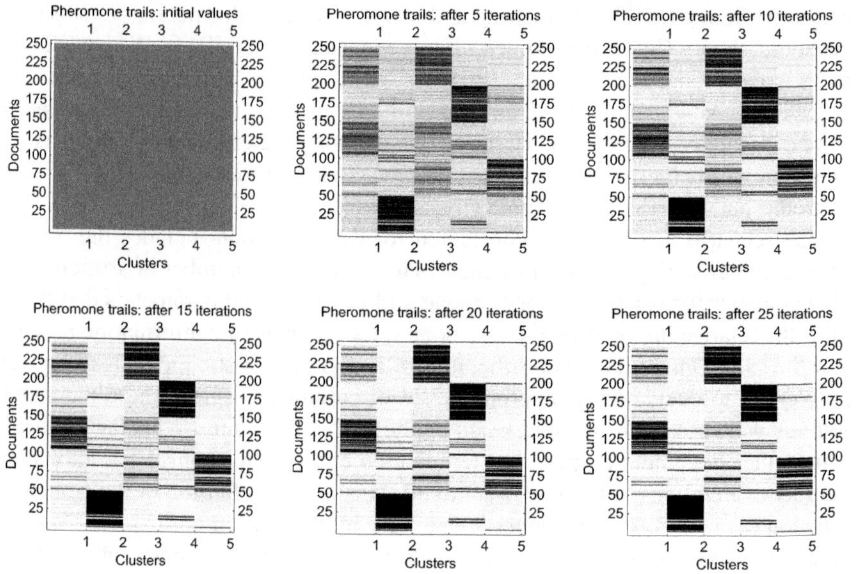

Fig. 1 Density plots of pheromone matrix.

where n is the number of documents in the corpus. The optimal F-measure value is 1. Finally, the entropy tells us how homogeneous a cluster is. Its optimal value is 0 and is defined by the following expression:

$$E = \frac{1}{n} \sum_j n_j E_j \quad \text{where} \quad E_j = -\sum_i n_{ij}/n_j \log(n_{ij}/n_j) \tag{8}$$

Table 1 Clustering results on the bilingual corpus of 250 documents

	Number of documents					Dominant	%
	ACC	MKT	HRM	ECO	IS	category	purity
Cluster 1	0	3	34	1	10	HRM	70.83%
Cluster 2	45	2	6	2	0	ACC	81.82%
Cluster 3	0	9	4	0	40	IS	75.47%
Cluster 4	4	4	6	45	0	ECO	76.27%
Cluster 5	1	32	0	2	0	MKT	91.43%
Quality measures							
Entropy: 0.6892; F-measure: 0.7813; Mean purity: 78.40%							

Table 2 Examples of document fuzzy classification

Document title	Journal	u_{i1}	u_{i2}	u_{i3}	u_{i4}	u_{i5}
Putting HR software to work	Workforce	0.071	0.012	0.893	0.015	0.008
Strategic Knowledge Management Tech.	Eur. J. of Information Systems	0.869	0.020	0.078	0.021	0.013
Integrating ERP using EAI	Eur. J. of Information Systems	0.249	0.017	0.709	0.015	0.010

The algorithm was run 100 times with the parameter settings described above. The averages over the 100 runs obtained for each of the quality measures using the proposed algorithm were entropy: 0.907; F-measure: 0.642; Purity: 0.65; and computational time (milliseconds[4]): 7649. The degree of purity of the clusters is quite fine. It is worth observing the difficulty of a clustering process in a corpus of documents in very close areas, as happens in our corpus. The documents belong to related areas and there are documents that can be assigned to different thematic categories, the pheromone information is very useful to discover them. In Table 1 the documents have been assigned to the category number j_s, with $j_s = \arg\max_j u_{ij}$, however they could be assigned to other related thematic categories and the values of the quality measures would be better. The membership levels in matrix (u_{ij}) obtained using the pheromone trails offer us very interesting information. For instance, Table 2 shows three documents with the membership levels returned by the algorithm. The first document was extracted from a human resource management journal, however the algorithm has associated it to information systems category; obviously this document has a very close relation with technology and software systems. In fact, the highest membership level is $u_{i3} = 0.893$, that associates the document to cluster 3, with information systems as dominant category. Something similar happens with the second document, despite having been extracted from a journal about information systems, it focus on planning, use, control and benefits of IT to support knowledge work in the organisation, these topics are also studied in human resource management. Finally, the third document belongs to information systems with level 0.709, and to human resource management with level 0.249.

Obviously, the circumstances in which algorithm performance has been analyzed are rather restricted and would require more experiments. In particular the number and size of the documents and the languages presented in the corpus could be varied. Algorithm behaviour could also be studied with different number of clusters and with other parameter settings.

4 Conclusions

In summary, in this work we have presented an ACO algorithm to solve fuzzy clustering problems. This algorithm can be applied when the number of clusters is known a priori and the pheromone matrix can be used to calculate membership levels that indicate the strength of the association between documents and clusters. The suggested clustering approach is meant for off-line document management, due to the computational efforts of pre-processing, similarity calculation and application of ACO technique, specially with very large document sets with unwieldy similarity matrices. The algorithm has been implemented and tested on a document corpus of research papers in different economic areas and in two languages (Spanish and English). The documents in this corpus were represented using four types of features, and the similarity between documents was calculated using a convex linear combination of similarities between groups of features. The automatic clustering of

[4] Hardware: Pentium(R) Dual-Core 2.00GHz, RAM memory 4Gb.

documents of very close areas is a very difficult task. The classic techniques of text mining have convincingly shown their worth and SI techniques have left their mark in recent years and are demonstrating their efficiency for solving a large number of complex problems. The integration of these new methodologies into technological tools for the management of documents allows optimising knowledge generation processes. The experimental results show a good performance of this methodology, given the difficulties of the problem addressed.

Acknowledgements

We gratefully acknowledge the support of the Office for Official Publications of the European Communities and the Language Services of the International Monetary Fund, because the use of editions of Eurovoc and IMF Glossary, respectively.

References

1. Bonabeau, E., Dorigo, M., Theraulaz, G.: Swarm Intelligence: From Natural to Artificial Systems. Oxford University Press, Oxford (1999)
2. Deneubourg, J., Goss, S., Franks, N., Sendova-Franks, A., Detrain, C., Chretien, L.: The dynamic of collective sorting robot-like ants and ants-like robots. In: First Int. Conf. on Simulation of Adaptive Behavior, pp. 356–363. MIT Press, Cambridge (1990)
3. Dorigo, M.: Optimization, learning and natural algorithms. Ph.D. thesis, Milano (1992)
4. Dorigo, M., Sttzle, T.: Ant Colony Optimization. MIT Press, Bradford (2004)
5. Handl, J., Meyer, B.: Improved ant-based clustering sorting in a document retrieval interface. In: Guervós, J.J.M., Adamidis, P.A., Beyer, H.-G., Fernández-Villacañas, J.-L., Schwefel, H.-P. (eds.) PPSN 2002. LNCS, vol. 2439, pp. 913–923. Springer, Heidelberg (2002)
6. Jursie, M., Lavrae, N.: Fuzzy clustering of documents. In: 11th Int Multi-conference on Information Society. Conf. on Data Mining and Data Warehouses (2008)
7. Lumer, E., Faieta, B.: Diversity and adaptation in population of clustering ants. In: Proc. 3rd Int. Conf. Simulation of Adaptive Behaviour, pp. 501–508 (1994)
8. Raju, G., Thomas, B., Tobgay, S., Kumar, T.: Fuzzy clustering methods in data mining: A comparative case analysis. In: Int. Conf. Adv. Comp. Theory and Engineering, pp. 489–493 (2008)
9. Runkler, T.: Ant colony optimization of clustering models. Int. J. Intell. Syst. 20, 1233–1251 (2005)
10. Shelokar, P., Jayaraman, V., Kulkarni, B.: An ant colony approach for clustering. Anal. Chim. Acta 509, 187–195 (2004)
11. Vizine, A., de Castro, L., Gudwin, R.: Text document classification using swarm intelligence. In: IEEE Int. Conf. Integration of Knowledge Intensive Multi-Agent Systems, pp. 134–139 (2005)
12. Wang, Z., Zhang, Q., Zhang, D.: A PSO-based Web document classification algorithm. In: 8th ACIS Int. Conf. Software Eng., Artificial Intelligence, Networking, and Parallel/Distributed Computing, pp. 659–664 (2007)
13. Zhang, F., Ma, Y., Hou, N., Liu, H.: An ant-based fast text clustering approach using pheromone. In: Fifth Int. Conf. on Fuzzy Systems and Knowledge Discovery, pp. 385–389 (2008)

Support Vector Machine Ensemble Based on Feature and Hyperparameter Variation for Real-World Machine Fault Diagnosis

Estefhan Dazzi Wandekoken, Flávio M. Varejão, Rodrigo Batista, and Thomas W. Rauber

Abstract. The support vector machine (SVM) classifier is currently one of the most powerful techniques for solving binary classification problems. To further increase the accuracy of an individual SVM we use an ensemble of SVMs, composed of classifiers that are as accurate and divergent as possible. We investigate the usefulness of SVM ensembles in which the classifiers differ among themselves in both the feature set and the SVM parameter value they use, which might increase the diversity among the classifiers and therefore the ensemble accuracy. We propose a novel method for building an accurate SVM ensemble. First we perform complementary feature selection methods to generate a set of feature subsets, and then for each feature subset we build a SVM classifier which uses tuned SVM parameters. The experiments show that this method achieved a higher estimated prediction accuracy in comparison to well-established approaches for building SVM ensembles, namely using a Genetic Algorithm based search to vary the classifier feature sets and using a predefined set of SVM parameter values to vary the classifier parameters. We work in a context of

Estefhan D. Wandekoken
Department of Informatics, PPGI, Federal University of Espírito Santo, 29060-970, Vitória, Brazil
e-mail: `estefhan@inf.ufes.br`

Flávio M. Varejão
Department of Informatics, PPGI, Federal University of Espírito Santo, 29060-970, Vitória, Brazil
e-mail: `fvarejao@inf.ufes.br`

Rodrigo Batista
Espírito Santo Business Unit. Petróleo Brasileiro S.A. PETROBRAS, Vitória, ES, Brazil

Thomas W. Rauber
Department of Informatics, PPGI, Federal University of Espírito Santo, 29060-970, Vitória, Brazil
e-mail: `thomas@inf.ufes.br`

A. Gaspar-Cunha et al. (Eds.): Soft Computing in Industrial Applications, AISC 96, pp. 271–282.
springerlink.com

real-world industrial machine fault diagnosis, using 2000 examples of vibrational signals obtained from operating faulty motor pumps installed on oil platforms.

1 Introduction

The *support vector machine* [16] (SVM) is currently considered one of the most powerful techniques for solving binary classification problems. Aiming to further increase the accuracy of an individual SVM, research in *classifier ensembles* [7] indicates that a better generalization power might be achieved when the class of an input pattern is predicted by a set of accurate SVMs that collectively disagree on their predictions as much as possible.

Previous work indicated two useful approches for building a SVM ensemble: training each SVM using a different SVM parameter value [14] [12]; and training each SVM using a different feature set [15] [11]. By now very few papers have investigated SVM ensembles based on the variation of both features and parameters. It can be expected that using different feature subsets and parameter values might increase the divergence among SVMs in an ensemble, therefore increasing the ensemble accuracy.

We propose a novel method for composing an optimized SVM ensemble based on feature and parameter variation. This method is based on feature selection to build a set of feature subsets and based on using a parameter tuning technique to vary the parameters of individual SVMs. We compare this proposed method to alternative approaches for building SVM ensembles: for varying the feature sets, we evaluated the Genetic Algorithm (GA) based GEFS [11] method; and for varying the SVM parameters, we evaluate an approach in which predefined SVM parameter values are used to replicate an ensemble [14] [12]. The experiments show that the proposed ensemble method achieved a higher estimated prediction accuracy.

We work in a context of machine fault diagnosis. The detection and diagnosis of faults in industrial machines is advantageos for economical and security reasons, as it allows damaged components to be repaired during planned maintanence, which minimizes machinery standstill besides providing more secure operations. An important novelty of our research is the use of data from real-world operating industrial machines instead of using data from a controlled laboratory environment which is almost always found in the literature [17]. This is highly desirable, as laboratory hardware cannot realistically represent intricate real-world fault occurrences. We work with 2000 examples of vibrational signals obtained from operational faulty motor pumps, acquired from 25 oil platforms off the Brazilian coast during five years. After extensive analysis, human experts provided a label for every fault present in each acquired example.

2 Model-Free Approach to Motor Pump Fault Diagnosis

Two principal approaches to the machine fault diagnosis problem exist: model-based techniques and model-free techniques. The former approach relies on an analytical model of the studied process, involving time dependent differential equations. However in real-world complex processes the availability of an analytical model is often unrealistic or inaccurate. In this case model-free techniques are an alternative approach [1]. We present a model-free method based on the supervised learning [2] classification paradigm. So it presents as a general advantage the requirement of a minimum of a priori knowledge about the plant, as the fault predictor is automatically defined based on the training data.

Several faults can simultaneously occur in a motor pump, and such a high diversity of defects has a direct impact on the subsequent classifier. We formulate the fault diagnosis problem as a multi-label classification task in which several labels (fault classes) may be simultaneously assigned to an example. Each fault category is represented by a distinct binary predictor. So a pattern $\mathbf{x}$ is considered as belonging to the positive class ω_{pos} if it presents this considered fault and as belonging to the negative class ω_{neg} if this considered fault is not present in $\mathbf{x}$. Therefore each fault category is detected in $\mathbf{x}$ by a distinct ensemble of SVM classifiers. We build a predictor for detecting each of the following faults: rolling element bearing failures; pump blade unbalance; hydrodynamic fault (due to blade pass and vane pass, cavitation or flow turbulence); shaft misalignment; mechanical looseness; and structural looseness.

We work with horizontal motor pumps with extended coupling between the electric motor and the pump. Accelerometers are placed at strategic positions along the main directions to capture specific vibrations of the main shaft. To diagnose a motor pump, the first step is to *extract* a global feature vector G to describe relevant aspects of the current motor pump condition. The cardinality of a feature vector G is 95 regardless of the fault under consideration, composed of features defined by different feature extraction methods, namely the Fourier transform, envelope analysis based on the Hilbert transform [10] and median filtering. So the features correspond to the energy in predetermined frequency bands of the vibrational signal, from the Fourier and the envelope spectrum.

3 The Support Vector Machine Classifier

The support vector machine (SVM) [16] is currently considered one of the most powerful techniques for solving binary classification problems. We compared SVMs with Multi-layer Perceptron (MLP) neural networks [2] and found that SVMs achieved a higher prediction accuracy, besides the MLPs being computationally much more expensive during training.

The objective of the SVM training is to define a maximum-margin separating hyperplane that lies in a transformed feature space defined implicitly by a kernel mapping. The hyperplane splits the mapped space into two regions, one associated to the positive class ω_{pos} and the other to the negative class ω_{neg}; a pattern is considered as positive class if it presents the fault considered by the SVM. The distance of a pattern $\mathbf{x}$ to the separating hyperplane, followed by a logistic discrimination, is used to estimate the a posteriori probability $\hat{P}_{\text{pos}}(\mathbf{x})$ that $\mathbf{x}$ belongs to the positive class ω_{pos}.

We use a widely adopted SVM model, namely a Radial Basis Function (RBF) kernel and the C-SVM architecture [2]. So we work with two parameters, namely the regularization parameter C which controls the model complexity and the kernel parameter γ which controls the nonlinear mapping of the features.

The performance of a SVM classifier strictly depends on its parameters. We use an effective, simple method to tune the SVM parameters, namely the grid search on the log-scale of the parameters in combination with cross-validation on each candidate parameter vector. Basically, pairs (C_i, γ_j) from a set of predefined values are tried by evaluating SVMs which use them, and the pair that provides the highest cross-validation accuracy defines the best parameters.

We use the `libsvm` library [3] to implement the SVM classification.

4 Feature Selection

Feature selection [6] is the process of choosing an optimized subset of features for classification from a larger set. It is composed of two ingredients: the selection criterion and the search strategy. The former is used to estimate the performance of a feature subset. A suboptimal search strategy is needed since an exhaustive search is not feasible to investigate the space of feature sets.

We estimate the criterion $J(X_k)$ of a candidate feature set X_k as the Area Under the ROC Curve (AUC) [5] achieved by a SVM classifier which uses X_k, estimating by cross-validation on the training data.

We use the *Sequential Backward Selection* (SBS) [6] search strategy, which allows accurate feature subsets to be found. The SBS method starts with every feature (from the global pool G) included in the set of selected features, and at each step one feature is removed from this set. Consider that k features are included in the set of selected features X_k. To remove one feature from X_k, each currently selected feature ξ_j must be tested by being individually excluded from X_k and ranked according to the criterion J, so that the feature ξ_h which provided the highest criterion $J(X_k \setminus \{\xi_h\})$ with its exclusion is removed from X_k. The exclusion process stops when the desired number of features is selected.

5 Classifier Ensembles

Combining decisions of divergent classifiers, that give erroneous answer in different regions of the global feature space, is becoming one of the most important techniques to improve classification accuracy [17]. Creating a so-called *classifier ensemble* entails addressing two issues: the construction of the base classifiers which constitute the ensemble and the combination of their individual decisions. To combine the classifiers decision we average the classification scores. So an ensemble $\mathscr{E}$ estimates the probability $\hat{P}_{pos}(\mathbf{x})$ that a a pattern $\mathbf{x}$ belongs to the positive class ω_{pos} as the average of the $\hat{P}_{pos}(\mathbf{x})^{c_h}$ value that the classifiers c_h in $\mathscr{E}$ output for $\mathbf{x}$ (for $h = 1,\ldots,|\mathscr{E}|$). Thus $\mathbf{x}$ is predicted as ω_{pos} if $\hat{P}_{pos}(\mathbf{x}) > 0.5$ or as ω_{neg} otherwise.

The classical approach to build a set of classifiers to compose an ensemble relies on using a different training data set for each classifier. See for instance bagging [2], in which the training patterns are sampled with replacement. However such an approach is not very well suited to SVM ensembles [4] because a small variation on the training data set tends to cause a small variation on the SVM decision function. On the other hand varying the value of the kernel parameter of a SVM classifier does decisively change its decision function, which is useful for building an ensemble as the divergence among the produced SVMs increases [9]. As an example of SVM ensemble based on parameter variation see [12] in which each SVM uses a different predefined SVM parameter value.

Another useful approach for building an ensemble is to use a different feature set for each classifier [17]. As a state-of-the-art ensemble method based on feature variation see the GEFS [11] method which relies on a Genetic Algorithm (GA) based search to investigate the space of feature sets. In GEFS, a member of the population (a chromosome) represents the feature set of a single classifier. Starting with randomly defined feature sets, the genetic operators (selection, cross-over and mutation) are used to evolve the population aiming to increase the classifiers fitness. The fitness of a classifier is estimated as a linear combination of its accuracy and its diversity, the latter being defined as the average difference between the prediction of this classifier and the prediction of the current ensemble. In the end of every generation the algorithm outputs the feature subsets of the classifiers in the current ensemble, thus the last generation defines the final produced feature subsets. Previous work shows that the GEFS method achived a higher prediction accuracy in comparison to other approaches [13].

6 Best Selected Feature Subsets Ensemble Method

We propose a method for composing SVM ensembles based on feature and parameter variation, which we call Best Selected Feature Subsets (BSFS). We use complementary SBS feature selection searches combined with the grid-search parameter optimization technique to build a large set $\mathscr{L}$ of classifiers that are candidates to constitute the ensemble. Further we use a sequential forward search to select just a reduced, optimized subset $\mathscr{E}$ of them to compose the final ensemble. This approach of building a large classifier set $\mathscr{L}$ and further searching for a subset $\mathscr{E}$ is known as *overproduce-and-choose* [7].

6.1 The Classifier Overproduction Stage

To build the set $\mathscr{L}$ of candidate classifiers to compose the ensemble, we first build a set Ξ of feature sets composed of several promising feature sets. The feature sets use features from the global pool G and differ on their cardinality. To define Ξ we perform m distinct SBS feature selection searches, $\{\mathscr{S}_1,\ldots,\mathscr{S}_i,\ldots,\mathscr{S}_m\}$, which differ among themselves on the SVM parameter value they use to build SVMs to estimate the selection criterion; this allows feature subsets to be found using complementary kernel mapped feature spaces. We require each search $\mathscr{S}_i$ to select a total of 1 feature, which is equivalent to require $\mathscr{S}_i$ to exclude $|G|-1$ features; each feature exclusion produces a new feature subset. So the feature sets in Ξ are determined by taking each produced feature set $X_k^{\mathscr{S}_i}$ which uses k features selected by the search $\mathscr{S}_i$, considering every k and i i. e. $k=1,2,3,\ldots,|G|$ and $i=1,2,\ldots,m$ (thus $|\Xi|=m\times|G|$).

Then the classifier set $\mathscr{L}$ is defined by building, for each feature set $X_k^{\mathscr{S}_i}$ in the set Ξ, a classifier $c_k^{\mathscr{S}_i}$ that uses this feature set, and we use the grid-search method to tune the SVM parameters of every classifier $c_k^{\mathscr{S}_i}$ aiming to increase its accuracy. Thus $\mathscr{L}$ is composed of every produced $c_k^{\mathscr{S}_i}$, each of which associated to a feature set $X_k^{\mathscr{S}_i}$ and to a tuned SVM parameter value (C',γ').

6.2 The Ensemble Classifier Selection Stage

After building the set $\mathscr{L}$ of candidate classifiers, we use the *Sequential Forward Selection* (SFS) search to select an optimized set of $|\mathscr{E}|$ classifiers to compose the final ensemble $\mathscr{E}$, selecting from $\mathscr{L}$. The SFS search operates in a similar way as SBS, but SBS removes objects, while SFS includes objects.

SFS starts with an empty set of selected classifiers, and at each step one classifier is included in this set, namely the one that provided the highest criterion with its individual inclusion in the current set of selected classifiers. We define the criterion J of a candidate ensemble (a subset of classifiers from $\mathscr{L}$) to be the AUC on training data achieved by this candidate ensemble. The score that each classifier in $\mathscr{L}$ gives to a training pattern $\mathbf{x}$ is previously estimated by cross-validation. Thus to obtain the criterion J of a candidate ensemble the score $\hat{P}_{pos}(\mathbf{x})$ assigned by this ensemble to each training pattern $\mathbf{x}$ must be obtained, by averaging the scores $\hat{P}_{pos}(\mathbf{x})^{c_j}$ assigned to $\mathbf{x}$ by the classifiers c_j in the candidate ensemble.

The first selected classifier $c_k^{\mathscr{S}_i}$ (from $\mathscr{L}$) is the one with the highest individual cross-validation AUC. In the following, each next selected classifier is the currently non-selected one which enabled the highest criterion J achieved by an ensemble composed of the currently selected classifiers and also this new selected one; thus the second selected classifier is the one which provided the highest criterion for an ensemble of two classifiers, namely the first and the second selected ones. When the desired number $|\mathscr{E}|$ of classifiers are selected, the inclusion process stops, so the ensemble $\mathscr{E}$ is finally built.

7 Experimental Results

To access the effectiveness of the studied classification approaches we performed a stratified 5×2 cross-validation [7]. So in the experiments we performed five replications of a 2-fold cross-validation. In each replication, the complete database of 2000 examples was randomly partitioned, in a stratified manner, into two sets each one with approximately 1000 examples (the stratification process preserves the distribution of the six fault categories between both sets). Then in each replication each considered classification model for building the predictor of a fault was trained on a set and tested on the remaining one; after the five replications we averaged the ten distinct test accuracies.

7.1 Studied Classification Approaches

For each of the six considered fault categories, we studied seven different classification models for building the predictor of that fault. We investigated the use of a single SVMs, a SVM ensemble based on feature variation, and a SVM ensemble based on feature and parameter variation.

We use a general three-stage approach to build SVM ensembles. First we produce a set Ξ of feature subset. Then we build a set $\mathscr{L}$ of SVM classifier. Finally we take from $\mathscr{L}$ a classifier set $\mathscr{E}$ to constitute the final ensemble.

7.1.1 The `SVM-s` Classification Model

This classification model is a single SVM. We used the global pool of features G as the feature set, and used the grid-search method to tune the SVM parameters as explained in section 3.

7.1.2 The `GEFS-not` Classification Model

This classification model is a SVM ensemble based on feature variation, built using the classical GEFS approach to generate ensembles based on feature variation. So the feature sets of the classifiers in the ensemble $\mathscr{E}$ are defined by the last generation of classifiers of the GEFS algorithm. Every SVM used the SVM parameter value $(C = 8.0, \gamma = 0.5)$ which usually provided accurate SVMs, according to parameter tuning experiments. We set the GEFS ensemble size parameter as 20, so the final produced ensemble $\mathscr{E}$ was composed of $|\mathscr{E}| = 20$ classifiers.

The GEFS algorithm has several parameters [11], and we performed preliminary experiments aiming to find parameter values that provided more accurate ensembles. We set the initial value of the GEFS λ parameter as 1.0, to estimate the initial value of the fitness of a classifier. We set 20 classifiers in the ensemble; in each generation, starting from the ensemble of 20 classifiers, we produce 10 new classifiers by mutation (randomly changing 10% of the features of each classifier) and more 10 new classifiers by cross-over (the two parents of each classifier are randomly selected from the current ensemble, proportionally to the fitness), and from these 40 classifiers the 20 more fit ones are selected to compose the ensemble in the end of this generation. The population evolved for 200 generations. We use 5-folds cross-validation to estimate the accuracy and fitness of each classifier.

7.1.3 The `BSFS-not` Classification Model

This classification model is a SVM ensemble based on feature variation, built using the BSFS method but without tuning the SVM parameters.

To build the set Ξ of feature sets we ran four SBS feature selection searches $\{\mathscr{S}_1, \ldots, \mathscr{S}_4\}$, each of which using a different SVM parameter value to build SVMs to estimate the selection criterion (which was the 5-fold cross-validation AUC). According to preliminary experiments, we used the SVM parameter value $(C = 8.0, \gamma = 0.5)$ which provided accurate SVMs. For the other three values, we focused on varying the γ parameter in order to introduce diversity among the SBS searches, allowing the selection of accurate feature subsets from different mapped feature spaces. So we used a low, a medium and a high value for the γ parameter; for the C parameter we used a high value, as it may cause some overfitting which is useful for increasing the diversity. So the four SVM parameter values $(C = 8.0, \gamma = 0.5)$, $(C = 30000.0, \gamma = 0.002)$, $(C = 30000.0, \gamma = 2.0)$ and $(C = 30000.0, \gamma = 36.0)$ were used to run the four SBS feature selection searches.

After performing the four SBS searches, to obtain the feature sets $X_k^{\mathscr{S}_i}$ to form Ξ, we use, for each SBS search $\mathscr{S}_i$, the feature subsets defined by using each number of selected features from $k = 1$ to $k = 60$; we do not used $k > 60$ as we observed that feature subsets with many features were less likely to be selected to compose the final ensemble $\mathscr{E}$. Thus $|\Xi| = 4 \times 60 = 240$ feature sets.

Then for each feature set in Ξ we built a SVM classifier using the SVM parameter value $(C = 8.0, \gamma = 0.5)$, including this classifier in the set $\mathscr{L}$ of candidates to compose the ensemble. In order to finally select a subset $\mathscr{E}$ of classifier from $\mathscr{L}$ we used the SFS search as explained in section 6.2. We set the desired ensemble size $|\mathscr{E}|$ as $|\mathscr{E}| = 10$ as we observed a general tendency of an AUC decrease with the usage of a larger set.

7.1.4 The `GEFS-rep` Classification Model

This classification model is a SVM ensemble based on feature and parameter variation. The feature subsets were defined by a genetic search, and to vary the value of the SVM parameter of the classifiers we used a predefined set of SVM parameter values [12].

First we performed the `GEFS-not` method to obtain an ensemble $\mathscr{E}'$ of $|\mathscr{E}'| = 20$ SVMs, based on feature variaton. In the following, we built n_{rep} replications of the ensemble $\mathscr{E}'$, with every SVM in a replication using the same SVM parameter value;

each replication used a different predefined parameter value. So the final ensemble $\mathscr{E}$ was composed of every classifier produced by every replication, in a total of $|\mathscr{E}| = |\mathscr{E}'| \times n_{rep}$ SVMs.

We used $n_{rep} = 20$ different SVM parameter values, chosen according to preliminary experiments as providing accurate, divergent SVMs (see the `BSFS-not` method). Each replication used the parameter $C = 30000.0$ and a different value for the γ parameter; the 20 values of γ were 0.03, 0.07, 0.12, 0.2, 0.4, 0.6, 0.8, 1.15, 1.5, 2.0, 3.0, 4.5, 6.0, 8.0, 11.0, 15.0, 20.0, 24.0, 29.0, 36.0.

So an ensemble built by the `GEFS-rep` method is composed of $|\mathscr{E}| = |\mathscr{E}'| \times n_{rep} = 20 \times 20 = 400$ SVMs. We evaluated approaches for removing some of those SVMs, similarly to the classifier selection stage of the BSFS method showed in section 6.2, but we found that in this `GEFS-rep` method the ensembles composed of all 400 produced classifiers achieved a higher prediction accuracy.

7.1.5 The BSFS-rep Classification Model

In this classification model we performed the `BSFS-not` method to obtain an ensemble $\mathscr{E}'$ of $|\mathscr{E}'| = 10$ SVMs, based on feature variation. In the following, we built $n_{rep} = 20$ replications of $\mathscr{E}'$ using in each replication a different SVM parameter value, as explained for the `GEFS-rep` method. So the final ensemble $\mathscr{E}$ was composed of every classifier produced by every replication, in a total of $|\mathscr{E}| = |\mathscr{E}'| \times n_{rep} = 200$ SVMs.

7.1.6 The GEFS-tun Classification Model

This classification model is a SVM ensemble based on feature and parameter variation, in which the feature subsets are defined by a genetic search and the classifiers use tuned parameters.

First, we used the GEFS algorithm as for the `GEFS-not` method. As the GEFS method produces 20 feature sets (classifiers in an ensemble) per generation and it evolved for a total of 200 generations, for producing 240 feature sets to compose Ξ (as for the BSFS method) we take the ensemble produced by each generation starting from the generation number 189 and finishing in the last generation (number 200). So the set Ξ of feature sets was composed of 240 feature sets.

Then for each feature set in Ξ we built a SVM classifier using the grid-search method to tune its SVM parameters, including this classifier in the set $\mathscr{L}$ of candidates to compose the ensemble. Then we used the SFS search to select an optimized ensemble $\mathscr{E}$ of $|\mathscr{E}| = 10$ classifiers (as explained in section 6.2).

7.1.7 The BSFS-tun Classification Model

We evaluated ensembles built by BSFS as showed in section 6, based on feature and parameter variation. So the difference to the `BSFS-not` method is that `BSFS-tun` used the grid-search method to tune the SVM parameters of every classifier in the set $\mathscr{L}$. The final ensemble $\mathscr{E}$ was composed of $|\mathscr{E}| = 10$ SVMs.

7.2 5 × 2 Cross-Validation Estimation Results

Table 1 presents, for each considered fault, the AUC estimated on test data by the 5 × 2 cross-validation estimation process, achieved by each considered classification model for building the predictor of that fault.

The results showed that SVM ensembles based on feature and parameter variation achieved higher prediction accuracies in comparison to SVM ensembles based on feature variation or to single SVMs.

Comparing the two studied approaches for varying the SVM parameters in an ensemble, the experiments showed that the approach of tuning the parameters of each classifier usually provided a higher estimated accuracy in comparison to the approach of using predefined different SVM parameter values.

Finally the experiments showed that the BSFS based methods (`BSFS-not`, `BSFS-rep` and `BSFS-tun`) achieved a higher prediction accuracy in comparison to the GEFS based methods (`GEFS-not`, `GEFS-rep` and `GEFS-tun`). Also, varying the SVM parameters of ensembles built by BSFS, specially the `BSFS-tun` method, provided an effective accuracy gain.

We performed the statistical testing to be used with the 5 × 2 cross-validation process [7], aiming to determine whether there is a significant difference on the accuracy achieved by using a single SVM, a SVM ensemble built by BSFS or a SVM ensemble built by GEFS. The level of significance of the statistical test is 0.05. For the `GEFS-tun` method, for the hydrodynamic fault category it was possible to accept that the accuracy achieved by the ensemble had a significant difference of the accuracy achieved by a single SVM. For the `BSFS-tun` method, for the misalignment, structural looseness and hydrodynamic fault the statistical test confirmed that there is a significant difference on the accuracy achieved by an ensemble in comparison to the accuracy achieved by a single SVM. Therefore this result confirmed the higher accuracy achieved by the BSFS method.

Table 1 AUC estimated on test data by 5 × 2 cross-validation.

Fault classifier	`SVM-s`	`GEFS -not`	`BSFS -not`	`GEFS -rep`	`BSFS -rep`	`GEFS -tun`	`BSFS -tun`
Misalignment	0.829	0.852	0.859	0.849	0.873	0.863	**0.876**
Bearing	0.909	0.934	0.929	0.935	0.939	0.938	**0.942**
Unbalance	0.836	0.866	0.870	0.853	0.879	0.869	**0.883**
Hydrodynamic	0.912	0.929	0.932	0.926	0.933	0.934	**0.936**
Structural L.	0.873	0.874	0.902	0.864	**0.918**	0.891	**0.918**
Mechanical L.	0.878	0.892	**0.902**	0.878	0.892	0.898	0.901

8 Conclusions and Future Work

Aiming to diagnose faults in industrial machines, we studied ensembles of SVM classifiers in which the classifiers differ on their feature set and SVM parameters.

To further increase the prediction accuracy, we plan to study more powerful approaches for tuning the parameters of SVM classifiers. This would produce SVMs that are more accurate and also more divergent in comparison to SVMs produced by the grid-search method. Specially, Particle Swarm Optimization (PSO) based techniques have been succesfully used to tune SVM parameters [8].

We plan to acquire more real-world data, from different machines and from more sources than just vibrational signals. Thus we plan to develop a multiparametric diagnostic system, which uses vibration signals complemented with electrical signals such as current, power and torque. This might increase the prediction accuracy as more information sources will be available.

Acknowledgements. We would like to thank the Brazilian energy company PETROBRAS S.A. for the financial support and for providing the examples of oil rig motor pump vibration signals.

References

[1] Bellini, A., Filippetti, F., Capolino, G.A.: Advances in diagnostic techniques for induction machines. IEEE Transactions on Industrial Electronics 55 (2008)
[2] Bishop, C.M.: Pattern Recognition and Machine Learning. Springer, Berlin (2007)
[3] Chen, P.H., Lin, C.J., Schölkopf, B.: A tutorial on ν-support vector machines. Applied Stochastic Models in Business and Industry 21, 111–136 (2005)
[4] Evgeniou, T., Pontil, M., Elisseeff, A.: Leave-one-out error, stability, and generalization of voting combinations of classifiers. Machine Learning (2002)
[5] Fawcett, T.: An introduction to ROC analysis. Pattern Recog. Letters 27 (2006)
[6] Kudo, M., Sklansky, J.: Comparison of algorithms that select features for pattern classifiers. Pattern Recognition 33, 25–41 (2000)
[7] Kuncheva, L.I.: Combining Pattern Classifiers. Springer, Heidelberg (2004)
[8] Li, S., Tan, M.: Tuning SVM parameter by using a hybrid CLPSO-BFGF algorithm. Neurocomputing 73, 2089–2096 (2010)
[9] Li, X., Wang, L., Sung, E.: Adaboost with SVM-based component classifiers. Engineering Applications of Artificial Intelligence 21 (2008)
[10] Mendel, E., Rauber, T.W., Varejao, F.M.: Automatic bearing fault pattern recognition using vibration signal analysis. In: Proc. of the IEEE Int. Symp. on Ind. Electronics ISIE 2008 (2008)
[11] Opitz, D.W.: Feature selection for ensembles. In: Proc. of the Sixteenth National Conf. on Artificial Intelligence AAAI 1999 (1999)
[12] Sun, B.Y., Zhang, X.M., Wang, R.J.: On constructing and pruning SVM ensembles. In: Proc. of the 2007 IEEE Conf. on Signal-Image Technologies (2007)
[13] Tsymbal, A., Pechenizkiy, M., Cunningham, P.: Diversity in search strategies for ensemble feature selection. Information Fusion 6, 83–98 (2005)

[14] Valentini, G., Dietterich, T.G.: Bias-variance analysis of support vector machines for the development of SVM-based ensemble methods. The Journal of Machine Learning Research (2000)
[15] Verikas, A., Gelzinis, A., Kovalenko, M., Bacauskiene, M.: Selecting features from multiple feature sets for SVM committee-based screening of human larynx. Expert Systems with Applications 37, 6957–6962 (2010)
[16] Widodo, A., Yang, B.S.: Support vector machine in machine condition monitoring and fault diagnosis. Mechanical Systems and Signal Processing 21 (2007)
[17] Zio, E., Baraldi, P., Gola, G.: Feature-based classifier ensembles for diagnosing multiple faults in rotating machinery. Applied Soft Computing 8, 1365–1380 (2008)

Application of Data Mining Techniques in the Estimation of Mechanical Properties of Jet Grouting Laboratory Formulations over Time

Joaquim Tinoco, António Gomes Correia, and Paulo Cortez

Abstract. Sometimes, the soil foundation is inadequate for constructions purpose (soft-soils). In these cases there is need to improve its mechanical and physical properties. For this purpose, there are several geotechnical techniques where Jet Grouting (JG) is highlighted. In many geotechnical structures, advance design incorporates the ultimate limit state (ULS) and the serviceability limit state (SLS) design criteria, for which uniaxial compressive strength and deformability properties of the improved soils are needed. In this paper, three Data Mining models, i.e. Artificial Neural Networks (ANN), Support Vector Machines (SVM) and Functional Networks (FN), were used to estimate the tangent elastic Young modulus at 50% of the maximum stress applied ($E_{tg50\%}$) of JG laboratory formulations over time. A sensitivity analysis procedure was also applied in order to understand the influence of each parameter in $E_{tg50\%}$ estimation. It is shown that the data driven model is able to learn the complex relationship between $E_{tg50\%}$ and its contributing factors. The obtained results, namely the relative importance of each parameter, were compared with the predictive models of elastic Young modulus at very small strain (E_0) as well as the uniaxial compressive strength (Q_u). The obtained results can help to understand the behavior of soil-cement mixtures over time and reduce the costs with laboratory formulations.

Joaquim Tinoco
Department of Civil Engineering/C-TAC, University of Minho, 4710, Guimarães, Portugal
e-mail: jabtinoco@civil.uminho.pt

António Gomes Correia
Department of Civil Engineering/C-TAC, University of Minho, 4710, Guimarães, Portugal
e-mail: agc@civil.uminho.pt

Paulo Cortez
Department of Information systems/R&D Algoritmi Centre, University of Minho, 4710, Guimarães, Portugal
e-mail: pcortez@dsi.uminho.pt

A. Gaspar-Cunha et al. (Eds.): Soft Computing in Industrial Applications, AISC 96, pp. 283–292.
springerlink.com

1 Introduction

Given the growth of the human population and the finite resources of the planet Earth, we are forced to use soft-soils as a soil foundation. In these cases, there is need to improve its physical and mechanical properties. For this purpose, the Jet Grouting (JG) technology is widely used [6, 12], given it great versatility. This technology allows the improvements of the mechanical and physical properties of several types of soil, since grain to fine soils, and different shapes of treatment (columns, panels, etc.) can be obtained. In few words, this technology consists in injecting high speed grouting of water-cement mixture into the subsoil with or without other fluids (air or water). The fluids are injected through small nozzles placed at the end of a rod which is inserted until the intended depth. This rod is continually rotated and slowly removed up to the surface. At the end, a soil-cement mixture with better properties is obtained. Currently adopted JG methods can be classified according to the number of fluids injected into the subsoil: water-cement grout → Jet 1; air + grout → Jet 2 and water + air + grout → Jet 3.

This paper will focus the JG initial stage, where a set of laboratory formulations, which are function of the soil type to be treated and the design properties, are used to set the soil-cement mixture that will be used in the construction works. In particular, this study allows the definition of the grout water/cement ratio, the amount of cement for cubical meter of treated soil and the cement type, needed to satisfy the design and economical requirements. The remaining parameters that control the final characteristics of the JG elements (e.g. the speed of withdrawal and rotation of the rods) will be evaluated with the execution of test columns, given the difficulty to simulate such parameters in laboratory. However, despite of all advantages of JG technology, its design is essentially based on empirical methods that are often too conservative [5, 11]. As a result, the economy and the quality of the treatment can be affected. Thus, given the high potential of JG technology, is very important to develop more accurate and rational models to estimate the effects of the different parameters involved in JG process. This will allow reducing field tests, optimizing all the constructive process and obtaining a higher technical and economical efficiency.

In the other hand, powerful tools have emerged that allow extract useful information from large databases, i.e. Data Mining (DM) techniques. These techniques enable the exploration of complex relationships between several inputs and the target variable. Hence, given the high complexity inherent to JG process, due to the number of parameters involved and the heterogeneity of the soil, DM techniques are an interesting tool to explore JG data. Thus, in order to develop rational models and satisfy the current project requirements, three DM techniques were applied to estimate the mechanical properties of JG laboratory formulations. So, we started by developing models to predict the uniaxial compressive strength (Q_u) of JG laboratory formulations over time [15]. However, to deal with the serviceability state of the structure, deformability properties of the improved ground are also necessary. In this context, predictive models for deformability moduli, namely elastic Young modulus at very small strain (E_0), of JG laboratory formulations were also developed [16].

Yet, the best parameter to analyze the deformability properties of soil-cement mixtures is the tangent elastic young modulus at 50% of the maximum applied stress ($E_{tg50\%}$). Thus, DM techniques were also applied to estimate this property over time.

In this work, the performance of the predictive models of E_0 and Q_u are summarized and the predictive capacity of $E_{tg50\%}$ by application of DM techniques, namely Artificial Neural Network (ANN), Support Vector Machine (SVM) and Functional Networks (FN) are exposed. Moreover, the key parameters in Q_u, E_0 and $E_{tg50\%}$ estimation were identified, compared and discussed.

2 Materials and Jet Grouting Laboratory Data

Three datasets were used to train and test the predictive models of each studied property. All data were prepared at University of Minho, under a huge laboratory experimental program, to analyze the influence of several parameters in Q_u, E_0 and $E_{tg50\%}$ of JG material. These mechanical properties were obtained in unconfined compression tests with on sample strain instrumentation [3]. In a non-linear stress-strain relationship different moduli can be defined. For this work, tangent elastic young modulus at 50% of the maximum applied stress was adopted since, is a key geotechnical parameter that better defines the deformability properties of soil-cement mixtures. Table 1 shows the number of records of each dataset used during the training phase of each predictive model as well as the number of different formulations in each dataset.

Table 1 Number of records and different formulations of each dataset used during the training of each predictive model

	Q_u	E_0	$E_{tg50\%}$
Number of records	175	188	49
Number of formulations	35	9	8

Based on expert knowledge about soil-cement mixtures [13] and after some experiments, the following input parameters were selected : Water/Cement ratio - W/C; Age of the mixture - t; Coefficient related with the cement type - s; Relation between the mixture porosity and the volumetric content of cement - $n/(C_{iv})^d$; Cement content of the mixture - %C; Percentage of sand - %Sand; Percentage of silt - %Silt; Percentage of clay - %Clay and Percentage of organic matter - %OM.

The basic statistics of the numerical parameters used in Q_u and E_0 datasets are described in [15, 16] respectively. Table2 shows the basic statistics for $E_{tg50\%}$. The geotechnical soil properties were evaluated using laboratory tests. While all of the soils were classified as fine grained soils they have different percentages of sand, silt, clay and organic matter. A detailed classification of soils can be found in [15]. All formulations were prepared with cement CEM I 42.5R, CEM II 42.5R and CEM IV/A 32.5R.

Table 2 Synopsis of the numerical input parameters in $E_{tg50\%}$

Soil	Parameter	Minimum	Maximum	Mean	Standard Deviation
	W/C	0.69	1.11	0.98	0.12
	t(days)	28.00	84.00	64.57	19.13
	$n/(C_{iv})^d$	38.73	73.81	61.46	6.55
clay	%C	24.19	64.86	44.39	11.79
	%Sand	0.00	39.00	14.10	13.68
	%Silt	33.00	57.00	50.00	8.27
	%Clay	22.50	45.00	35.71	7.45
	%OM	0.40	8.30	3.71	2.43

3 Data Mining Models and Evaluation Measures

3.1 Data Mining Models

Three Data Mining models where trained to estimate Q_u [15], E_0 [16] and $E_{tg50\%}$, of JG laboratory formulations over time.

ANN mimic some basic aspects of brain functions [9], which processes information by means of interaction among several neurons. We adopted the most popular model, the multilayer perceptron that contains only feedforward connections, with one hidden layer with H processing units. The general model of the ANN is:

$$\hat{y} = W_{o,0} + \sum_{j=I+1}^{o-1} f\left(\sum_{i=1}^{I} X_i \cdot W_{j,i} + W_{j,0}\right) \cdot W_{o,i} \tag{1}$$

where $W_{j,i}$ represents the weight of the connection from neuron j to unit i, f is a logistic function $1/(1+e^{(-x)})$, and I is the number of input neurons. To chose the best value of H, we used a grid search within $\{2, 4, ..., 10\}$ [8].

SVM was initially proposed for classification tasks [7]. After the introduction of the ε-insensitive loss function, it was possible to apply SVM to regression tasks [14]. SVM has theoretical advantages over ANN, such as the absence of local miminima in the learning phase. The main idea of the SVM is to transform the input data into a high-dimensional feature space by using a nonlinear mapping. For this purpose, the popular Gaussian kernel was adopted:

$$k(x,x') = e^{\left(-y \cdot \|x-x'\|^2\right)}, y > 0 \tag{2}$$

Under this setup, performance of the regression is affected by three parameters: γ - the parameter of the kernel, C - a penalty parameter, and ε - the width of a ε-insensitive zone. To reduce the search space, the first two values will be set using the heuristics of [2]: $C = 3$ and $\varepsilon = \hat{\sigma}/\sqrt{N}$, where $\hat{\sigma} = 1.5/N \cdot \sum_{i=1}^{N} (y_i - \hat{y}_i)^2$ and

$\hat{y}_i$ is the value predicted by a 3-nearest neighbor algorithm. To optimize the kernel parameter γ, we adopted a grid search of $\{2^{-15}, 2^{-13}, ..., 2^3\}$.

FN are a general framework useful for solving a wide range of problems [1], where the functions of the neurons can be multivariate, multi-argument and it is also possible to use different learnable functions, instead of fixed functions. Moreover, there is no need to associate weights to connections between nodes, since the learning is achieved by the neural functions. When compared with ANN, there are some advantages [18]. For example, unlike ANN, FN can reproduce certain physical characteristics that lead to the corresponding network in a natural way. Also, the estimation of the network parameters can be obtained by resolving a linear system of equations, which returns a fast and unique solution, i.e. the global minimum is always achieved. These two types of networks have a similar structure, but they also have important differences. In FN the selection of the initial topology is normally based on the properties of the problem at hand and can be further simplified using functional equations and its neural functions (normally functions from a given family, such as polynomial or exponential) can be multidimensional and set during the learning phase. Furthermore, outputs neurons can be connected, which is not the case of standard ANN. In this work we use the FN to solve the following generic expression:

$$\hat{y} - \beta_0 \cdot \prod_{i=1}^{N} x_i^{\alpha_i} \quad (3)$$

where, $\{x_1, ..., x_i\}$ are the input parameters and $\{\beta_0, \alpha_1, ..., \alpha_i\}$ are the coefficients to be adjusted.

To learn the coefficients in equation (3), the following minimization problem was used:

$$MinimizeQ = \sum_{S=1}^{S} \delta_s^2 = \sum_{S=1}^{S} \left(y_s - \beta_0 \cdot \prod_{i=1}^{N} x_i^{\alpha_i} \right)^2 \quad (4)$$

We also tested the classic multiple regression (using the R tool). Yet, the poor results achieved (when compared with ANN, SVM and FN) are not reported here due to space limitations.

The ANN and SVM models were training using rminer library [4], which facilitates the application of DM techniques in the R tool. The formulation and resolution of the FN was implemented in the free version of the GAMS [17].

3.2 Evaluation Measures

To assess and compare the performance of each predictive model, three evaluation measures were calculated: Mean Absolute Deviation - MAD; Root Mean Square Error - RMSE and the coefficient of determination - R^2:

$$MAD = \frac{\sum_{i=1}^{N} |y - \hat{y}|}{N}; RMSE = \sqrt{\frac{\sum_{i=1}^{N} (y - \hat{y})^2}{N}} \quad (5)$$

$$R^2 = \left(\frac{\sum_{i=1}^{N} (y - \bar{y}) \cdot (\hat{y} - \bar{\hat{y}})}{\sqrt{\sum_{i=1}^{N} (y - \bar{y})^2 \cdot \sum_{i=1}^{N} (\hat{y} - \bar{\hat{y}})^2}} \right)^2 \quad (6)$$

When compared with MAD, RMSE metric is more sensitivity to extreme errors. In a model with good performance, both MAD and RMSE should present lower values and R^2 should be close to unit value.

The Leave-One-Out scheme was adopted for measuring the predictive capability of each model, where sequentially one example is used to test the model and the remaining data is used for fitting the model. Under this scheme, there is need around N (the number of data samples) times more computation, since N models are fitted. The final generalization estimate is evaluated by computing the MAD, RMSE and R^2 metrics for all N test samples. To understand better the behavior of the JG material, the influence of each parameter was also quantified by applying a sensitivity analysis procedure [10]. This procedure determines the most important variables by successively holding all but one input constant and varying the other over its range of values to observe its effect on the system. A high variance observed in the outputs denotes a high input relevance.

4 Results of the Different Predictive Models

Similarly to the previous works, where DM techniques were used to study the behavior of JG laboratory formulation [15, 16], in the present study, a high performance was also reached in $E_{tg50\%}$ estimation. This performance is proved by lower values of MAD and RMSE metrics and the R^2 value close to the unit (see table 3).

Figure 1 shows the relation between $E_{tg50\%}$ measured versus predicted by SVM model. The same relation for the remaining models (ANN and FN) is very similar. As one can see in figure 1, all points are very close to the diagonal line that represents the perfect prediction. Despite of the better values of the metrics MAD, RMSE and R^2 in FN model, SVM is more consistent and interesting in terms of relative importance of each parameter. According to FN model the $E_{tg50\%}$ of JG laboratories formulations is almost only controlled by clay percentage of soil (55.92%), and we know that is not truth. Based on empirical knowledge, the mechanical properties of soil-cement mixtures are also affected by cement content.

Table 3 Comparison of the performance between the three models: ANN, SVM and FN, in $E_{tg50\%}$ estimation

	ANN	FN	SVM
MAD	0.27	0.18	0.19
RMSE	0.50	0.24	0.28
R^2	0.74	0.93	0.91

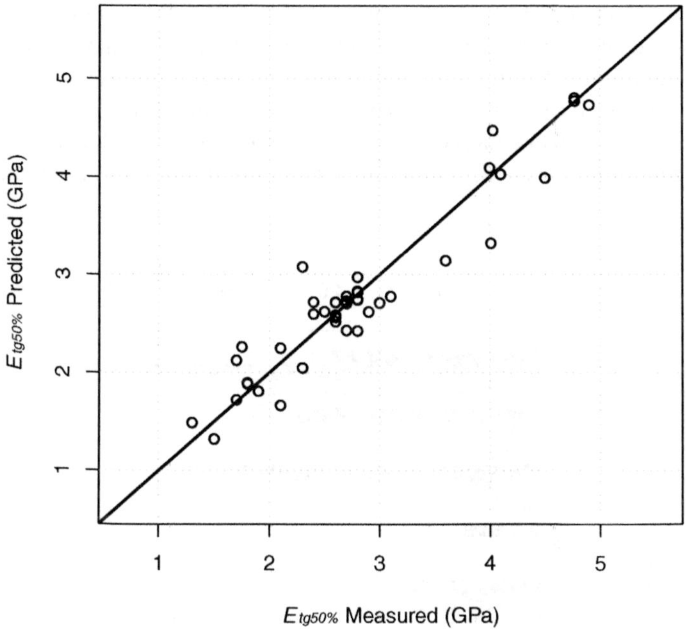

Fig. 1 Predicted versus desired $E_{tg50\%}$ of JG laboratory formulations using the SVM model

The performance reached by SVM model in $E_{tg50\%}$ prediction is very similar to the performance reached by the same model in Q_u [15] and E_0 [16] prediction. Table 4 summarizes the values of the coefficient correlation (R^2) of SVM model in Q_u, E_0 and $E_{tg50\%}$ estimation over time.

Table 4 Performance of each SVM predictive model in Q_u, E_0 and $E_{tg50\%}$

	SVM-Q_u	SVM-E_0	SVM-$E_{tg50\%}$
R^2	0.93	0.96	0.91

In spite of the high performance, assessed by metric MAD, RMSE and R^2, as well as the high relation between $E_{tg50\%}$ measured versus predicted, it is also important to quantify and analyze the relative relevance of each parameter in the model. Observing figure 2, which shows the importance of each input parameter in SVM predictive model of Q_u, E_0 and $E_{tg50\%}$, we can see that the relation between porosity and the volumetric content of cement $(n/(C_{iv})^d)$, the water/cement ratio (W/C) and the soil properties, namely the percentage of clay (%Clay), are the key parameters in $E_{tg50\%}$ prediction. Moreover, in Q_u estimation the age of the mixture and the percentage of cement should be included. In the other hand, it is interesting to observe that the soil properties are more relevant in deformability properties estimation (E_0 and $E_{tg50\%}$) than in strength prediction (Q_u). This observation makes some sense if we take into account that for low deformations, the grain size is the responsible for the main resistance capacity of the material. After the grains broke, the cohesion is sustained by soil-cement matrix. So, from this time, the age of the mixture and the percentage of cement take the main role in the strength capacity of the soil-cement mixture.

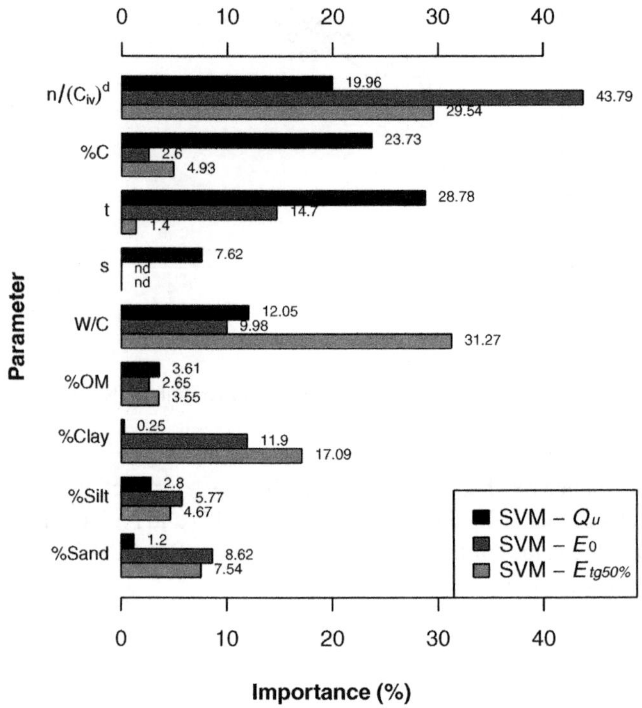

Fig. 2 Comparison of the relative importance of each parameter in each SVM predictive model of Q_u, E_0 and $E_{tg50\%}$

5 Conclusions and Future Works

A rational approach, based in soft computing tools, namely Artificial Neural Networks (ANN), Support Vector Machines (SVM) and Functional Networks (FN), was successfully applied to improve the knowledge about mechanical properties of Jet Grouting (JG) laboratory formulations. The uniaxial compressive strength (Q_u), the elastic Young modulus at very small strain (E_0) and the tangent elastic young modulus at 50% of the maximum applied stress ($E_{tg50\%}$) of Jet Grouting (JG) laboratory formulations can be estimated over time with high accuracy.

The high performance reached in mechanical properties prediction of JG laboratory formulations, allows us to conclude that Data Mining techniques are a useful tool (in particular, the SVM model) to better understand the behavior of soil-cement mixtures over time.

The sensitivity analysis carried out in this study allows the conclusion that the relation between porosity and the volumetric content of cement $(n/(C_{iv})^{d})$, the water/cement ratio (W/C) and the soil properties, are the key parameters in deformability properties prediction. Furthermore, in Q_u estimation, the age of the mixture and the percentage of cement should be included. It was also possible to observe, that the soil properties are more relevant in deformability properties than in ultimate strength estimation of soil-cement mixtures.

The knowledge obtained from this study allows understanding better the mechanical behavior of soil-cement mixtures and will reduce the number of laboratory formulations carried out. As a result, the quality control process of JG columns is improved and the costs of laboratory material formulations are reduced. In future works, we intend to apply Data Mining techniques to develop predictive models for final diameter of real JG columns, as well as its mechanical properties. A sensitivity analysis procedure will be also applied in order to analyze the influence of each parameter.

Acknowledgements. The authors wish to thank to Portuguese Foundation for Science and Technology (FCT) the support given through the doctoral grant SFRH/BD/45781/2008. Also, the authors would like to thank the interest and financial support by Tecnasol-FGE and Tiago Valente for the dataset gathered from the laboratory formulations.

References

1. Castillo, E., Cobo, A., Gutirrez, J., Pruneda, R.: Functional networks with applications: a neural-based paradigm. Springer, Heidelberg (1999)
2. Cherkassky, V., Ma, Y.: Selection of SVM parameters and noise estimation for SVM regression. Neural Networks 17(1), 113–126 (2004)
3. Gomes Correia, A., Valente, T., Tinoco, J., Falção, J., Barata, J., Cebola, D., Coelho, S.: Evaluation of mechanical properties of jet grouting columns using different test methods. In: Procedings of 17th International Conference on Soil Mechanics and Geotechnical Engineering, pp. 2179–2171 (2009)
4. Cortez, P.: Data Mining with Neural Networks and Support Vector Machines Using the R/rminer Tool. In: Perner, P. (ed.) ICDM 2010. LNCS(LNAI), vol. 6171, pp. 572–583. Springer, Heidelberg (2010)
5. Croce, P., Flora, A.: Analysis of single-fluid jet grouting. Geotechnique 50(6), 739–748 (2001)
6. Falco, J., Pinto, A., Pinto, F.: Case histories of ground improvement solutions using jet grouting. Tecnasol FGE: Fundacoes e Geotecnia S. A (2000)
7. Goh, A.T.C., Goh, S.H.: Support vector machines: Their use in geotechnical engineering as illustrated using seismic liquefaction data. Computers and Geotechnics 34(5), 410–421 (2007)

8. Hastie, T., Tibshirani, R., Friedman, J.: The Elements of Statistical Learning II: Data Mining, Inference, and Prediction. Springer, Heidelberg (2009)
9. Kenig, S., Ben-David, A., Omer, M., Sadeh, A.: Control of properties in injection molding by neural networks. Engineering Applications of Artificial Intelligence 14(6), 819–823 (2001)
10. Kewley, R.H., Embrechts, M.J., Breneman, C.: Data strip mining for the virtual design of pharmaceuticals with neural networks. Neural Networks 11(3), 668–679 (2000)
11. Kirsch, F., Sondermann, W.: Ground improvement and its numerical analysis. In: Proceedings of the International Conference on Soil Mechanics and Geotechnical Engineering, vol. 3, pp. 1775–1778 (2001)
12. Padura, A.B., Sevilla, J.B., Navarro, J.G., Bustamante, E.Y., Crego, E.P.: Study of the soil consolidation using reinforced jet grouting by geophysical and geotechnical techniques. Construction and Building Materials 23(3), 1389–1400 (2009)
13. Shibazaki, M.: State of practice of jet grouting. In: Grouting and Ground Treatment, pp. 198–217 (2004)
14. Smola, A.J., Schölkopf, B.: A tutorial on support vector regression. Statistics and computing 14(3), 199–222 (2004)
15. Tinoco, J., Gomes Correia, A., Cortez, P.: A data mining approach for Jet Grouting Uniaxial Compressive Strength prediction. In: World Congress on Nature & Biologically Inspired Computing, NaBIC 2009, pp. 553–558 (2009)
16. Tinoco, J., Gomes Correia, A., Cortez, P.: Applications of Data Mining Techniques in the estimations of Jet Grouting Laboratory Formulations over Time. In: 1st International Conference on Information Technology in Geo-Engineering, ICITG-Shanghai 2010, pp. 92–100 (2010)
17. GAMS Development Corporation. Online Documentation, Welcome to the GAMS Home Page, `http://www.gams.com/docs/document.htm` (accessed May 17, 2010)
18. Zhou, Y.Q., He, D.X., Nong, Z.: Application of Functional Network to Solving Classification Problems. Proceedings of the World Academy of Science, Engineering and Technology 7, 390–393 (2005)

Hybrid Intelligent Intrusion Detection Scheme

Mostafa A. Salama, Heba F. Eid, Rabie A. Ramadan, Ashraf Darwish, and Aboul Ella Hassanien

Abstract. This paper introduces a hybrid scheme that combines the advantages of deep belief network and support vector machine. An application of intrusion detection imaging has been chosen and hybridization scheme have been applied to see their ability and accuracy to classify the intrusion into two outcomes: normal or attack, and the attacks fall into four classes; R2L, DoS, U2R, and Probing. First, we utilize deep belief network to reduct the dimensionality of the feature sets. This is followed by a support vector machine to classify the intrusion into five outcome; Normal, R2L, DoS, U2R, and Probing. To evaluate the performance of our approach, we present tests on NSL-KDD dataset and show that the overall accuracy offered by the employed approach is high.

Mostafa A. Salama
Department of Computer Science, British University in Egypt, Cairo, Egypt
e-mail: mostafa.salama@gmail.com

Heba F. Eid
Faculty of Science, Al-Azhar University, Cairo, Egypt
e-mail: heba.fathy@yahoo.com

Rabie A. Ramadan
Cairo University, Faculty of Engineering, Computer Engineering Department, Cairo, Egypt
e-mail: rabieramadan@gmail.com

Ashraf Darwish
Faculty of Science, Helwan University, Cairo, Egypt
e-mail: amodarwish@yahoo.com

Aboul Ella Hassanien
Faculty of Computers and Information, Cairo University
e-mail: aboitcairo@gmail.com

A. Gaspar-Cunha et al. (Eds.): Soft Computing in Industrial Applications, AISC 96, pp. 293–303.
springerlink.com

Keywords: Deep Belief Network (DBN), Network Intrusion detection system, Support Vector Machines (SVMs), Dimensional reduction.

1 Introduction

The Internet and online procedures is an essential tool of our daily life today. They have been used as an important component of business operation [1]. Therefore, network security needs to be carefully concerned to provide secure information channels. Intrusion detection (ID) is a major research problem in network security, where the concept of ID was proposed by Anderson in 1980 [2]. ID is based on the assumption that the behavior of intruders is different from a legal user [3]. The goal of intrusion detection systems (IDS) is to identify unusual access or attacks to secure internal networks [4] Network-based IDS is a valuable tool for the defense-in-depth of computer networks. It looks for known or potential malicious activities in network traffic and raises an alarm whenever a suspicious activity is detected. In general, IDSs can be divided into two techniques: misuse detection and anomaly detection [5, 6]. Misuse intrusion detection (signature-based detection) uses well-defined patterns of the malicious activity to identify intrusions [7,8]. However, it may not be able to alert the system administrator in case of a new attack. Anomaly detection attempts to model normal behavior profile. It identifies malicious traffic based on the deviations from the normal patterns, where the normal patterns are constructed from the statistical measures of the system features [9]. The anomaly detection techniques have the advantage of detecting unknown attacks over the misuse detection technique [10]. Several machine-learning techniques including neural networks, fuzzy logic [11], support vector machines (SVM) [9, 11] have been studied for the design of IDS. In particular, these techniques are developed as classifiers, which are used to classify whether the incoming network traffics are normal or an attack. In this paper, we propose an anomaly intrusion detection scheme using Deep Belief Network (DBN) based on Restricted Boltzmann Machine (RBM) [12, 13]. DBN is used as feature reduction method [14] that is followed by SVM classifier. We evaluate the effectiveness of the proposed DBN-SVM scheme by conducting several experiments on NSL-KDD dataset. We examine the performance of the DBN-SVM scheme in comparison with standalone DBN and standalone SVM classifier. Also, DBN as a feature reduction method is compared with other known feature reduction methods. The rest of this paper is organized as follows: Section 2 gives an overview of RBM architecture and DBN. Section 3 describes DBN classifier and the proposed DBN-SVM intrusion detection scheme. The experimental results and conclusions are presented in Section 4 and 5 respectively.

2 An Overview

This section discusses the deep belief network structure including the exploration of the restricted Boltzmann machine.

2.1 *Restricted Boltzmann Machine*

RBM is an energy-based undirected generative model that uses a layer of hidden variables to model a distribution over visible variables [14, 15]. The undirected model for the interactions between the hidden and visible variables is used to ensure that the contribution of the likelihood term to the posterior over the hidden variables is approximately factorial which greatly facilitates inference [16]. Energy-based model means that the probability distribution over the variables of interest is defined through an energy function. It is composed from a set of observable variables $V = \{v_i\}$ and a set of hidden variables $H = \{h_j\}$, i node in the visible layer, j node in the hidden layer. It is restricted in the sense that there are no visible-visible or hidden-hidden connections. The steps of the RBM learning algorithm can be declared as follows:

1. Due to the conditional independence (no connection) between nodes in the same layer (Property in RBM), the conditional distributions are given in Equations (1) and (2).

$$\begin{cases} P(H|V) = \prod_j p(h_j|v) \\ p(h_j = 1|v) = f(a_i + \sum_i w_{ij} v_i) \\ p(h_j = 0|v) = 1 - p(h_j = 1|v); \end{cases} \tag{1}$$

And

$$\begin{cases} P(H|V) = \prod_i p(v_i|h) \\ p(v_i = 1|h) = f(b_j + \sum_j w_{ij} h_j) \\ p(v_i = 0|h) = 1 - p(v_i = 1|h); \end{cases} \tag{2}$$

Where f is a sigmoid function (σ) which takes the form $\sigma(z) = 1/1+e^{-z}$ for binary data vector.

2. The likelihood distribution between hidden and visible units is defined as:

$$P(v,h) = \frac{e^{-E(v,h)}}{\Sigma_i e^{-E(v_i,h)}} \tag{3}$$

Where $E(x,h)$=$-\bar{h}wv - \bar{b}v - \bar{c}h$,
And $\bar{h}$,$\bar{b}$, $\bar{c}$ are the transposes of matrices h, b and c.

3. The average of the log likelihood with respect to the parameters is given by

$$\begin{aligned}\Delta w_{ij} &= \varepsilon^*(\delta \log p(v)/\delta w_{ij}) \\ &= \varepsilon(< x_i h_j >_{data} - < v_i h_j >_{model})\end{aligned} \tag{4}$$

$$\Delta v_i = \varepsilon(< v_i^2 >_{data} - < v_i^2 >_{model}) \tag{5}$$

$$\Delta h_i = \varepsilon(< h_i^2 >_{data} - < h_i^2 >_{model}) \tag{6}$$

4. The term $<>_{model}$ takes exponential time to compute exactly so the Contrastive Divergence (CD) approximation to the gradient is used instead [6]. Contrastive divergence is a method that depends on the approximation that is to run the sampler for a single Gibbs iteration, instead until the chain converges. In this case the term $<>_1$ will be used such that it represents the expectation with respect to the distribution of samples from running the Gibbs sampler initialized at the data for one full step, the new update rule will be.

$$\Delta w_{ij} = \varepsilon(< v_i h_j >_{data} - < v_i h_j > 1) \tag{7}$$

$$\Delta v_i = \varepsilon(< v_i^2 >_{data} - < v_i^2 > 1) \tag{8}$$

$$\Delta h_i = \varepsilon(< h_i^2 >_{data} - < h_i^2 > 1) \tag{9}$$

The Harmonium RBM is an RBM with Gaussian continuous hidden nodes [6]. Where f is normal distribution function which takes the form shown in Equation (10)

$$P(h_j = h|x) = N(c_j + w_j.x, 1) \tag{10}$$

Harmonium RBM is used for a discrete output in the last layer of a deep belief network in classification.

2.2 Deep Belief Network

The key idea behind training a deep belief network by training a sequence of RBMs is that the model parameters θ , learned by an RBM define both $p(v \mid h, \theta)$ and the prior distribution over hidden vectors, $p(h \mid \theta)$, so the probability of generating a visible vector, v, can be written as:

$$p(v) = \Sigma_h p(h \mid \theta).p(v \mid h, \theta) \tag{11}$$

After learning θ, $p(v \mid h, \theta)$ is kept while $p(h \mid \theta)$ can be replaced by a better model that is learned by treating the hidden activity vectors $H = h$ as the training data (visible layer) for another RBM. This replacement improves a variation lower bound on the probability of the training data under the composite model. The study in [17] proves the following three rules:

1. Once the number of hidden units in the top level crosses a threshold; the performance essentially flattens at around certain accuracy.
2. The performance tends to decrease as the number of layers increases.
3. The performance increases as we train each RBM for an increasing number of iterations.

In case of not using class labels and back-propagation in the DBN Architecture (unsupervised training) [14], DBN could be used as a feature extraction method for dimensionality reduction. On the other hand, when associating class labels with feature vectors, DBN is used for classification. There are two general types of DBN classifier architectures which are the Back-Propagation DBN (BP-DBN) and the Associate Memory DBN (AM-DBN) [8]. For both architectures, when the number of possible classes is very large and the distribution of frequencies for different classes is far from uniform, it may sometimes be advantageous to use a different encoding for the class targets than the standard one-of-K softmax encoding.

3 Hybrid Intelligent Intrusion Detection Scheme

This section shows DBN as a standalone classifier and the proposed DBN-SVM hybrid scheme.

3.1 DBN Classifier

In the paper, the Constructed DBN will be composed of two RBMs, lower and higher RBM layers. The number of visible nodes of lower RBM is attribute number and the number of hidden nodes of the higher RBM is the available class number. While the number of hidden nodes in the lower RBM layer and number of visible nodes in the higher RBM layer are the same and equal to a random number, e.g. 13. Each hidden node in the higher RBM represents one of the classes under testing, such that if "0" is the class label associated with the input, then the first node in the hidden nodes in the higher RBM is 1, and the rest of nodes will be of value 0. e.g. If the output in the first hidden node is 0.6 and if the class label is 0, which means that the expected output in this node is 1, then there is an error of a value of 0.4. Algorithm 1 shows the steps of DBN classifier.

The training of the two restricted Boltzmann machines is required to initialize the weights of the deep belief network. So the network may have a better performance than using random weights.

Algorithm 1. DBN classifier

```
1: Use the training dataset to train the lower RBM layer.
2: Used output of the lower RBM layer to train the higher RBM layer.
3: Test the output in the higher RBM layer hidden nodes according to the
   output class label.
4: Back-propagate the error to fix the weights of the network.
5: Run the complete dataset through the network to produce a reduced
   output of the data.
6: for each object in the testing dataset do
7:    Run the input through the trained DBN network to produced an output
      in the hidden nodes of the higher RBM layer.
8:    Get the node of the maximum output value.
9:    Assign a class label that correspond to this node (of maximum output).
10:   if Assigned class label is equal to actual class label then
11:      object is classifier correctly
12:   end if
13: end for
14: Calculate the sum of the correctly classified object to find the classifica-
    tion accuracy.
```

3.2 DBN-SVM Hybride Scheme

The proposed hybrid intelligent intrusion detection network system is composed of three main phases; Preprocessing phase, DBN feature reduction phase and classification phase. Figure 1 describes the structure of the hybrid intelligent intrusion detection network system.

3.2.1 NSL-KDD Dataset Preprocessing

Pre-processing of NSL-KDD dataset contains three processes; (1) Mapping symbolic features to numeric value, (2) Data scaling, since the data have significantly varying resolution and ranges. The attribute data are scaled to fall within the range [0, 1]. and (3) Assigning attack names to one of the five classes, 0 for *normal*, 1 for *DoS* (Denial of Service), 2 for *U2R* (User to Root), 3 for *R2L* (Remote to Local) , and 4 for *Probe*.

3.2.2 DBN Feature Reduction

In this paper, DBN has been used as dimensionality reduction method with back-propagation to enhance the reduced data output. The DBN Network has the BP-DBN structure that is constructed of 2 RBM layers, the first RBM layer efficiently reduces the data(e.g. from 41 to 13 feature and the second from 13 features to 5 output features based on NSL-KDD data).

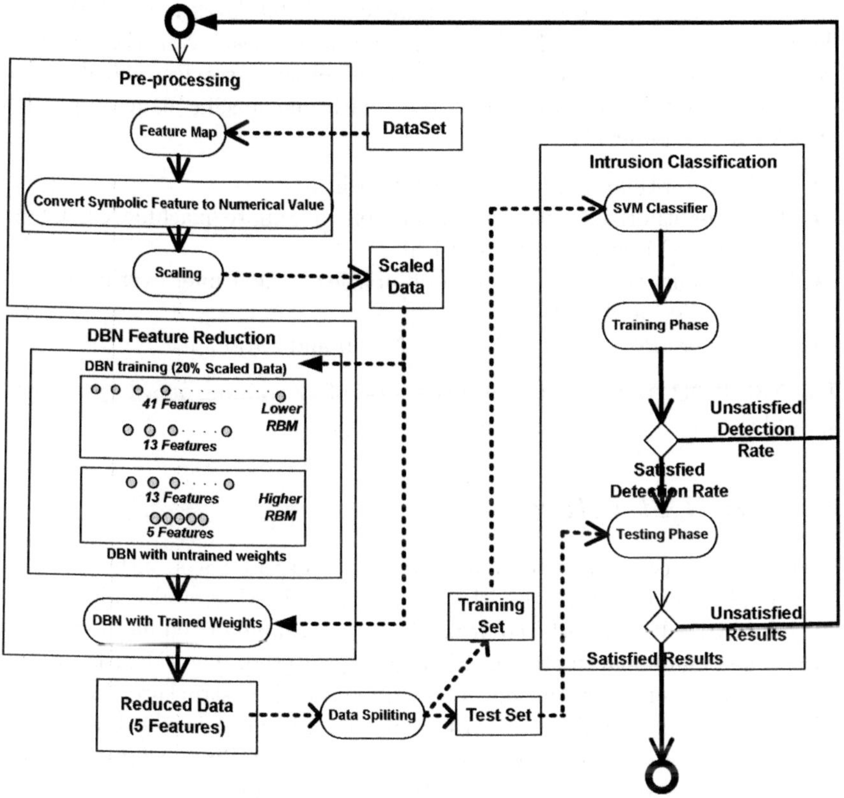

Fig. 1 Hybrid Intelligent Intrusion Detection Network Scheme

3.2.3 Intrusion Classification

The 5 features output from the DBN where pass to the SVM classifier to be classified. SVM is a classification technique based on Statistical Learning Theory (SLT). It is based on the idea of a hyper plane classifier, where it first maps the input vector into a higher dimensional feature space and then obtains the optimal separating hyper plane. The goal of SVM is to find a decision boundary (i.e. the separating hyper-plane) so that the margin of separation between the classes is maximized [4].

4 Experimental Results and Discussion

4.1 Dataset Characteristics

The data used in classification is NSL-KDD, which is a new dataset for the evaluation of researches in network intrusion detection system. NSL-KDD consists of selected records of the complete KDD'99 dataset [18]. NSL-KDD dataset solve the issues of KDD'99 benchmark [19]. Each NSL-KDD connection record contains 41 features (e.g., protocol type, service, and flag) and is labeled as either normal or an attack, with one specific attack type. The attacks fall into four classes:

- DoS e.g Neptune, Smurf, Pod and Teardrop.
- R2L: unauthorized access to local from a remote machine e.g Guess-password, Ftp-write, Imap and Phf.
- U2R: unauthorized access to root privileges e.g Buffer-overflow, Load-module, Perl and Spy.
- Probing eg. Port-sweep, IP-sweep, Nmap and Satan.

The NSL-KDD training set contains a total of 22 training attack types, with an additional 17 types in the testing set only.

4.2 DBN Structure

Deep Belief network has been used in two different ways, either as a dimensionality reduction method before applying SVM as a classifier or as a classifier by itself. Support Vector machine is a parameterized method, in this paper the default parameters of SVM has been used. The RBM training is considered as weights initializer. The number of RBM structures in the used DBN is two. The number of features are 41, 13 and 4 in the first, second and last layer in the DBN Network. The number of Gipps iteration is 150. Classification is applied on different training percentage.

4.3 Experiments and Analysis

The NSL- KDD dataset are taken to evaluate the proposed DBN-SVM intrusion detection scheme. All experiments have been performed using Intel Core 2 Duo 2.26 GHz processor with 2 GB of RAM and weka software [21].

4.3.1 Case 1: DBN vs. SVM vs. DBN-SVM Scheme

A comparison between SVM, DBN and the proposed DBN-SVM scheme is shown in Table 1. The classification accuracy achieved using DBN as dimensional reduction method before SVM is improved than using SVM or DBN as standalone classifier. Also the testing speed of DBN-SVM scheme is improved which is important for real time network applications. One of the conclusions that appear during experiment is that the accuracy starts to increase, when number of Gipps methods is 100 and reaches high performance then starts to deteriorate again.

Table 1 SVM, DBN AND THE HYRIDE DBN-SVM Scheme TESTING ACCURACY AND SPEED

Training percentage	SVM	DBN	DBN-SVM
20%	82.30 (10.4 Sec)	89.63 (0.31 Sec)	90.06 (2.54Sec)
30%	87.6 (10.4 Sec)	89.44 (0.26 Sec)	91.50 (2.54Sec)
40%	88.33 (16.67Sec)	89.54 (0.24 Sec)	92.84 (3.07 Sec)

4.3.2 Case 2: DBN as Feature Reduction Method vs. Different Feature Reduction Methods

We compared the DBN as a feature reduction method with other feature reduction methods like PCA, Gain Ratio and chi square. Using DBN, PCA, Gain Ratio and chi square the 41 features of the NSL- KDD dataset is reduced to 13 features. Table 2 gives the testing performance accuracy of the reduced data using SVM classifier. Table 2 illustrate that DBN gives better performance than the other reduction methods.

Table 2 Performance accuracy of DBN with different feature reduction methods

Training percentage	PCA	Gain-Ratio	Chi-Square	DBN
20%	68.72	65.83	66.0	90.06
30%	68.98	65.88	65.68	91.50
40%	71.01	70.99	65.82	92.84

5 Conclusion

Deep Belief network has proved a good addition to the field of network intrusion classification. In comparison with known classifier and feature reduction methods, DBN provides a good result as a separate classifier and as a feature reduction method. In this paper we proposed a hybrid DBN-SVM intrusion detection scheme, where DBN is used as a feature reduction method and SVM as a classifier. We examine the performance of the proposed DBN-SVM scheme by reducing the 41-dimensional of NSL-KDD dataset to approximately 87% of its original size and then classify the reduced data by SVM. The DBN-SVM scheme shows higher percentage of classification than SVM and enhances the testing time due to data dimensions reduction. Also, we compare the performance of the DBN as a feature reduction method with PCA, Gain Ratio and Chi-Square feature reduction method.

References

1. Shon, T., Moon, J.: A hybrid machine learning approach to network anomaly detection. Information Sciences 177, 3799–3821 (2007)
2. Anderson, J.P.: Computer security threat monitoring and surveillance.Technical Report, James P. Anderson Co., Fort Washington (April 1980)
3. Stallings, W.: Cryptography and network security principles and practices. Prentice Hall, USA (2006)
4. Tsai, C., Hsu, Y., Lin, C., Lin, W.: Intrusion detection by machine learning: A review. Expert Systems with Applications 36, 11994–12000 (2009)
5. Biermann, E., Cloete, E., Venter, L.M.: A comparison of intrusion detection Systems. Computer and Security 20, 676–683 (2001)
6. Verwoerd, T., Hunt, R.: Intrusion detection techniques and approaches. Computer Communications 25, 1356–1365 (2002)
7. Ilgun, K., Kemmerer, R.A., Porras, P.A.: State transition analysis:A rule-based intrusion detection approach. IEEE Trans. Software Eng. 21, 181–199 (1995)
8. Marchette, D.: A statistical method for profiling network traffic. In: proceedings of the First USENIX Workshop on Intrusion Detection and Network Monitoring (Santa Clara), CA, pp. 119–128 (1999)
9. Mukkamala, S., Janoski, G., Sung, A.: Intrusion detection: support vector machines and neural networks. In: Proceedings of the IEEE International Joint Conference on Neural Networks (ANNIE), St. Louis, MO, pp. 1702–1707 (2002)
10. Lundin, E., Jonsson, E.: Anomaly-based intrusion detection: privacy concerns and other problems. Computer Networks 34, 623–640 (2002)
11. Wu, S., Banzhaf, W.: The use of computational intelligence in intrusion detection systems: A review. Applied Soft Computing 10, 1–35 (2010)
12. Mohamed, A.R., Dahl, G., Hinton, G.E.: Deep belief networks for phone recognition. In: NIPS 22 Workshop on Deep Learning for Speech Recognition (2009)
13. Hinton, G.E.: A fast learning algorithm for deep belief nets. Neural Computation 18, 1527–1554 (2006)
14. Noulas, A.K., Krse, B.J.A.: Deep Belief Networks for Dimensionality Reduction. In:Belgian-Dutch Conference on Artificial Intelligence, Netherland (2008)
15. Larochelle, H., Bengio, Y.: Classification using discriminative restricted boltzmann machines. In: Proceedings of the 25th International Conference on Machine learning, vol. 307, pp. 536–543 (2008)
16. McAfee, L.: Document Classification using Deep Belief Nets, CS224n, Sprint (2008)
17. Larochelle, H., Bengio, Y., Louradour, J., Lamblin, P.: Exploring Strategies for Training Deep Neural Networks. Journal of Machine Learning Research 10, 1–40 (2009)

18. Cohen, I., Tian, Q., Zhou, X.S., Huang, T.S.: Feature Selection Using Principal Feature Analysis. In: Proceedings of the 15th International Conference on Multimedia, Augsburg, Germany, September 25-29 (2007)
19. KDD 1999 dataset Irvine, CA, USA (July 2010), http://kdd.ics.uci.edu/databases
20. Tavallaee, M., Bagheri, E., Lu, W., Ghorbani, A.A.: A Detailed Analysis of the KDD CUP 99 Data Set. In: Proceeding of the IEEE Symposium on Computational Intelligence in security and defense application, CISDA (2009)
21. Weka. Data Mining Software in java, http://www.cs.waikato.ac.nz/ml/weka/

Multi-Agent Association Rules Mining in Distributed Databases

Walid Adly Atteya, Keshav Dahal, and M. Alamgir Hossain

Abstract. In this paper, we present a collaborative multi-agent based system for mining association rules from distributed databases. The proposed model is based on cooperative agents and is compliant to the Foundation for Intelligent Physical Agents standard. This model combines different types of technologies, namely the association rules as a data mining technique and the multi-agent systems to build a model that can operate on distributed databases rather than working on a centralized database only. The autonomous and the social abilities of the model agents provided the ability to operate cooperatively with each other and with other different external agents, thus offering a generic platform and a basic infrastructure that can deal with other data mining techniques. The platform has been compared with the traditional association rules algorithms and has proved to be more efficient and more scalable.

Keywords: Multi-Agent Systems, Distributed Data Mining, Association Rules.

1 Introduction

Nowadays, both Data Mining Technology and Agent Technology have reached an acceptable level of maturity, and each one alone has its own scope and applicability [5]. Integration between the two technologies has been proposed to combine the

Walid Adly Atteya
School of Computing, Informatics and Media, Bradford University, United Kingdom
e-mail: waaabdo@bradford.ac.uk

Keshav Dahal
School of Computing, Informatics and Media, Bradford University, United Kingdom
e-mail: k.p.dahal@bradford.ac.uk

M. Alamgir Hossain
School of Computing, Informatics and Media, Bradford University, United Kingdom
e-mail: M.A.Hossain1@bradford.ac.uk

A. Gaspar-Cunha et al. (Eds.): Soft Computing in Industrial Applications, AISC 96, pp. 305–314.
springerlink.com

benefits of both worlds and to help computer scientists for building more sophisticated software systems. In other words, the discovered knowledge nuggets may constitute the building blocks of agent intelligence.

Due to the overwhelming amount of data and the complexity of data mining algorithms, researchers have tried to adopt distributed or parallel approaches [6]. There are many reasons for developing these approaches, one of which is that in most of the cases the data itself is distributed in different sites. Loading these data into a centralized location for mining interesting rules is not a good approach as it violates common issues like data privacy and it imposes network overhead. This will not be acceptable especially in real time problems. The other reason for developing distributed approaches is the increased complexity of the algorithms which made the researches present other ways to divide the data into independent subsets and apply the required algorithm on each subset and then combine their partial results.

Some researchers have presented various parallel data mining algorithms for association rules [21]. The problem of applying such approaches is scalability especially when number of distributed data sources is large and heterogeneous. In this context, we believe that a general Distributed Data Mining framework based on multi-agent systems can enable and support and accelerate the deployment of solutions that can solve the distributed computational problems.

Several algorithms have been proposed by many researchers based on multi-agent systems for context based Distributed Data Mining [11][14]. A few researchers have presented multi-agent systems for mining association rules as a data mining technique [20], although even this technique do not comply with global standards like the Foundation for Intelligent Physical Agents (FIPA) [10] or any other standards thus reducing the chance of integration and interoperability with other existing agent based systems. In this paper, we present a cooperative multi-agent system model for mining association rules from distributed databases. The model is based on our previously implemented association rules algorithm [7]. The previously implemented algorithm was proven to outperform Apriori algorithm in terms of performance and computational overhead for a considered case study of a centralized database. The reason we based our implementation on enhanced versions of Apriori like algorithm is that the main functionalities of these algorithms can be parallelized and allocated for agents in the MAS environment. Moreover, these algorithms scan the database once. Thus, improving the computational time and cost. Apriori like algorithms also have smaller computational complexity compared to other algorithms. The agent model developed in this paper complies with FIPA global standards in communication between agents, thus enabling the ability to cooperate with other standard agents also the future extension for the proposed model.

The rest of the paper is organized as follows. The next section explains a brief overview about basic techniques we are combining, namely, the distributed data mining, multi agent systems and FIPA as a global standard for agents communication. Section (3) describes the motivation for implementing our model. Our proposed algorithm, the solution architecture and the message negotiation between agents are described in Section (4). Section (5) describes the model experiments and verification. Section (6) describes the conclusion and the future work.

2 Overview of Basic Techniques Used in the Model

2.1 Distributed Data Mining and Multi-Agent Systems (MAS)

The last decade has seen an ever increasing demand for Data mining techniques which can reveal the valuable knowledge hidden in data. However, these techniques often require high performance approaches in order to cope with the large amounts of rough data and the complexity of algorithms [6]. Many such approaches fall into the area of distributed systems, where a number of entities work together to cooperatively solve complex problems. Multi-Agent systems (MAS), emphasizes the joint behaviours of agents with some degree of autonomy and the complexities arising from their interactions [16]. This agent paradigm presents a new way for analysing data mining systems especially if the data sources are distributed in different nodes (Distributed Data Mining). The purpose is to combine different algorithms of data mining to prepare elements for decision makers, benefiting from the possibilities offered by the multi-agent systems [22]. Several researchers have attempted to provide a various and meaningful classification of the attributes that agents might have. Common agent attributes are Autonomy, Cooperation, Continuity, Reactivity, Proactiveness, Inferential capability, Adaptivity and Trust Worthy [18] [17]. MAS is an emerging subfield of distributed artificial intelligence, which aims at providing both: principles for the construction of complex systems involving multiple agents and mechanisms for the coordination of independent agents behaviour [4].

2.2 FIPA Agent Management Reference Model

FIPA is an IEEE Computer Society standards organization that promotes agent based technology and the interoperability of its standards with other technologies; moreover, it was officially accepted by the IEEE as its eleventh standards committee on 8 June 2005 [9]. FIPA has developed generic agent specifications. FIPA agent management reference model is the framework in which FIPA agents exist and operate. The main components of this framework are the Agent Platform (AP), the Directory Facilitator (DF), the Agent Management System (AMS) and the Message Transport Service (MTS). Agent Communication Languages (ACL) have been proposed based on the speech-act theory which is derived from the linguistic analysis of human communication [19]. Two most popular declarative agent languages are KQML (Knowledge Query and Manipulation Language) and FIPA ACL. KQML was conceived both as a message format and a message handling protocol to support runtime knowledge sharing among agents but the support for KQML has been discontinued in favour of FIPA ACL [3].

3 Motivation

Frequent pattern mining plays an important role in several Data Mining tasks including association rules [1] [2]. In our previous work [7], we presented an enhanced algorithm for mining association rules from large databases. This algorithm was proved to have a better performance over the Apriori Algorithm. The major drawback of Apriori is that it leads to Input/Output overhead, which reduces the performance of the algorithm. The interesting feature of the previously proposed algorithm is that it scans the database once at k=1. At the following iterations, the database is not used for counting the support as it uses a certain data structure to calculate the candidate itemsets on the fly from the previous iterations. An important result of the enhanced approach was the achievement of the I/O improvement over Apriori Algorithm, which was one of the best previous algorithms. Rules were generated using both algorithms. The experiments showed how the enhanced algorithm outperformed the Apriori Algorithm.

The following is a brief description for the previously proposed Algorithm:

1. Initially, the algorithm finds the frequent 1-itemsets from the database.

2. The supports of these itemsets are checked. If the itemset support is greater or equal to the minimum support threshold given by the user, it is counted as large itemset, otherwise it is not taken in consideration at the next iteration.

3. The algorithm loops while there are existing candidate itemsets and at every iteration the following steps occur:

3.1 The algorithm generates on the fly from the previous iteration the candidate frequent k-itemsets for the next phase.

3.2 The support of these itemsets is incremented.

3.3 If the support of the candidate itemsets is greater than the minimum support, these itemsets are considered as frequent itemsets

4. When there are no more candidate itemsets, the list of final frequent itemsets is then presented to the user.

The problem of the previously proposed algorithm is that it can be applied on centralized database only and not on distributed databases. This is because some of the required variables and calculations like counting the itemsets support and generating the candidate itemsets cannot be calculated at each local site, thus comes the need for a main agent with a global scope to accept some information from local agents, do some calculations then return the results to all the local agents at sites.

4 Multi-Agent Based Enhanced Algorithm

This paper proposes reconstructing the previously proposed algorithm using the multi-agent systems (MAS). The main advantage of this model is the ability to apply association rules on distributed databases utilizing the benefit from the multi agent systems flexibility. The agents can be easily modified or reconstructed to change an existing methodology without the need for detailed rewriting for the application. Moreover, this model is based upon the notion of interacting agents to achieve a certain level of automation between agents [13]. We provide to this platform the ability to operate automatically thanks to autonomous and intelligent agents. This new methodology combines the powerful capabilities of the agent approach with the powerful capabilities of data mining techniques in order to discover the hidden association rules inside the huge number of distributed data. The model had taken in consideration a number of fundamental tools as proposed by Jennings [13] for helping to manage the algorithms complexity like decomposition, abstraction and organization. Our development framework acts as an integrated GUI based environment that facilitates the design process of a data mining system [22].

4.1 Model Compliance to Global Standards

Any implemented MAS model should comply to a standard to ensure that the model agents can interact and cooperate with each other and with other systems. We have investigated the two main standards for the MAS architectures which are the FIPA (Foundation for Intelligent Physical Agents) [8] and the OMG MASIF [15]. Our model complies with the FIPA standard and complies with its communication language namely the FIPA-ACL (Agent Communication Language). The reason for choosing FIPA standard is that while OMG MASIF does not define any standard about communication between agents, FIPA , on the other side have became a strong standard in MAS development due its richness in not only the agent management, but also language specifications, conversations, personal assistants, etc [12].

4.2 Solution Architecture

The model shown in Fig. 1 consists of three types of cooperative agents that work together to achieve the required goals. The first kind of agents is the ***Interface Agent.*** This agent can cooperate with the human and accepts the user required support. This agent sends a message containing the value of the support to the second agent which is the ***Main Agent*** , which in turn sends it to the third type of agents namely the ***Local Agent*** . Each ***Local Agent*** calculates the local support of the 1-itemsets according to the number of data records in his local database and sends the results back to the ***Main Agent*** . The ***Main Agent*** has a global view of all ***Local Agents*** . When this agent receives all messages from all ***Local Agents*** , it starts summing up all itemsets local supports from all agents and compares it with the minimum support supplied by the ***Interface Agent*** . If the summation of the local supports for an itemset is greater than or equal to the minimum support threshold, this itemset is counted as a large itemset and is placed at the central site, else this itemset is considered as small and will not be taken in consideration at the next iteration. The ***Main Agent*** also calculates the candidate k+1 itemsets from the set of large previous k itemsets, and send a message to all ***Local Agents*** containing the small itemsets that are to be pruned from each local site and the k+1 candidate itemsets. Each ***Local Agent*** eliminates the received small itemsets from its local site and uses the received candidate itemsets to start calculating the support. These cooperative messages between the ***Main Agent*** and the ***Local Agent*** continue until no more candidate itemsets are generated by the ***Main Agent*** . Finally, the ***Main Agent*** sends the set of frequent itemsets back to the ***Interface Agent*** which is responsible for presenting the results to the user using graph outputs.

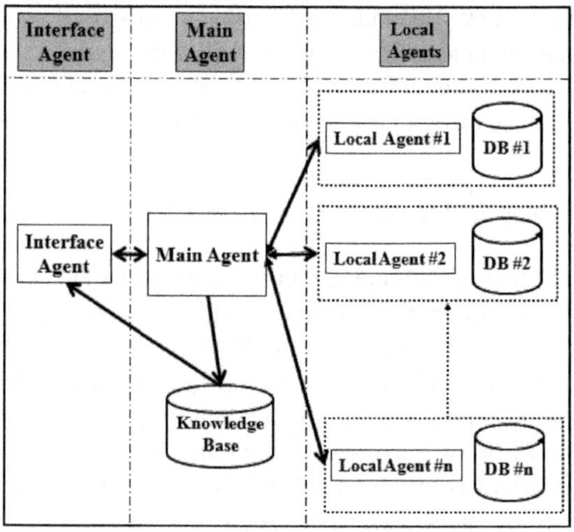

Fig. 1 Solution architecture for the proposed multi-agent based model

4.3 Proposed MAS Algorithm and FIPA Messages between Agents

The proposed algorithm is described as follows:

1. The Interface Agent accepts the support from the user.
2. The Interface Agent sends the required support to the main agent.
3. Main Agent sends a "propose performative" FIPA message to Local Agents:
   (**Propose**
   :sender (agent-identifier :name main_agent)
   :receiver (set (agent-identifier :name local_agent))
   :content "Start mining with support = minsupp"
   :reply-with start_mining_proposal)
4. Local Agents reply with an "agree performative" to Main Agent as follows:
   (**Agree**
   :sender (agent-identifier :name local_agent)
   :receiver (set (agent-identifier :name main_agent))
   :content "proposal approved and mining started at k=1"
   :in-reply-to start_mining_proposal)
5. Each Local Agent starts counting the local supports for all 1- candidate itemsets in its local database according to its local number of records.
6. Local Agent replies with "inform performative" to Main Agent as follows:
   (**Inform**
   :sender (agent-identifier :name local_agent)
   :receiver (set (agent-identifier :name main_agent))
   :content "finished counting candidate 1-itemsets")

7. Main Agent compares the summation of the local supports sent from all agents for 1-candidate itemsets with the min support supplied by the user.
8. Main Agent finds the 1-large itemsets and save it in the database in the list of frequent itemsets.
9. Main Agent sends an "Inform performative" FIPA message to Local Agents:
(**Inform**
:sender (agent-identifier :name main_agent)
:receiver (set (agent-identifier :name local_agent))
:content "frequent itemsets at k=1 are successfully generated")
10. Main Agent generates the k-candidate itemsets
11. Main Agent sends the generated k-candidate itemsets to all local agents.
12. Main Agent sends a "Request performative" FIPA message to all Local Agents:
(**Request**
:sender (agent-identifier :name main_agent)
:receiver (set (agent-identifier :name local_agent))
:content "candidates are generated at iteration K, please count the support"
:reply-with iteration_k)
13. Each Local Agent calculates the k-candidate itemsets in its local databases
14. Local Agents send an "Inform performative" FIPA message to Main Agent:
(**Inform**
:sender (agent-identifier :name local_agent)
:receiver (set (agent-identifier :name main_agent))
:content "finished counting candidate itemsets for the current iteration K"
:in-reply-to iteration_k)
15. The Main Agent considers any k-candidate itemset as frequent if the summation of all local supports for this itemset from all local agents is greater than the min global support
16. Frequent itemsets are saved in the central database in the list of k-frequent itemsets while small itemsets are not considered in the next iteration.
17. Steps (10) to (16) are iterative and finish when there are no more k+1 candidate itemsets.
18. Main Agent sends an "Inform performative" message to all Local Agent :
(**Inform**
:sender (agent-identifier :name main_agent)
:receiver (set (agent-identifier :name local_agent))
:content "Finished mining of frequent itemsets")
19. Main Agent sends all frequent itemsets to Interface Agent for representation.

5 Model Experiments and Verification

Experiments using the proposed multi-agent based algorithm were done on medical data to investigate the functional and the non functional requirements (sometimes referred to as system correctness and completeness) of the proposed model. Experiments using the proposed multi-agent based algorithm and the previously proposed algorithm were repeated three times against different database sizes. Results of this comparison are explained as follows:

5.1 *Satisfying the Non Functional Requirements*

The previously proposed algorithm and the proposed multi-agent based algorithm were tested on medical databases with different sizes. The data used contains patients symptoms related to the diseases Inflammation of urinary bladder and Nephritis of renal pelvis origin. Fig. 2 shows that the proposed multi-agent based algorithm outperformed the previously proposed algorithm. The reason for this increase in performance is that the support count and the candidate generation processes which are the most time consuming processes in Apriori like algorithms are parallelized and distributed among the model cooperative agents. This parallelism results in the decrease in time required to accomplish these processes.

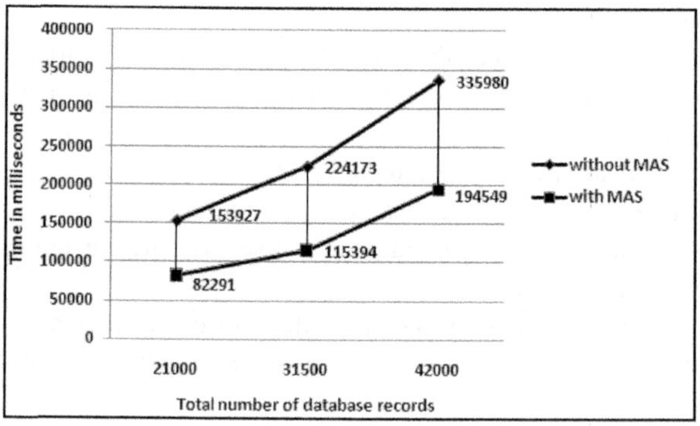

Fig. 2 A time comparison between the two implementations on different database sizes

From Fig. 2 we can deduce that the time needed for mining frequent itemsets using the proposed multi-agent based algorithm is less than that needed using traditional Apriori like algorithms. Moreover, the time difference between the two implementations increases when the number of database records increases.

5.2 *Satisfying the Functional Requirements*

5.2.1 Generating the Same Frequent Itemsets

The proposed multi-agent based algorithm was found to satisfy the functional requirements (functions, services and tasks that the system or system components must be able to perform) of the previously proposed algorithm, in which the proposed multi-agent based algorithm generated the same exact frequent itemsets as those generated by the Apriori like algorithms.

5.2.2 Satisfying the MAS Features

Moreover, the agent in the model satisfies the MAS agent features as follows:

1. Autonomy: All agents in the model satisfy the MAS autonomous feature, as all agents either main or local agents operate without the direct intervention of humans, and have some kind of control over their actions and internal state.

2. Sociable ability: All agents in this model interact with each other through a negotiation mechanism, thus our model satisfies the cooperative feature of MAS.

3. Continuous: Agents are continuously running processes, thus the continuous feature is satisfied in our model.

4. Reactivity: Our model agents perceive their environment and respond to changes that occur in it, for instance the main agent accepts the itemsets support from the local agents in each site, then generates the candidate itemsets and send them back to the local agents, thus our model satisfies the reactiveness feature.

5. Proactiveness: Our agents have a goal directed behaviour which is generating the frequent itemsets according to a specified minimum threshold.

6. Trustworthy: Model agents are trustworthy as the experiments have proved that the frequent itemsets generated are exactly the same like those generated by the algorithm proposed in[7] . The correct and exact frequent itemsets generated in different database size gives a trust to these agents.

6 Conclusion

This paper has presented a model gathering a number of techniques, namely, the distributed data mining, multi-agents and association rules. The objective of this paper is to extend the association rules techniques to be applied on distributed data bases. The model is an improvement for the previously proposed and implemented association rules algorithm [7]. The main achievement of the model is the compliance with global standards namely FIPA thus having the ability to interact with other standard external agents. Another important feature is the model flexibility, since agents can be added to, modified and reconstructed, without the need for detailed application rewriting. The proposed model was applied and verified on medical data and was proved to outperform the previously implemented association rules algorithm. The reason for this increase in performance is that the support count and the candidate generation processes which are the most time consuming processes in Apriori like algorithms are parallelized and distributed among the model cooperative agents. This parallelism results in the decrease in time required to accomplish these processes.

References

1. Agrawal, R., Imieliński, T., Swami, A.: Mining association rules between sets of items in large databases. ACM SIGMOD Record 22(2), 207–216 (1993)
2. Agrawal, R., Srikant, R.: Fast algorithms for mining association rules in large databases. In: Proceedings of the 20th International Conference on Very Large Data Bases, pp. 487–499. Morgan Kaufmann Publishers Inc., San Francisco (1994)
3. Ahmad, H.: Multi-agent systems: overview of a new paradigm for distributed systems. In: Proceedings of 7th IEEE International Symposium, pp. 101–107. IEEE, Los Alamitos (2003)
4. Alhajj, R., Kaya, M.: Multiagent association rules mining in cooperative learning systems. In: Advanced Data Mining and Applications, pp. 75–87 (2005)
5. Cao, L.: Data Mining and Multi-agent Integration. Springer, Heidelberg (2009)
6. Di Fatta, G., Fortino, G.: A customizable multi-agent system for distributed data mining. In: Proceedings of the 2007 ACM symposium on Applied computing, pp. 42–47. ACM Press, New York (2007)
7. Fakhry, M., Atteya, W.A.: An Enhanced Algorithm for Mining Association Rules. In: First International Conference on Intelligent Computing and Information Systems (2002)
8. FIPA. FIPA Abstract Architecture Specification, Technical Report, SC00001L (2002), `http://www.fipa.org/`
9. FIPA. FIPA Specification (2002), `http://www.fipa.org/`
10. FIPA. FIPA Agent Management Specification Technical Report, SC00023k (2004), `http://www.fipa.org/specs/fipa00023/SC00023K.html`
11. Fortino, G., Russo, W., Frattolillo, F., Zimeo, E.: Mobile Active Object for Highly Dynamic Distributed Computing. In: ipdps, p. 118. IEEE Computer Society, Los Alamitos (2002)
12. González, E., Hamilton, A.: Hamilton: Software experience when using ontologies in a multi-agent system for automated planning and scheduling. Software-Practice and Experience 36, 667–688 (2006)
13. Jennings, N.: An agent-based approach for building complex software systems. Communications of the ACM 44(4), 35–41 (2001)
14. Luck, M., McBurney, P., Preist, C.: Agent technology: Enabling next generation computing. Citeseer (2003)
15. OMGA. Technical Report agent/00-09-01 (2000), `http://www.omg.org/mda/`
16. Panait, L., Luke, S.: Cooperative multi-agent learning: The state of the art. Autonomous Agents and Multi-Agent Systems 11(3), 387–434 (2005)
17. Regli, W.: Development and specification of a reference model for agent-based systems. IEEE Transactions on Systems, Man, and Cybernetics 39(5), 572–596 (2009)
18. Russell, S., Norvig, P.: Artificial intelligence: a modern approach. Prentice-Hall, Englewood Cliffs (2009)
19. Smith, B.: John Searle: From speech acts to social reality. John Searle, pp. 1–33 (2003)
20. Wang, F., Helian: A distributed and mobile data mining system. In: Proceedings of the Fourth International Conference on Parallel and Distributed Computing, Applications and Technologies, pp. 916–918. IEEE, Los Alamitos (2003)
21. Zaki, M.: Parallel and distributed association mining: A survey. In: Concurrency, IEEE, Los Alamitos (2002)
22. Zghal, H., Faiz, S., Ghezala, H.: A framework for data mining based multi-agent: An application to spatial data. World Academy of Science, Engineering and Technology (2005)

Part VI: Soft Computing for Pattern Recognition

A Novel Initialization for Quantum Evolutionary Algorithms Based on Spatial Correlation in Images for Fractal Image Compression

Mohammad H. Tayarani N., Adam Prugel Bennett, Majid Beheshti, and Jamshid Sabet

Abstract. Quantum Evolutionary Algorithm (QEA) is a novel optimization algorithm proposed for class of combinatorial optimization problems. While Fractal Image Compression problem is considered as a combinatorial problem, QEA is not widely used in this problem yet. Using the spatial correlation between the neighbouring blocks, this paper proposes a novel initialization method for QEA. In the proposed method the information gathered from the previous searches for the neighbour blocks is used in the initialization step of search process of range blocks. Then QEA starts searching the search space to find the best matching domain block. The proposed algorithm is tested on several images for several dimensions and the experimental results shows better performance for the proposed algorithm than QEA and GA. In comparison with the full search algorithm, the proposed algorithm reaches comparable results with much less computational complexity.

1 Introduction

Fractal Image Compression, proposed by Barnsley has, recently become one of the most promising encoding technologies in the generation of image compression [1]. The high compression ratio and the quality of the retrieved images attract many of researchers, but the high computational complexity of the algorithm is its main drawback. One way of decreasing the time complexity is to move from full search method to some optimization algorithms like Genetic Algorithms. From this point of view, several works try to improve the performance of fractal image compression algorithms using Genetic algorithm. In [2] a new method for finding the IFS

Mohammad H. Tayarani N.
University of Southampton
e-mail: mhtn1g09@ece.soton.ac.uk

Adam Prugel Bennett
University of Southampton
e-mail: apb@ecs.soton.ac.uk

A. Gaspar-Cunha et al. (Eds.): Soft Computing in Industrial Applications, AISC 96, pp. 317–325.
springerlink.com

code of fractal image is developed and the influence of mutation and the crossover is discussed. The low speed of fractal image compression blocks its way to practical application. In [3] a genetic algorithm approach is used to improve the speed of searching process in fractal image compression. A new method for genetic fractal image compression based on an elitist model in proposed in [4]. In the proposed approach the search space for finding the best self similarity is greatly decreased. Reference [5] makes an improvement on the fractal image coding algorithm by applying genetic algorithm. Many researches increase the speed of fractal image compression but the quality of the image will decrease. In [6] the speed of fractal image compression is improved without significant loss of image quality. Reference [7] proposes a genetic algorithm approach which increases the speed of the fractal image compression without decreasing of the quality of the image. In the proposed approach a standard Barnsley algorithm, the Y. Fisher based in classification and the genetic compression algorithm with quad-tree partitioning are compared. In GA based algorithm a population of transformations is evolved for each range block. In order to prevent the premature convergence of GA in fractal image compression a new approach is proposed in [8], which controls the parameters of GA adaptively. A spatial correlation genetic algorithm is proposed in [9], which speeds up the fractal image compression algorithm. In the proposed algorithm there are two stages, first the spatial correlations in image for both the domain pool and the range pool is performed to exploit local optima. In the second stage if the local optima were not certifiable, the whole of image is searched to find the best self similarity. A schema genetic algorithm for fractal image compression is proposed in [10] to find the best self similarity in fractal image compression.

Using spatial correlation between the neighbor range and domain blocks this paper proposes a novel initialization method for QEA. In the proposed method, based on the information gathered from the search process for the neighbor range blocks, the q-individuals are initialized to represent the better parts of the search space with higher probability. Performing this new method the q-individuals have more chance to find better solutions in less time. The proposed algorithm is tested on several images and experimental results shows better performance for the proposed algorithm than GA and original form of QEA.

The rest of the paper is organized as follows. Section 2 introduces QEA, in section 3 the new algorithm is proposed and in section 4 is experimented on several images and finally section 5 concludes the paper.

2 Quantum Evolutionary Algorithm

QEA is inspired from the principles of quantum computation, and its superposition of states is based on qubits, the smallest unit of information stored in a two-state quantum computer. A qubit could be either in state "0" or "1", or in any superposition of the two as described below:

$$|\psi\rangle = \alpha|0\rangle + \beta|1\rangle \tag{1}$$

Where α and β are complex numbers, which denote the corresponding state appearance probability, following below constraint:

$$|\alpha|^2 + |\beta|^2 = 1 \tag{2}$$

This probabilistic representation implies that if there is a system of m qubits, the system can represent 2^m states simultaneously. At each observation, a qubits quantum state collapses to a single state as determined by its corresponding probabilities.

Consider $i-th$ individual in $t-th$ generation defined as an m-qubit as below:

$$\begin{bmatrix} \alpha_{i1}^t \; \alpha_{i2}^t \cdots \alpha_{ij}^t \cdots \alpha_{im}^t \\ \beta_{i1}^t \; \beta_{i2}^t \cdots \beta_{ij}^t \cdots \beta_{im}^t \end{bmatrix} \tag{3}$$

Where $\left|\alpha_{ij}^t\right|^2 + \left|\beta_{ij}^t\right|^2 = 1$, $j = 1,2,,m,m$ is the number of qubits, i.e., the string length of the qubit individual, $i = 1,2,,n,n$ is the number of possible solution in population and t is generation number of the evolution. If there is, for instance, a three-qubits ($m = 3$) individual such as 4:

$$q_i^t = \begin{bmatrix} \frac{1}{\sqrt{2}} & \frac{1}{\sqrt{3}} & \frac{1}{2} \\ \frac{1}{\sqrt{2}} & \frac{\sqrt{2}}{\sqrt{3}} & \frac{\sqrt{3}}{2} \end{bmatrix} \tag{4}$$

Or alternatively, the possible states of the individual can be represented as:

$$q_i^t = \frac{1}{2\sqrt{6}}|000\rangle + \frac{1}{2\sqrt{2}}|001\rangle + \frac{1}{2\sqrt{3}}|010\rangle + \frac{1}{2}|011\rangle +$$

$$\frac{1}{2\sqrt{6}}|100\rangle + \frac{1}{2\sqrt{2}}|101\rangle + \frac{1}{2\sqrt{3}}|110\rangle + \frac{1}{2}|111\rangle \tag{5}$$

In QEA, only one qubit individual such as 4 is enough to represent eight states, whereas in classical representation eight individuals are needed. Additionally, along with the convergence of the quantum individuals, the diversity will gradually fade away and the algorithm converges.

2.1 QEA Structure

In the initialization step of QEA, $[\alpha_{ij}^t \;\; \beta_{ij}^t]^T$ of all q_i^0 are initialized with $\frac{1}{\sqrt{2}}$. This implies that each qubit individual q_i^0 represents the linear superposition of all possible states with equal probability. The next step makes a set of binary instants; x_i^t by observing $Q(t) = \{q_1^t, q_2^t, ..., q_n^t\}$ states, where $X(t) = \{x_1^t, x_2^t, ..., x_n^t\}$ at generation t is a random instant of qubit population. Each binary instant, x_i^t of length m, is formed by selecting each bit using the probability of qubit, either $|\alpha_{ij}^t|$ or $|\beta_{ij}^t|$ of q_i^t. Each instant x_i^t is evaluated to give some measure of its fitness. The initial best solution $b = max_{i=1}^n\{f(x_i^t)\}$ is then selected and stored from among the binary instants of $X(t)$. Then, in 'update' $Q(t)$, quantum gates U update this set of qubit individuals

$Q(t)$ as discussed below. This process is repeated in a while loop until convergence is achieved. The appropriate quantum gate is usually designed in accordance with problems under consideration.

2.2 Quantum Gates Assignment

The common mutation is a random disturbance of each individual, promoting exploration while also slowing convergence. Here, the quantum bit representation can be simply interpreted as a biased mutation operator. Therefore, the current best individual can be used to steer the direction of this mutation operator, which will speed up the convergence. The evolutionary process of quantum individual is completed through the step of "update $Q(t)$". A crossover operator, quantum rotation gate, is described below. Specifically, a qubit individual q_i^t is updated by using the rotation gate $U(\theta)$ in this algorithm. The $j-th$ qubit value of $i-th$ quantum individual in generation t, $[\alpha_{ij}^t \;\; \beta_{ij}^t]^T$ is updated as:

$$\begin{bmatrix} \alpha_{ij}^t \\ \beta_{ij}^t \end{bmatrix} = \begin{bmatrix} cos(\Delta\theta) & -sin(\Delta\theta) \\ sin(\Delta\theta) & cos(\Delta\theta) \end{bmatrix} \begin{bmatrix} \alpha_{ij}^{t-1} \\ \beta_{ij}^{t-1} \end{bmatrix} \tag{6}$$

Where $\Delta\theta$ is rotation angle and controls the speed of convergence and determined from Table 1. Reference [11] shows that these values for $\Delta\theta$ have better performance.

Table 1 Lookup Table of $\Delta\theta$, the rotation angle.

x_i	b_i	$f(x) \geq f(b)$	$\Delta\theta$
0	0	false	0
0	0	true	0
0	1	false	0.01π
0	1	true	0
1	0	false	0.01π
1	0	true	0
1	1	false	0
1	1	true	0

3 Proposed Method

Statistical studies on fractal image compression problems show for the range blocks, the potential quasi-affine matched domain centralized around its vicinity [1]. It means that the neighbor range blocks have similar fractal codes. Using this idea, [1] proposes a two stage fractal image coding method. The first stage searches around the best matched domain blocks of neighbor range blocks. If the result is not satisfying, the search process starts to find a good domain block. Inspiring this idea, this paper proposes a novel initialization method for Quantum Evolutionary Algorithm in solving fractal image compression problem. In solving fractal image compression problem using genetic algorithm, for each range block, GA searches among the domain pool to find the best matched domain block [10]. Basically the search process for each range block is performed independently and the results of the previous searches are not used in future searches for other range blocks. This paper proposes a novel method which uses the information from previous searches to help the evolutionary algorithm finding better solutions. In the proposed method, based on the information gathered from the previous searches for neighbor range blocks, the q-individuals are initialized to represent the better parts of the search space with higher probability. The new algorithm is proposed as follows:

```
The Proposed Algorithm
begin
    t = 0
1.  Initialize Q(0) based on H^b, H^m and H^w
1.  Make X^0 by observing the states of Q(0)
3.  Evaluate X(0)
4.  Store X(0) into B(0). Store the best solution among X(0) into b
5.  While not termination condition do
    begin
        t=t+1
6.      Make X^t by observing the states of Q(t-1)
7.      Evaluate X(t)
8.      Update Q(t) using Q-Gates
9.      Store the best solutions among B(t-1) and X(t) into B(t)
10.     Store the best solution among B(t) into b
        end
end
```

QEA has a population of quantum individuals $Q(t) = \{q_1^t, q_2^t, ..., q_n^t\}$, where t is generation step and n is the size of population.

A comprehensive description of QEA can be found in [11]. The QEA procedure is described as:

1. In QEA the possible solutions are initialized with the values of $\frac{1}{\sqrt{2}}$ to represent the whole search space with the same probability. Giving these values to q-gates means a complete random initialization. Random initialization works well when we do not have previous information about the search space. Knowing where to search, helps QEA searching better parts of the search space and finding better solutions with less searching time. Here we propose a novel initialization approach using the information from previous searches. In the proposed method the best, worst and median solutions of each iteration in the search process for each range block are

stored in $\mathcal{H}^b$, $\mathcal{H}^m$ and $\mathcal{H}^w$ respectively. Where $\mathcal{H}$ is the history of X, containing the best, median and worst possible solutions in previous searches. The initialization step in the proposed algorithm is performed as follows:

$$\mathcal{B}_{l,i} = \frac{1}{T}\sum_{j=1}^{T} \mathcal{H}^b_{l,ij}, \quad \mathcal{M}_{l,i} = \frac{1}{T}\sum_{j=1}^{T} \mathcal{H}^m_{l,ij}, \quad \mathcal{W}_{l,i} = \frac{1}{T}\sum_{j=1}^{T} \mathcal{H}^w_{l,ij} \tag{7}$$

Where $\mathcal{H}^b$, $\mathcal{H}^m$ and $\mathcal{H}^w$ are the history of the best, median and worst possible solutions respectively, $\mathcal{H}^b_{l,ij}$ is the $i-th$ bit of $j-th$ possible solution in $\mathcal{H}^b$, and the index l shows the $l-th$ neighbor range block, $\mathcal{B}_{l,i}$ is the average of $i-th$ bit among all the possible solutions in $\mathcal{H}^b$ of $l-th$ neighbor range block. Here $l = 1,2,...,8$ shows the $l-th$ neighbor of the range block. Each range block has 8 neighbor range blocks. Here $\mathcal{B}_{l,i}$, $\mathcal{M}_{l,i}$ and $\mathcal{W}_{l,i}$ contain the percentage of ones in the best, median and worst possible solutions during the previous searches respectively. The value of $\mathcal{B}_i$ shows the importance of the $i-th$ bit in possible solutions, if $\mathcal{B}_i$ is near 1, it means that in most of better possible solutions, this bit has the value of 1, and it is better to give a value to this q-bit which represents 1 with more probability. Analogously, if $\mathcal{W}_i$ is near 1, it means that in most of worse possible solutions, this bit is 1, therefore it is better to give a value of to this q-bit which represents 0 with more probability. Accordingly, this paper proposes the following method for reinitialization step:

$$\theta^t_{ij} = \frac{\pi}{4} + [2\mathcal{B}_i + \mathcal{M}_i - 3\mathcal{W}_i] \times \frac{\pi}{16} \tag{8}$$

Where $j = 1,2,...,l$, l is the number of neighbor blocks, $i = 1,2,,m$, m is the number of q-bits in the q-individuals, i. e. the dimension of the problem. In order to prevent the q-individuals getting stuck in local optima of previous searches the other q-individuals are initialized randomly:

$$\theta^t_{ij} = \frac{\pi}{4} \tag{9}$$

For $j = l+1, l+2, ..., n$ where n is the size of the population.

The proposed initialization operator gathers information from the previous searches and initializes the q-individuals with the values representing better parts of search space.

2. In this step the binary solutions $X(0) = \{x^0_1, x^0_2, ..., x^0_n\}$ at generation $t = 0$ are created by observing $Q(0)$. Observing x^t_{ij} from qubit $[\alpha^t_{ij} \quad \beta^t_{ij}]^T$ is performed as below:

$$x^t_{ij} = \begin{cases} 0 & if\ U(0,1) < |\alpha^t_{ij}|^2 \\ 1 & otherwise \end{cases} \tag{10}$$

Where $U(.,.)$, is a uniform random number generator.

3. All solutions in $X(t)$ are evaluated with fitness function.

4. Store $X(0)$ into $B(0)$. Select best solution among $X(0)$ and store it to b.

5. The while loop is running until termination condition is satisfied. Termination condition can be considered as maximum generation condition or convergence condition.

6. Observing $X(t)$ from $Q(t-1)$.

7. Evaluate $X(t)$ by fitness function.

8. Update $Q(t)$.

9, 10. Store the best solutions among $B(t-1)$ and $X(t)$ to $B(t)$. If the fittest solution among $B(t)$ is fitter than b then store the best solution into b.

3.1 Coding

In the proposed algorithm for each range block, QEA searches among all the domain pool to find the best match domain block. The domain blocks are coded by their horizontal and vertical address in the image. Therefore a solution is a binary string having 3 parts, p_x, p_y, p_T, representing the horizontal and vertical location of domain block in the image and the transformation respectively. The length of the possible solution for a $M \times N$ image is calculated as follows:

$$m = \lceil log_2(M) \rceil + \lceil log_2(N) \rceil + 3 \tag{11}$$

Where m is the size of the possible solutions. Here 8 ordinary transformation are considered: rotate $0°, 90°, 180°, 270°$, flip vertically, horizontally, flip relative to $45°$, and relative to $135°$.

4 Experimental Results

This section experiments the proposed algorithm and compares the proposed algorithm with the performance of GA and original version of QEA in fractal image compression. The proposed algorithm is examined on images Lena, Pepper and Baboon with the size of 256×256 and gray scale. The size of range blocks is considered as 8×8 and the size of domain blocks is considered as 16×16. In order to compare the quality of results, the PSNR test is performed:

$$PSNR - 10 \times log \left(\frac{255^2}{\frac{1}{M \times N} \sum_{i=1}^{N} \sum_{j=1}^{M} (f(i,j) - g(i,j))^2} \right) \tag{12}$$

Where M×N is the size of image.

The crossover rate in GA is 0.8 and the probability of mutation is 0.003 for each allele. Table 2 shows the experimental results using proposed algorithm and GA. The number of iterations for GA, QEA and the proposed algorithm for all the experiments is 200. According to Table 2 the proposed algorithm improves the performance of fractal image compression for all the experimental results.

Table 2 Experimental results on Lena, Pepper, and Baboon

Picture	Method	Pop Size	MSE Computations	PSNR
Lena	Full Search	-	59,474,944	28.85
	QEA	30	6,144,000	28.49
		25	5,120,000	28.28
		20	4,096,000	28.95
		15	3,072,000	27.43
	Proposed Method	30	6,144,000	28.58
		25	5,120,000	28.39
		20	4,096,000	29.02
		15	3,072,000	27.56
	GA	30	6,144,000	28.11
		25	5,120,000	28.04
		20	4,096,000	27.55
		15	3,072,000	27.27
Pepper	Full Search	-	59,474,944	29.85
	QEA	30	6,144,000	29.55
		25	5,120,000	29.09
		20	4,096,000	28.87
		15	3,072,000	28.12
	Proposed Method	30	6,144,000	29.62
		25	5,120,000	29.28
		20	4,096,000	28.93
		15	3,072,000	28.51
	GA	30	6,144,000	29.14
		25	5,120,000	28.92
		20	4,096,000	28.64
		15	3,072,000	28.11
Baboon	Full Search	-	59,474,944	20.04
	QEA	30	6,144,000	19.28
		25	5,120,000	19.18
		20	4,096,000	18.95
		15	3,072,000	18.62
	Proposed Method	30	6,144,000	19.63
		25	5,120,000	19.25
		20	4,096,000	19.09
		15	3,072,000	18.77
	GA	30	6,144,000	19.17
		25	5,120,000	19.02
		20	4,096,000	18.65
		15	3,072,000	18.41

5 Conclusion

Using spatial correlation between the neighbor blocks in images, this paper proposes a novel initialization method for QEA in fractal image compression. In the proposed method, during the search process for each range block, some information about the best, worst and median possible solutions is gathered. Using this information in the initialization step in the search process of neighbor range blocks, this paper proposes a novel method in solving fractal image compression problem. Several experiments on Lena, Pepper, and Baboon pictures show an improvement on evolutionary algorithms solving fractal image compression.

References

1. Xing-yuan, W., Fan-ping, L., Shu-guo, W.: Fractal image compression based on spatial correlation and hybrid genetic algorithm. Journal of vis. commun. image R, 505–510 (2009)
2. Xuan, Y., Dequn, L.: An improved genetic algorithm of solving IFS code of fractal image. In: IEEE 3rd international conference on signal processing (1996)
3. Chen, X., Zhu, G., Zhu, Y.: Fractal image coding method based on genetic algorithms. In: International Symposium on Multispectral Image Processing (1998)
4. Mitra, S.K., Murthy, C.A., Kundu, M.K.: Technique for fractal image compression using genetic algorithm. IEEE Trans. on Image Processing 7(4), 586–593 (1998)
5. Xun, L., Zhongqiu, Y.: The application of GA in fractal image compression. In: 3rd IEEEWorld Congress on Intelligent Control and Automation (2000)
6. Gafour, A., Faraoun, K., Lehireche, A.: Genetic fractal image compression. In: ACS/IEEE International Conference on Computer Systems and Applications (2003)
7. Mohamed, F.K., Aoued, B.: Speeding Up Fractal Image Compression by Genetic Algorithms. Springer Journal of Multidimention Systems and Signal processing 16(2) (2005)
8. Xi, L., Zhang, L.: A Study of Fractal Image Compression Based on an Improved Genetic Algorithm. International Journal of Nonlinear Science 3(2), 116–124m (2007)
9. Wu, M., Teng, W., Jeng, J., Hsieh, J.: Spatial correlation genetic algorithm for fractal image compression. Journal of Chaos, Solitons and Fractals 28(2), 497–510 (2006)
10. Wu, M., Jeng, J., Hsieh, J.: Schema genetic algorithm for fractal image compression. Elsevier Journal of Engineering Applications of Artificial Intelligence 20(4), 531–538 (2007)
11. Han, K., Kim, J.: Quantum-inspired evolutionary algorithm for a class of combinatorial optimization. IEEE Transactions on Evolutionary Computing 6(6) (2002)

5 Conclusion

References

Identification of Sound for Pass-by Noise Test in Vehicles Using Generalized Gaussian Radial Basis Function Neural Networks

María Dolores Redel-Macías, Francisco Fernández-Navarro, Antonio José Cubero-Atienza, and Cesar Hervás-Martínez

Abstract. The sound of road vehicles plays a major role in providing quiet and comfortable rides. Automotive companies have invested a great deal over the last few decades to achieve this goal and attract customers. Engine noise has become one of the major sources of passenger car noise today and the demand for accurate prediction models is high. The purpose of this paper is to develop a novel noise prediction model in vehicles using a Pass-by noise test based on Artificial Neural Networks at high frequencies. The artificial neural network used in the experiments was the Generalized Gaussian Radial Basis Function Neural Network (GRBFNN). This type of RBF can reproduce different RBFs by updating a real τ parameter and allowing different shapes of RBFs in the same Neural Network. At low frequencies the system behaves linearly and therefore the proposed method improves the accuracy of the system in frequencies over 2.5 kH, obtaining a Mean Squared Error (MSE) of $0.018 \pm 3 \times 10^{-4}$, enough for our noise prediction aim.

1 Introduction

The successful development of new products depends on the capability of assessing the performance of conceptual design alternatives in an early design phase. In recent years, major progress has been made in this area, based on the extensive use of prediction models, particularly in the automotive industry. Noise and vibration in vehicles are topics of increasing interest, owing to the fact that new regulations have

Francisco Fernández-Navarro · Cesar Hervás-Martínez
Department of Computer Science and Numerical Analysis,
University of Córdoba, Spain
e-mail: i22fenaf@uco.es, chervas@uco.es

María Dolores Redel-Macías · Antonio José Cubero-Atienza
Area de Proyectos de Ingeniería. Dpto. de Ingeniería Rural.
University of Córdoba
e-mail: mdredel@uco.es, ir1cuata@uco.es

A. Gaspar-Cunha et al. (Eds.): Soft Computing in Industrial Applications, AISC 96, pp. 327–336.
springerlink.com

been issued to restrict the emission of exterior noise. Automobile manufacturers are supervised to certify that their vehicles comply with noise emission standards by measuring noise levels according to procedures defined by international standards, commonly known as Pass-by and Coast-by noise tests. The noise generated by a passenger vehicle is the result of several different sources like the engine, tires, exhaust, etc, but the major source is engine noise, so the demand for accurate prediction models is high. For these reasons, it is imperative to establish noise prediction models which are as simple as possible to identify the experience of an observer in the position of the receptor with sufficient accuracy. The efficiency of sound prediction models in passenger vehicles can improve the results of the Pass-by test because of the design phase which has to fulfill the international review standard, like ISO 362 and ISO 362:2005. The Pass-by Noise test examines the noise generated by several sources in the vehicle, changing operation conditions at different receptor positions.

Auralization is the process which renders the sound field of a physical sound source audible in a space, in such a way as to simulate the listening experience at a given position in the modeled space. The process of auralization is carried out by modeling the physical source in a sound synthesis model. Although there are some synthesis procedures that work purely in the time domain, there are numerous advantages of working partially in the frequency domain. The sound synthesis procedure starts with a source-transmission path-receiver model for the prediction of the time-frequency spectrum of the sound field produced by the active source at the receiver location. The next step is to synthesize the signal of the sound within the time period. This first part of the procedure starts by identifying different contributing noise sources, such as engine noise, tire noise or exhaust noise, amongst others. The second step is often very difficult, since the noise sources identified need to be quantified. Next, the transfer functions between the source and receiver must be determined experimentally or numerically. Finally, the sound spectrum at the receiver position is calculated by using the results from the first two steps. These prediction methods are based on linear models [13, 16]. The main advantage of these methods is their robustness, while the drawback is their resolution limitations at high frequencies since the system has nonlinear behavior. Various soft computing methods have been used to improve accuracy in the noise signal prediction in different areas. Monophonic sound source separation systems based on neural networks can be found in [12]. Hu [6] introduced a method based on back-propagation artificial neural networks to obtain individual head-related transfer functions (HRTF). By using HRTF, a sound source can be defined more accurately than with a non- individual HRTF.

On the other hand, various methods, such as Artificial Neural Networks (ANNs) and Wavelet Networks have been developed in recent years for analyzing and solving identification problems [18]. Radial Basis Function Neural Networks (RBFNNs) are a well-established tool for approximating multidimensional functions in industrial applications [1]. An important advantage of an RBFNN is an understandable interpretation of the functionality of basis functions [8]. There are several common types of functions used as transfer functions, for example, the standard Gaussian (SRBF), the Multiquadratic (MRBF), the Inverse Multiquadratic (IMRBF), and the

Cauchy (CRBF). This type of RBF can reproduce different RBFs, by updating a real τ parameter, and allowing different shapes of RBFs in the same Neural Network. This novel basis function was previously proposed for classification problems [3], and in this paper, has been adapted to regression problems.

This paper investigates the performance of the Generalized Gaussian RBFNN (GRBFNN) in an industrial application, the identification of sound for the Pass-by Noise test for vehicles. The transfer functions between the source and receiver are established by means of GRBFNNs using only the data corresponding to a frequency between 2.5-10 kHz since the system has linear behavior at low frequencies. Therefore, a novel strategy is proposed based on the auralization of sound to simulate the listening experience at a given position in modeled space at high frequencies.

2 Radial Basis Function Neural Networks for Industrial Applications

Radial Basis Function Neural Networks (RBFNNs) [2] are well suited for function approximation. A RBF is a function which has been built into a distance criterion with respect to a center. Let the number of nodes in the input and hidden layer be p and m, respectively. For any sample $\mathbf{x} = (x_1, x_2, \ldots, x_p)$, the output of the RBFNN is $y_{\text{RBFNN}}(\mathbf{x})$. The model of a RBFNN can be described by the following equation:

$$y_{\text{RBFNN}}(\mathbf{x}) = \beta_0 + \sum_{i=1}^{m} \beta_i \times \phi_i(d_i(\mathbf{x})) \tag{1}$$

where $\phi_i(d_i(\mathbf{x}))$ is a non-linear mapping from the input layer to the hidden layer, $\beta = (\beta_1, \beta_2, \ldots, \beta_m)$ is the connection weight between the hidden layer and the output layer, β_0 is the bias. The function $d_i(\mathbf{x})$ can be defined as:

$$d_i(\mathbf{x}) = \frac{\|\mathbf{x} - \mathbf{c}_i\|^2}{\theta_i^2} \tag{2}$$

where θ_i is the scalar parameter that defines the width for the i-th radial unit, $\|.\|$ represents the Euclidean norm and $\mathbf{c_i} = [c_1, c_2, \ldots, c_p]$ the centers of the RBFs. The standard RBF (SRBF) is the Gaussian function, which is given by: $\phi_{\text{SRBF}_\text{i}}(d_i(\mathbf{x})) = e^{-d_i(\mathbf{x})}$.

The radial basis function $\phi_i(d_i(\mathbf{x}))$ can take different forms, including the Cauchy RBF (CRBF) defined by: $\phi_{\text{CRBF}_\text{i}}(d_i(\mathbf{x})) = 1/(1 + d_i(\mathbf{x}))$, and the Inverse Multi-quadratic RBF (IMRBF), given by: $\phi_{\text{IMRBF}_\text{i}}(d_i(\mathbf{x})) = 1/(\sqrt{1 + d_i(\mathbf{x})})$.

This paper proposes the use of the Generalized Gaussian Function as RBF. This work, based on Generalized Gaussian Distribution (GGD) [9, 14], defines a novel RBF by removing the constraints of a probability function from the GGD. This basis function is called the Generalised Gaussian Radial Basis Function (GRBF), which is defined using the following expression:

$$\phi_{\mathrm{GRBF_i}}(\mathbf{x}) = \exp\left(-\frac{\|\mathbf{x}-\mathbf{c_i}\|^{\tau_i}}{\theta_i^{\tau_i}}\right) \tag{3}$$

where τ_i is the exponent of the i-th GRBF. Figure 1 presents the radial unit activation for the GRBF for different values of τ.

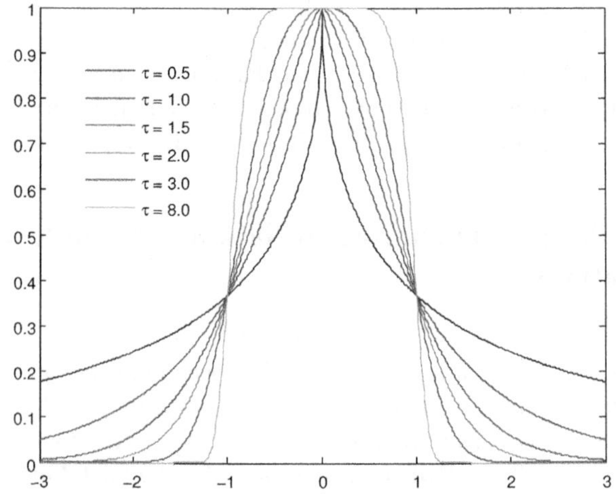

Fig. 1 Radial unit activation in one-dimensional space with $c = 0$ and $\theta = 1$ for the Generalized RBF (GRBF) with different values of τ

The error surface associated with the model is very convoluted. Thus, the parameters of the RBFNNs are estimated by means of a Hybrid Evolutionary Algorithm (HEA) (detailed in Section 3). The HEA was developed to optimize the error function given by the Mean Squared Error (MSE) for N observation s, which is defined for an individual g of the population:

$$MSE(g) = \frac{1}{N}\sum_{i=1}^{N}(y_i - \hat{y}_i)^2 \tag{4}$$

where $\hat{y}_i$ are the predicted values.

3 Hybrid Evolutionary Algorithm

The basic framework of the Evolutionary Algorithm (EA) is the following: the search begins with an initial population of Generalized Gaussian RBFNNs and, in each iteration, a population-update algorithm which evolves both its structure and weights is applied. The population is subject to the operations of replication, mutation and recombination.

We considered $MSE(g)$ defined in (Equation 4) as the error function of an individual g of the population. The fitness measure needed for evaluating the individuals is a strictly decreasing transformation of the error function $MSE(g)$ given by: $A(g) = \frac{1}{1+MSE(g)}; 0 < A(g) \leq 1$.

The severity of a mutation with respect to an individual RBFNN model is dictated by the temperature, $T(g)$ of the RBFNN model. $T(g)$ is related to $A(g)$ by means of the expression $T(g) = 1 - A(g), 0 \leq T(g) < 1$ and, for that reason, $T(g)$ is in descent throughout the evolutionary process, resulting in abrupt changes at the beginning (exploration) and slight changes at the end (exploitation).

Parametric mutation consists of a simulated annealing algorithm [10]. Structural mutation implies a modification in the structure of the RBFNNs and allows the exploration of different regions in the search space, helping to maintain the diversity of the population. There are four different structural mutations: hidden node addition, hidden node deletion, connection addition and connection deletion. These four mutations are applied sequentially to each network. More information about the genetic operators proposed can be seen in [4, 5].

With regard to the mutation of theτ parameter: if the structural mutator adds a new node in the Generalized Gaussian RBFNN, the τ parameter is assigned to a γ value, where $\gamma \in [1.75, 2.25]$, because when $\tau \rightarrow 2$, the Generalized Gaussian RBF reproduces the SRBF. The τ parameter is updated by adding a uniform ε value, where $\varepsilon \in [-0.25, 0.25]$, because the modification of the Generalized Gaussian RBFNN is very sensitive to τ variation.

Finally, the Hybrid Evolutionary Algorithm (HEA) applies the local optimization algorithm to the best solution obtained by the EA in the last generation. The local improvement procedure considered in this work was the *iRprop+* algorithm [7].

4 Experiments

4.1 Description of the Dataset and the Experimental Design

The set-up used for this research consisted of a rectangular box with outer dimensions of 902 mm x 602 mm x 190mm, with 24 loudspeakers separated from it at a distance of 150 mm. The nearest indicator microphones are positioned at a distance of 0.10 m from the loudspeaker cabinet, see Fig 2.

An LMS instrumentation series, consisting of a portable and multi channel SCADAS meter, a Bruel and Kjaer (BK) pre-polarized free-field half-inch microphone and a pre-polarized free-field quarter-inch microphone was utilized as the measuring device. LMS Test.Lab was the measurement software package and all the microphones were calibrated with a BK calibrator. All recordings were carried out inside a semi anechoic chamber. The measurements were taken with a sampling frequency of 2.56 kHz over a frequency span of 12.8 kHz. The frequency resolution was 1.5625 Hz and 100 spectral averages were implemented for analyses. Linear averaging was used to place equal emphasis on all spectra or time records. This type of averaging is helpful for the analysis of stationary signals. Hanning weighting

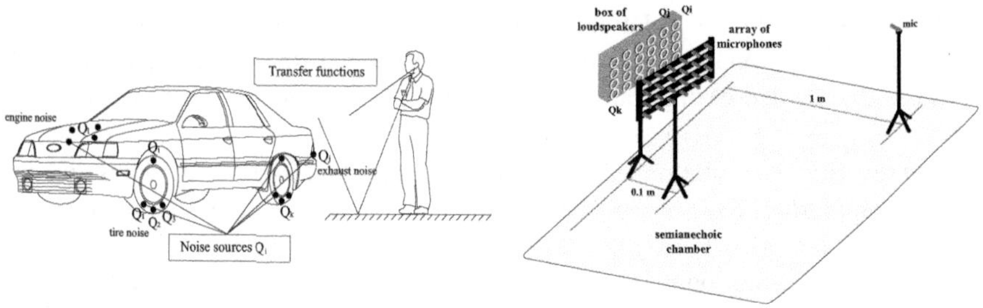

Fig. 2 Measurements set-up

with maximum overlap (up to 99%) was used as a general-purpose window for the continuous signal.

For the engine mock-up sound measurements, the height of the one free-field microphone above the ground was 1.65 m and the distance between the microphones and the loudspeakers is presented in Fig. 2. The microphone was pointed towards the source and situated parallel to the ground at $45 \pm 10°$. In addition, the sound emitted by the source was a random burst of noise between 100-10000 Hz.The sound produced by the source is recorded by the microphone array and by the microphone situated in the receptor position. The objective is to identify the sound source at the receptor position using the signals of the microphone array situated at 10 cm. The signal registered at receptor position is used to check the accuracy of the model. As the main drawback of the traditional methods is their low resolution at high frequency and their linear behavior at a low frequency, only high frequencies have been used ($f > 2.5kHz$). Figure 3 shows the condition number of the transfer function; under 2.5 kHz the system is well-conditioned while above the 2.5 kHz the function is ill-conditioned. Therefore, from 8193 registered data, for our proposal 6000 data corresponding to high frequencies have been utilized, , 4000 for training and 2000 for the generalization of the ANN model.

The proposed model (GRBF) is compared to other RBFs obtained with the same HEA (detailed in Section 3). In particular, the GRBF model has been compared to SRBF, CRBF and IMRBF models. For the other RBFs (CRBF, SRBF and IMRBF), the $iRprop^+$ algorithm was modified slightly taking into account which RBF was being used in the hidden layer. Furthermore, the performance of the best GRBF model is compared (obtained by the HA) to other state-of-the-art approaches (available in the WEKA machine learning tool [17]). In particular, the GRBF model is compared to:

- A Gaussian Radial Basis Function Network (RBFN), deriving the centre and width of hidden units using k-means, and combining the outputs obtained from the hidden layer using linear regression.
- A Multilayer Perceptron (MLP) with sigmoid units as hidden nodes, obtained by means of the backpropagation algorithm
- Ridged Linear Regression (RLR) where the Akaike Information Criterion (AIC) is used to select the variables to be included in the model.

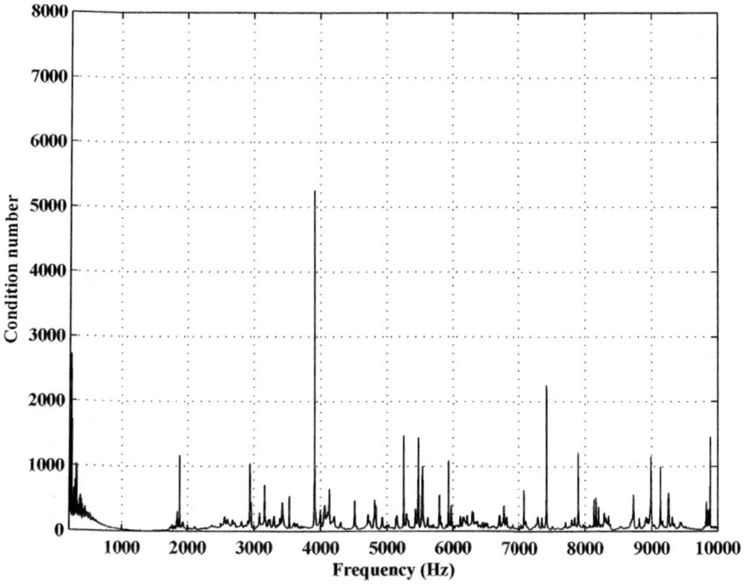

Fig. 3 Condition number of the system

For the selection of the maximum and minimum number of RBFs in the hidden layer ($[M_{\min}, M_{\max}]$) and the number of generations (#*Gen*), a grid search algorithm was applied with ten-fold cross-validation in the same way as for SVM, using the following ranges: $[M_{\min}, M_{\max}] \in \{[2,5],[4,7]\}$ and $\#Gen \in \{20,40,100,400\}$.

A simple linear rescaling of the input variables was done in the interval $[-2,2]$, with X_i^* as the transformed variables. The connections between the hidden and output layer were initialised in the $[-5,5]$ interval (i.e. $[-I,I] = [-5,5]$). The size of the population was $N = 200$. For the structural mutation, the number of nodes that could be added or removed was within the $[1,2]$ interval, and the number of connections to add or delete in the hidden and the output layer during structural mutations was within the $[1,7]$ interval.

The performance of each basis function was evaluated using the *MSE* and the Standard Error of Prediction (*SEP*) in the generalisation set. SEP is defined as: $SEP = (100/|\bar{y}|) \times \sqrt{MSE}$.

4.2 Comparison to Other Radial Basis Functions Neural Networks

Table 1 shows the results obtained with the different RBFs tested. The Generalized Gaussian Radial Basis Function (GRBF) model obtained the best result in terms of MSE_G and SEP_G out of all models compared both in EA and HEA. Furthermore, the GRBF models generated by the EA and HEA have fewer connections.

Table 1 Statistical results of the MSE_G and SEP_G, obtained using the different basis functions proposed and the EA and HEA algorithms. Ordered mean for the statistical multiple comparison Tukey test

	MSE_{EA_G}	SEP_{EA_G}	MSE_{HEA_G}	SEP_{HEA_G}
Model	Mean ± SD	Mean ± SD	Mean ± SD	Mean ± SD
SRBF	$0.021 \pm 4\times10^{-4}$	259.224 ± 2.865	$0.022 \pm 8\times10^{-3}$	263.924 ± 38.764
CRBF	*$0.020 \pm 3\times10^{-4}$*	*257.157 ± 1.728*	*$0.020 \pm 2\times10^{-4}$*	*255.576 ± 1.570*
IMRBF	$0.021 \pm 3\times10^{-4}$	258.203 ± 1.860	$0.020 \pm 3\times10^{-4}$	256.683 ± 2.007
GRBF	$\mathbf{0.019 \pm 1\times10^{-4}}$	$\mathbf{251.240 \pm 4.231}$	$\mathbf{0.018 \pm 3\times10^{-4}}$	$\mathbf{245.561 \pm 13.976}$

	Tukey HSD test	
	MSE_G	SEP_G
EA Ranking	$\mu_4 > \mu_3 \geq \mu_2 \geq \mu_1$	$\mu_4 > \mu_3 \geq \mu_2 \geq \mu_1$
HEA Ranking	$(\mu_4 \geq \mu_2 \geq \mu_3 \geq \mu_1)$ $(\mu_4 > \mu_1)$	$(\mu_4 \geq \mu_2 \geq \mu_3 \geq \mu_1)$ $(\mu_4 > \mu_1)$

	Comparison to other state-of-art approaches			
	RBFN	MLP	RLR	GRBF
MSE_G	0.023	0.027	*0.022*	**0.017**
SEP_G	269.536	297.486	*269.358*	**241.351**

(1): SRBF; (2): CRBF; (3): IMRBF; (4): GRBF
$\mu_A \geq \mu_B$: The differences were not significant; $\mu_A > \mu_B$: Significant differences were found.
The best quantitative result method is represented in bold face.

In order to determine the best methodology (in the sense of its influence on the MSE and on the SEP in the generalization set, MSE_G and SEP_G), an ANOVA statistical method test was carried out, based on a previous Normality Kolmogorov-Smirnov (K-S) test of the MSE_G and SEP_G values. The results of the ANOVA analysis for the MSE_G and SEP_G values show that the effect associated with the methodology was statistically significant at a level of signification of 5%.

Once this test guaranteed that there were significant differences between the results of the different methods, a multiple comparison test was performed on the MSE_G and SEP_G values in order to establish a ranking of the different methods. First, a Levene test [11] was carried out to evaluate the equality of variances. Then, a Tukey HSD test [15] was performed because the variances were equal (either for MSE_G or SEP_G) in order to rank the different methods.

Table 1 show the results obtained by the Tukey HSD test. On analyzing the average results in EA for MSE_G and SEP_G, the GRBF model was seen to obtain significantly better results than those obtained with other models, for a level of signification of 5%, over the other methodologies. On the other hand, the results of the average using the HEA methodology showed that the GRBF model obtained better performance both in MSE_G and SEP_G than those obtained with other models, although significant differences were found only with the SRBF model.

Finally, the best performing RBFs models were compared, the GRBF model to other state-of-art approaches. In particular, the GRBF model was compared to the RBFN, MLP and SLR methodologies. As the HEA employed to estimate the GRBF parameters is a stochastic method and the other state-of-art approaches are deterministic methods, the results provided by the best performing GRBF model over the 30 HEA executions were compared to the results of the other baseline approaches. Table 1 also shows the results obtained with different state-of-art approaches tested and the GRBF model. The GRBF model obtained the best result in terms of MSE_G and SEP_G. In general, these results showed that the proposed approaches based on GRBFNNs are robust to identify the sound in vehicles in the interval of frequency [2.5-10]kHz (since the system has linear behavior at a low frequency), obtaining better results than the remaining RBFs.

5 Conclusions

An engine noise at high frequencies was experimentally characterized by means of a novel method based on Generalized Gaussian Radial Basis Function Neural Networks for use in a Pass-by Test for vehicle noise. The data corresponding to noise at low frequencies show linear behavior; which is why they were not considered in the identification process.

A multiple comparison test on MSE_G and SEP_G values was performed in order to rank the different RBFs tested. Finally, the GRBF model obtained better performance both in MSE and SEP than those obtained with other models.

The accuracy achieved using GRBFNNs in the prediction of engine noise at high frequencies is considered acceptable for our purposes.

Acknowledgements. The section corresponding to the investigation and measurements about noise was supported by Research Group of Noise and Vibration from Katholieke Universiteit Leuven with the supervision of the Professor Paul Sas. This work has been partially subsidized by the TIN 2008-06681-C06-03 project of the Spanish Inter-Ministerial Commission of Science and Technology (MICYT), FEDER funds and the P08-TIC-3745 project of the "Junta de Andaluca" (Spain). The research of Francisco Fernndez-Navarro has been funded by the "Junta de Andalucia" Predoctoral Program, grant reference P08-TIC-3745.

References

1. Bauer, M., Buchtala, O., Horeis, T., Kern, R., Sick, B., Wagner, R.: Technical data mining with evolutionary radial basis function classifiers. Applied Soft Computing Journal 9(2), 765–774 (2009)
2. Broomhead, D.S., Lowe, D.: Multivariable functional interpolation and adaptive networks. Complex Systems 2, 321–355 (1988)
3. Castaño, A., Hervás-Martínez, C., Gutierrez, P.A., Fernández-Navarro, F., García, M.M.: Classification by evolutionary generalized radial basis functions. In: ISDA 2009: Proceedings of the 2009 Ninth International Conference on Intelligent Systems Design and Applications, Pisa, Italy, pp. 203–208 (2009)
4. Gutirrez, P.A., Hervs-Martnez, C., Carbonero, M., Fernndez, J.C.: Combined Projection and Kernel Basis Functions for Classification in Evolutionary Neural Networks. Neurocomputing 72(13-15), 2731–2742 (2009)

5. Gutirrez, P.A., Hervs-Martnez, C., Lozano, M.: Designing multilayer perceptrons using a guided saw-tooth evolutionary programming algorithm. In: Soft Computing (2009), (in press) doi:10.1007/s00500-009-0429-x
6. Hu, H.: Hrtf personalitation based on artificial neural network in individual virtual auditory space. Applied Acoustic 69, 163–172 (2008)
7. Igel, C., Hsken, M.: Empirical evaluation of the improved rprop learning algorithms. Neurocomputing 50(6), 105–123 (2003)
8. Jin, Y., Sendhoff, B.: Extracting interpretable fuzzy rules from rbf networks. Neural Processing Letters 17(2), 149–164 (2003)
9. Krupinski, R., Purczynski, J.: Approximated fast estimator for the shape parameter of generalized gaussian distribution. Signal Processing 86(2), 205–211 (2006)
10. Martnez-Estudillo, F.J., Hervs-Martnez, C., Gutirrez, P.A., Martnez-Estudillo, A.C.: Evolutionary product-unit neural networks classifiers. Neurocomputing 72(1-2), 548–561 (2008)
11. Miller, R.G.: Beyond ANOVA, Basics of App. Statistics. Chapman & Hall, London (1996)
12. Pichevar, R., Rouat, J.: Monophonic sound source separation with an unsupervised network of spiking neurones. Neurocomputing 71(1-3), 109–120 (2007)
13. Redel-Macías, M.D., Berckmans, D., Cubero-Atienza, A.: Model of identification of sound source. application to noise engine. Revista Iberoamericana de Automática e Informática Industrial 34(7) (2010)
14. Sharifi, K., Leron-Garcia, A.: Estimation of shape parameter for generalized gaussian distributions in subband decompositions of video. IEEE Transactions on Circuits and Systems for Video Technology 5(1), 52–56 (1995)
15. Tamhane, A.C., Dunlop, D.D.: Statistics and Data Analysis. Prentice-Hall, Englewood Cliffs (2000)
16. Williams, E.G., Maynard, J.D., Skudrzyk, E.: Sound source reconstructions using a microphone array. Journal of the Acoustical Society of America 68(1), 340–344 (1980)
17. Witten, I.H., Frank, E.: Data Mining: Practical Machine Learning Tools and Techniques, Data Management Systems, 2nd edn. Morgan Kaufmann (Elsevier), San Francisco (2005)
18. Yao, X., Wei, C.J., He, Z.Y.: Evolving wavelet neural networks for function approximation. Electron. Lett. 32(4), 360–361 (1996)

Case Study of an Intelligent AMR Sensor System with Self-x Properties

Muhammad Akmal Johar and Andreas Koenig

Abstract. Numerous research efforts have tried to mimic the capabilities of living organisms in performing self-monitoring and self-repairing denoted as self-x features to achieve robust and dependable systems. In sensor systems applications, self-x features carry the promise to deliver properties requested by standards organizations, e.g., the NAMUR[1], such as improved flexibility, better accuracy and reduced vulnerability to deviations and drift caused by manufacturing and the environmental changes. In this paper, the concept of self-x properties implemented on an Anisotropic Magnetoresistive AMR sensor system is investigated as a first case study to be carried on in MEMS implementations. The degradation of AMR sensor can occur when the sensor is exposed to the strong magnetic field shown by weak sensitivity of sensor and inaccurate measurement output. The self-x properties are required to monitor and recover the sensor performance by employing the compensating and flipping coils. The experimental result shows the recovering of sensor performance in terms of classification accuracy for vehicle recognition application by implementing the self-x features.

1 Introduction

The progress of micro and nano technologies along with advanced packaging technologies has opened the possibilities to design new sensor systems that have better performance, smaller size, affordable price and facilitated implementation. Sensors are now not only limited to electrical and mechanical domains but also chemical and biological domains are now amenable to be integrated in one system. MEMS sensor systems are particularly interesting platforms for intelligent system implementation. The emergence of new sensor systems with intelligent functionality commonly

Muhammad Akmal Johar · Andreas Koenig
Institute of Integrated Sensor System, University of Kaiserslautern,
67663 Kaiserslautern, Germany
e-mail: {johar,koenig}@eit.uni-kl.de

A. Gaspar-Cunha et al. (Eds.): Soft Computing in Industrial Applications, AISC 96, pp. 337–346.
springerlink.com

denoted as smart sensors have brought enlarged possibilities to extract more information from the testing environment which lead to produce more intelligent systems. However, this also increases the complexity of design systems, and the vulnerability of the outcome.

In critical applications, dependable sensor systems with no or tightly bounded minimum error rates are desired. As preventive actions have been taken at the design and manufacturing stage, proactive actions are required at the deployment stage. Once the sensor systems are deployed, they are now exposed to many sources of errors from environment. In order to deal with this, added intelligence and redundancy in sensors systems are required, e.g. by the NAMUR 2005-2015 roadmap[1]. The self-x concept has been introduced in this context for general computing systems. This term has been introduced by von der Malsburg et al.[2, 3], based on studies of self-organization systems in nervous systems. Inspired by the observed biological mechanisms, substantial research effort has been expended to mimic these self-x features[4, 5, 6, 7]. Most of the research are concentrated at the system level, e.g., evolvable hardware (EHW) focusing on the electronics level.

The next step, pursued also in our research, is to bring this capability to the component level, in other word to sensor component. Self-repairing capabilities has been done in MEMS accelerometer by having redundant structure in one integrated circuit[8, 9]. It is interesting to explore and generalize these self-x capabilities to other types of sensors and combine them with advanced EHW sensor electronics.

The design of an intelligent, integrated sensor systems is a demanding task for engineers nowadays in many applications, e.g., in automotive, biometrics, robotics and automation. It is a task that requires high skilled engineers, time consuming effort in selecting and optimizing the right sensor element to build an integrating sensor system (Fig.1a).There are several possible ways to provide intelligence in integrated sensor systems. Observation and optimization activities are one way to do it which at the same time can significantly increase the robustness of the system. It can be implemented at various stage of sensor system (Fig.1c). Previous work has been done in the area of signal conditioning and mixed signal electronics [10, 11]. Here the implementation of system reconfiguration and optimization is achieved at hardware electronics level in a mixed signal chip. Optimization processes have been done locally to find the best structure and parameter settings. Several works on the area from signal processing and feature computation until classification have been reported in [12, 13].

The intention of this case study is to expand the observation and optimization capabilities to the sensory level Fig.1b . These can be implemented by introducing the self-x features concept. Self-monitoring features provides the capability to monitor the state of sensor performance and ensures it to work at the optimum level. Without costly external test and calibration, the user will be unaware of a sensor malfunction and accept the output as actual measurement. Self-repairing provides the action to handle errors occurring at or in sensors in the limits of the available redundancy. The implementation can be in a simple way by having binary judgment or by more complex algorithm via continuous-valued features.

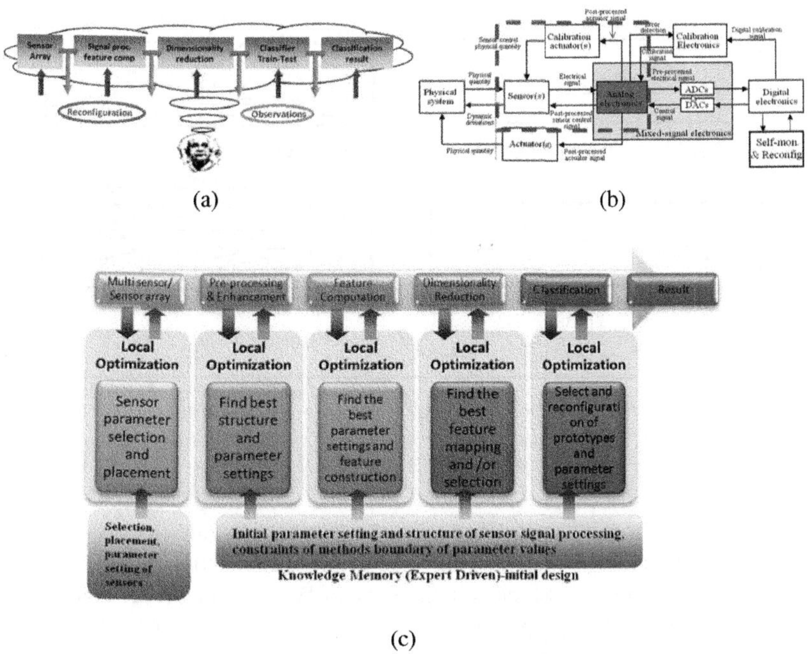

Fig. 1 (a) Issues in designing intelligent system, (b) self-x features implementation at the sensor front end of the system, (c) architectures for intelligent system design.

In this work, based on a simple intelligent sensor system for vehicle recognition by AMR sensors, self-x principles for error detection and repair will be investigated working interleaved with the recognition task. Section 2 will explain the sensor with its embedded actors for self-x features, section 3 will explain the intelligent system task and implementation, and section 4, before concluding, will present results of interleaved recognition and self-x phases.

2 AMR Sensor with Embedded Actuators

Unlike most of the sensor that directly measure the physical properties, magnetic sensors offer the capabilities to detect changes of the magnetic field. This can give information of presence, rotations, current and angle.

Anisotropic Magnetoresistive or AMR sensors can be realized when the thin film technology is available to develop the resistive layer of nickel-iron or Permalloy on silicon. During fabrication, a strong magnetic field (M vector) is applied to create a preferred axis also known as the easy axis in the thin film as shown in Fig.2a. The length of the thin film will determine the initial resistive value and the M vector is parallel with the length of the film. So the direction of the vector can be set

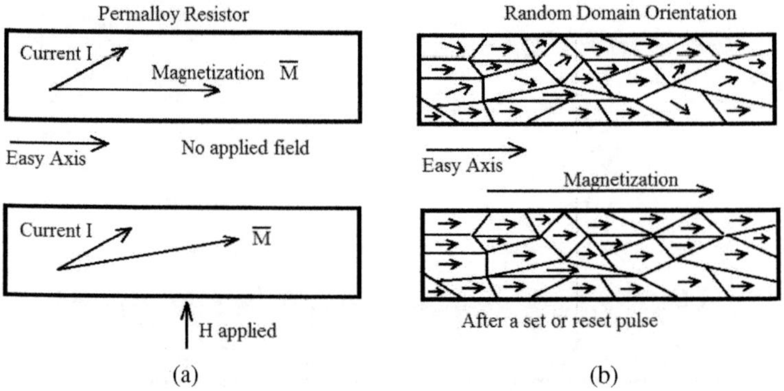

Fig. 2 (a) Magnetoresistive effect in Permalloy thin film, (b) degradation effect and recovery phase.

either to the left or right in the film. When external magnetic field is applied, the magnetization vector will deviate from the original easy axis and will change the resistance of the thin film which is in the form of Wheatstone bridge configuration.

AMR sensor technology possessed degradation effect (Fig.2b), when the magnetic domain is not aligned anymore due to the presence of strong magnetic field that causes the magnetic domain aligned to the random orientation. This will affect the sensitivity of the sensor and give the wrong output that lead to false in-formation extracted from the sensor. In order to counter this problem, strong magnetic fields need to be applied in the specific direction to the thin film. It reported from the sensor manufacturer [15] that a reset current of 150mA for 1s is required to aligned the magnetic domain back to their easy axis. Flipping can be done preemptive, i.e., before every measurement. Here, it will be investigated to monitor first and flip only on detected failure, which can also be interesting in the light of the energy budget of autonomous systems. Numerous applications of using AMR sensors, e.g., compass and navigation system, vehicle detection, and recently including 3D Localization [16] in our research group have been introduced. In this 3D Localization project triaxial AMR sensor module has been built for wireless sensor networks application. In this paper, the application of AMR sensor is focused on train recognition system. During the experiment we encounter several time of degradation effect in the sensors that lead to errors. With this experience the needs of self-x features will be implemented in order to have a robust sensor system.

Self-monitoring in sensor system can be defined as a capability to monitor the condition of the sensor whether it is in optimal condition or not. While self-repairing concept is the capability of the sensor to return to the optimal condition if degradation happens. In AMR sensors, both concepts can be applied to the sensor by having a coil at the specific directions.

In this project sensor module using AMR Sensor- AFF755B from Sensitec (Fig. 3b) were built. This sensor has one compensation coil and one flipping coil

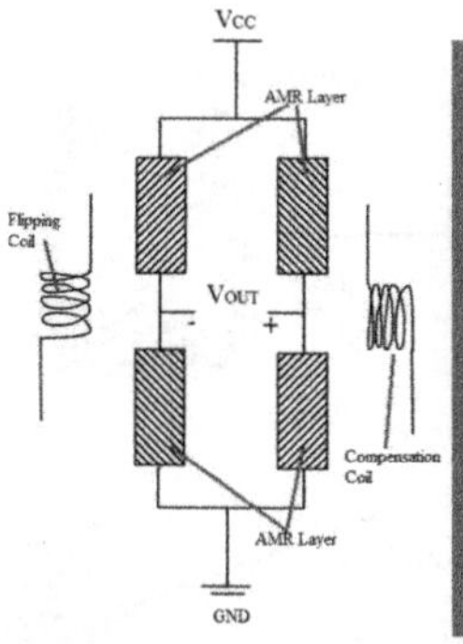

(a) Sensor layout

(b) AMR sensor module

Fig. 3 (a) A typical Wheatstone bridge configurations of AMR sensors with embedded coils, (b) AMR sensor module.

implemented inside the sensor chip and package. In implementing self-monitoring feature, an external controlled signal that provide on-off current state to the compensation coil of the sensor A test current with maximum value of 200mA can be supplied to the compensation coil during turn on cycle. Self-monitoring data are taken interleaved with the measurement and intelligent system operation for several cycles to determine the state of the sensor.

The implementation of self-repairing features also similar to self-monitoring circuit. DC current with the typical value of 150mA is required to align the magnetic domain on the Permalloy thin film. The self-repairing circuit will activate only when sensor degradation is detected.

3 Explanation of the Experimental Setup

This work studies implementation options of self-x features in an AMR sensor-based intelligent system for a vehicle identification task. A train system in N-scale is used as testing environment (Fig.4). It can distinguish between the train engine and two different wagons with different length via theirs magnetic signature. The plastic wagons label as Brown wagon and Iron wagon are preloaded with same amount of nails as a metal load to provide significant magnetization changes. The AMR sensor module is connected to a DAQ board DT9816 from Data Translation and connected to a PC via USB. Sets of data which consist of 10 rounds of measurements each for training and testing were taken at sampling rate of 5000 samples per seconds. The sensor is triggered to take measurement for two seconds which will give us 10000 samples for every measurement. For every measurement, the triggering point is set so that it will capture the whole magnetization changes produced by the train system. The train speed is fixed for all data acquisition.

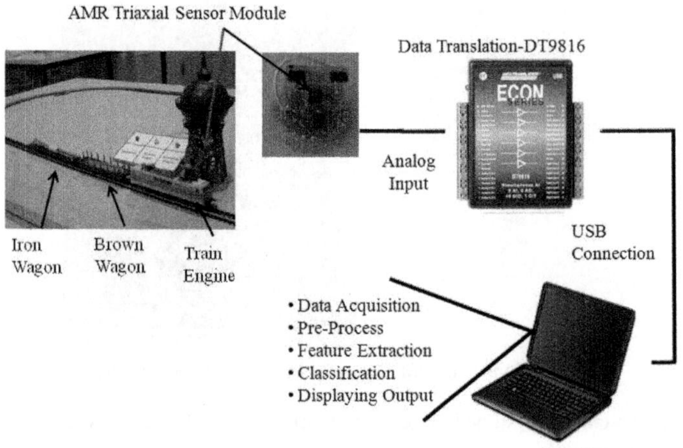

Fig. 4 Experimental setup for vehicle recognition using AMR sensors.

Identification classes are set to four different classes namely: no train class, train engine class, brown wagon class and iron wagon class. It should be mentioned here, that the first class could unknowingly be confused with a defect sensor output easily. Region of interest (ROI) of each class are manually selected from the training data set. After features extraction and selection are employed, k-Nearest Neighbor (k-NN) and Reduced Nearest Neighbor (RNN) classifiers are then applied for classification. Two modes in our experiments were tested. Offline mode classification is computed by using pre-recorded data while online mode classification is done directly with the train running continuously in a track circle.

The experiments of self-x concept are implemented as follows. There are two modes of operation which are self-monitoring operation and self-repairing operation. The process begins with resetting the sensor then a self-monitoring signal is activated for recording the reference value. Sensors aging are done by running the train around for 10, 30 40, 50 and 100 rounds. At each aging level, a self-monitoring signal is activated and the result is compared to the reference. For self-repairing process also start with the similar step for reference measurement. Dur-ing the idle time the self-monitoring signal is activated. A small fixed DC current is supplied to the compensation coil. The voltage output is measured and compared to the reference value. This is to ensure the sensor is working properly. Then the data of one round of the train are taken and the classification testing is done in both modes. After that a complete degradation effect is artificially introduced to the sensor via deliberately applying strong magnetic field on it. The data is taken once more and classification test is done once again. Self-monitoring signal is activated to check the sensor condition and now it will sense the error of the sensor. This will activate the self-repairing function and a DC current is applied to the flipping coil. Once again the data is taking and the classification test is done to prove the repairing concept. During this self-x experiment only k-NN classifier has been used for classification. SVM could be considered as future improvement.

4 Results and Discussion

4.1 Details of Intelligent System for Vehicle Recognition

A typical magnetization 'foot-print' with the corresponding event of a passing train is shown in Fig. 6. A set of training data was taken from several runs or rounds of the train passing the sensor. A set of manually selected Region-Of-Interest (ROI) data were selected for four different classes. The size of each ROI is same at 1000 samples per ROI. Fig. 7 shows one plotted data of ROI for each class. Several features are extracted from the ROI samples. The statistical features, namely mean, maximum, minimum and variance, value are heuristically used here for classification. From four extracted features, three features which are mean, minimum and variance have been selected. Other features have been considered, but the computationally simplest ones have prevailed for microcontroller based system implementation.

Using a k-NN classifier with k equal from 1 to 5, our training and test data sets (offline mode), were achieved 100% of classification accuracy were achieved. While online data testing with 10 runs has yielded 97.5% of classification accuracy. RNN classifier produced slightly lower accuracy with 99% during off line testing and 96% during online testing. The lower accuracy is understandable in this case as selecting weight from RNN is reduced thus increase higher probability of error. Confusion might occur when two reference points from different class have similar distance and lead to error in decision making.

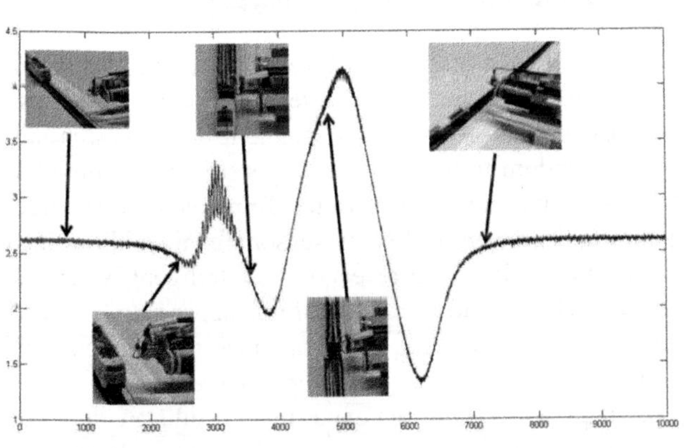

Fig. 5 Data recorded by AMR sensor with respect to train position.

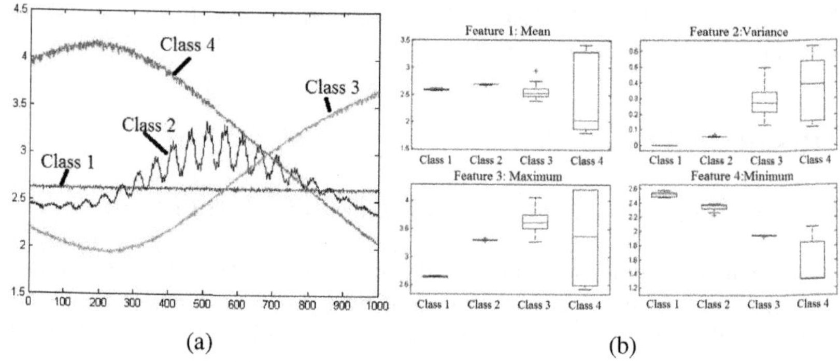

Fig. 6 (a) Plot of ROI's for each of class, (b) plot of features data.

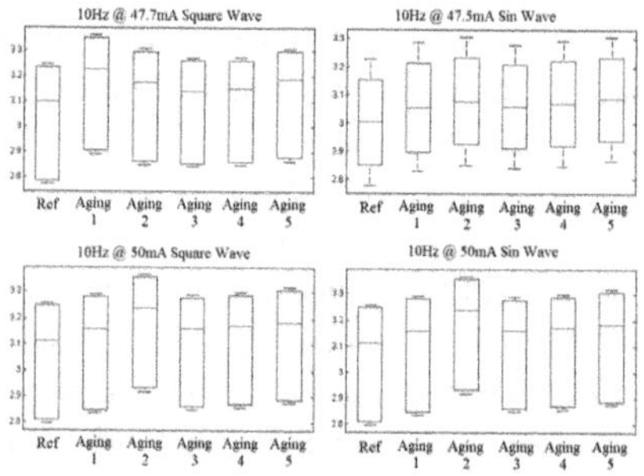

Fig. 7 Box plots of self-monitoring data with several monitoring signal.

4.2 The Implementation of Self-x Features

The efficacy of self-monitoring at sensor level has been proved in this experiment. Fig.7 shows varies level of output reading after every aging processes. It is very difficult to predict the degradation level as the magnetic domains are randomly affected and, thus produced random results. The sensor is still functioning but doesn't give the same output as the initial reference value. To recover the accuracy of the reading, the compensation here can be done in sensor signal conditioning or processing level, including adaptivity in higher levels of the intelligent system (s. Fig.1b), and not at the sensory level. Since the intention of this paper is only to prove the efficay of self-x features at the sensor level, the data concerning the classification performances with this concept has not yet been implemented.

The self-repairing feature study and experiment results are summarized in Fig.8. The experiment showing the classification results before, during and after sensor degradation. Good results before sensor degradation have been achieved, i.e.,100% correct classification during offline and 97.5% via online operation.

During the AMR sensor degradation phase, the classification result drops to only 40% correct classification, which is of course worse than guessing. This is due to the classifier only able to detect the first No train class. Without any diagnosis, this will go undetected by the user or following information processing systems. The output signal plot shows the distorted signal where virtually no magnetization changes recorded as the sensor is saturated. The next self-monitoring cycle will detect this error and trigger the self-repairing signal.

After applying self-repair, the sensor recovers the capability of recording magnetization changes. The classification results show the same results as before degradation phase. This simple experiment showed by implementing the self-repairing it can help to recover the classifications success rate back to optimum level again without supervisor interference. Similarly,continuous-valued defects or deviations, i.e., varying sensitivities of the AMR bridge, can be detected by a more sophisticated and precisely known stimulus signal to the compensation coil in a multiphase measurement to eliminate background field influences. Corresponding changes to, i.e., electronics gain or zoom can be done employing even commercially available reconfigurable sensor signal conditioners. This extension of the self-x features is currently under investigation for the same system.

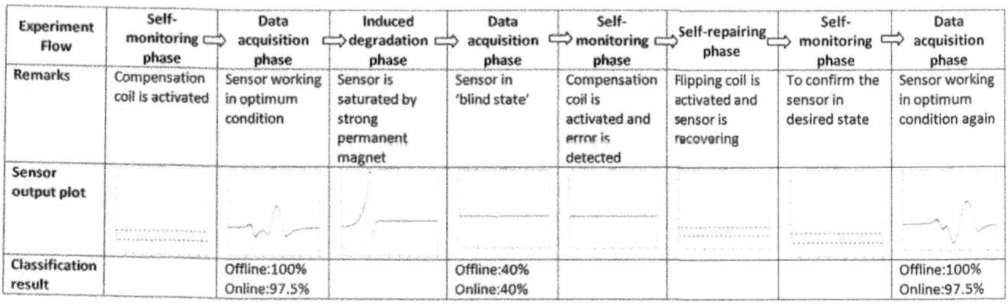

Experiment Flow	Self-monitoring phase ⇨	Data acquisition phase ⇨	Induced degradation phase ⇨	Data acquisition phase ⇨	Self-monitoring phase ⇨	Self-repairing phase ⇨	Self-monitoring phase ⇨	Data acquisition phase
Remarks	Compensation coil is activated	Sensor working in optimum condition	Sensor is saturated by strong permanent magnet	Sensor in 'blind state'	Compensation coil is activated and error is detected	Flipping coil is activated and sensor is recovering	To confirm the sensor in desired state	Sensor working in optimum condition again
Sensor output plot								
Classification result		Offline:100% Online:97.5%		Offline:40% Online:40%				Offline:100% Online:97.5%

Fig. 8 Self-repair experiment flow.

5 Conclusion

A plethora of available sensor technologies and their wider spread use in applications, e.g., implementing intelligent systems, requires for more and more sophisticated yet affordable means for test and correction within the sensory systems itself [1]. In this context, we carried out first research based on integrated AMR sensors with embedded actors in a nutshell recognition task of vehicle recognition in an N-scale train system. From our experiment in robust condition, a classification rate at 97.5% has been achieved. The extreme degradation in AMR sensor has been emulated by applying high external magnetic field to it. Since the sensor system has no knowledge of the perturbation by the incoming magnetic field from the environment, it is essential to continuously self-monitor the state of the sensor interleaved with the recognition operation. In the event of sensor degradation, this was detected during self-monitoring and the self-repairing features have been activated, restoring the sensor and system back to the optimum level of recognition. Thus, the research paves the way to dependable embedded and sensory system.

Our future work will expand the AMR sensor capability in conjunction with reconfigurable electronics to self calibration/trimming features by imposing a more sophisticated and well-defined control signal to the compensation coil in a multiphase measurement process to also cope with gradual variation or degradation of sensor parameters, e.g., sensitivity of the bridge in SoC and MEMS.

References

1. NAMUR,VDE/VDE, Final Report Technology Roadmap Process-Sensors 2005-2015, NAMUR Publication (2006)
2. Malsburg, V.D.: The Handbook of the Brain Theory and Neural Networks. In: Self-Organization in the Brain, 2nd edn., The MIT Press, Cambridge (2002)
3. Malsburg, V.D.: Self-organization of Orientation Sensitive Cells in the Striate Cortex. Kybernetik, 85–100 (1973)
4. Ghosh, D., et al.: Self-healing systems- Survey and Synthesis. Decision Support Systems, 2164–2185 (2007)
5. Carlson, J.A., English, J.M., Coe, D.J.: A flexible, self-healing sensor skin. Smart Materials and Structures, 129–135 (2006)
6. Obha, Y.: Intelligent sensor technology, p. 185. John Wiley & Sons, Chichester (1992) ISBN:0471934232
7. Gert, V.D.H., Huijsing, J.L.: Integrated Smart Sensor: Design and Calibrations. Kluwer Academic Publishers, Dordrecht (1998)
8. Xiong, X., Wu, Y.L.,Jone,W.B.: Reliability Analysis of Self-RepairableMEMS Accelerometer. In: 21st IEEE International Symposium Defect and Fault Tolerance in VLSI Systems, DFT 2006, (2006)
9. Xiong, X., Wu, Y.L., Jone, W.B.: Yield analysis for self-repairable MEMS devices. Analog Integrated Circuits and Signal Processing 56(1), 71–81 (2008)
10. Senthil, K.L., Koenig, A.: Statistical Analysis of Compensating properties of Reconfigurable Analog Circuits for Generic Self-X Sensor Interface, Sensor+Test Conf. (2009)
11. Tawdross, P., Koenig, A.: Mixtrinsic Multi-Objective Reconfiguration of Evolvable Sensor Electronics. In: Conf. on Adaptive Hardware and Systems, pp. 51–57 (AHS2007) (2007)
12. Iswandy, K., Koenig, A.: Methodology, Algorithms, and Emerging Tool for Automated Design of Intelligent Integrated Multi-Sensor Systems. J. Algorithm, Open Access 2, 1368–1409 (2009), doi:10.3390/a2041368
13. Stefanie, P., Koenig, A.: Optimized texture operators for the automated design of image analysis systems: Non-linear and oriented kernels vs. gray value co-occurrence matrices. Int. Journal of Hybrid Intelligent Systems, IJHIS (2007)
14. Caruso, M.J., Schneider, R., et al.: A new perspective on magnetic field sensing. Sensors Mag. (1998)
15. Magnetoresistive Field Sensor-AFF755B Data sheet, `http://www.sensitec.com`
16. Stefano, C., Koenig, A., et al.: 3D-Localization of low power wireless sensor nodes based on AMR-sensors in Industrial and AmI Applications. In: Conf. Sensors and Test (2010)

Part VII: Traffic and Transportation Systems

Application of Markov Decision Processes for Modeling and Optimization of Decision-Making within a Container Port

Mohamed Rida, Hicham Mouncif, and Azedine Boulmakoul

Abstract. In modern container terminals, efficiently managing the transit of the containers becomes more and more of a challenge. Due to the progressive evolution of container transport, traffic management within container ports is still an evolving problem. To provide adequate strategy for the increased traffic, ports must either expand facilities or improve efficiency of operations. In investigating ways in which ports can improve efficiency, this paper proposes a Markov Decision Process (MDP) for loading and unloading operations within a container terminal. The proposed methodology allows an easy modeling for optimizing complex sequences of decisions that are to be undertaken at each time. The goal is to minimize the total waiting time of quay cranes and vehicles, which are allocated to service a containership. In this paper, reinforcement learning, which consists of solving learning problems by studying the system through mathematical analysis or computational experiments, is considered to be the most adequate approach.

Keywords: Container port, Markov Decision Process (MDP), Optimization.

Mohamed Rida
Mathematics and Computer Sciences Department
Faculty of Sciences Ain Chock, University Hassan 2 Casablanca Morocco
e-mail: Mohamed_ridama@yahoo.fr

Hicham Mouncif
Mathematics and Computer Sciences Department
Faculty of Sciences Ain Chock, University Hassan 2 Casablanca Morocco
e-mail: hmouncif@yahoo.fr

Azedine Boulmakoul
LIST Lab., Computer Sciences Department
Mohammedia Faculty of Sciences and Technology (FSTM)
e-mail: boul@uh2m.ac.ma

A. Gaspar-Cunha et al. (Eds.): Soft Computing in Industrial Applications, AISC 96, pp. 349–358.
springerlink.com

1 Introduction

The container port provides the interface between railroads, ocean-going ships, and over the road trucks, and represents the critical node in the transport network. In container ports, the quay crane is the critical element of the container port that is served by all other port equipments. Because the quay crane is the only direct link between the storage yard and the ship, an improvement in its operations participates to the minimization of the time a ship requires to load or to unload. In attempting to improve port operations, managers must make decisions, regarding labor and equipment assignments that directly affect quay crane productivity.

For decades, many researchers in engineering, as well as in computer science, have approached the problem of container terminal management in different ways. Existing literatures report several approaches to manage a container terminal. The simulation is one of the most important approaches that have been adopted (Bruzzone, A., *et al.,* 1998) (Shabayek, A. *et al.*, 2002) [2] [1] [3] [8]. It aims to study and compare alternative layout plans, leasing policies, or handling equipments, and identify the best solution in terms of efficiency and cost-effectiveness. Most proposed approaches are based on deterministic optimization methods, although recently a stochastic optimization model was proposed. In our previous work [1], we have developed software for container terminal simulation that is based on object oriented paradigm and distributed discrete event simulation approach. The simulator is used as a test bench to evaluate management policies produced by the optimization modules, and it is charged with a realistic reproduction of the activities and flows that occur inside the terminal. The stochastic problem that we envisage for the container terminal is to achieve different goals in an uncertain and complex environment characterized by container arrivals using different transportation modes (trucks, trains, and ships) and decisions to be taken during each stage. Loading/unloading operations, resource allocations, and storage yard management are the main criteria that we seek to optimize. Some operations are to be achieved in sequence (loading and unloading a ship) or parallel (unloading containers from a ship and storing them in the yard) in order to attain the final goal (ship operations completed). It may also have to take into account some global environment variables (wind, visibility, etc.), which can disturb the entire system operations. Markov Decision Process (MDP) is a classical stochastic model for planning in the context of decision theory. It is suitable where the system states can be considered as uncertainties on the outcome of actions. A MDP is a Markov chain controlled by one agent. A control strategy associated to each state consists at choosing an action whose result is a stochastic state. The Markov property means that the probability of arriving in a particular state after completion of an action depends only on the previous state in the chain, not the entire historical of states of the chain. Formally, an MDP is a quadruple *(S, A, T, R)* where *S* is the set of state, *A* is the set of actions, *T* and *R* are respectively the probabilities and the transition rewards that are functions of the initial state, the state of arrival and the chosen action. The optimization criterion most used is to maximize over an infinite horizon the average $E(\sum_{t=0}^{\infty} \gamma^t . r_t)$, which is the sum of successive rewards r_t weighted by a discount factor $0 < \gamma < 1$ that ensures the convergence of the sum, but can also be interpreted as a probability of system failure (end of mission) between two moments of the process.

2 Review of Markov Decision Processes (MDP)

Markov decision processes (MDPs), named after Andrey Markov, provide a mathematical framework for modeling decision-making in situations where outcomes are partly random and partly under the control of a decision maker. It consists of a 4-tuple *(S; A; R; T)*:

(1) The set of states *S* is the finite set of all possible states of the system.
(2) The finite set of actions *A*.
(3) The reward function: *R(s; a)* depends on the state of the system and the taken action. We assume that the reward function is bounded.
(4) The Markovian transition model: It is represented by the probability of going from a state *s* to another state *s'* by doing the action *a*: *P(s'/s; a)=T(s',a,s)*.

If the probabilities or rewards are unknown, the problem is one of reinforcement learning [7]. The main task of reinforcement learning is finding a policy that optimizes the value of rewards. This policy can be represented by a map π:

$$\pi: S \rightarrow A$$
$$s \rightarrow \pi(s)$$

Where $\pi(s)$ is the action, which the agent (It can be a human, a robot, a part of a machine or anything susceptible to take a decision) takes at the state *s*.

Most MDP algorithms are based on estimating value functions. The state-value function according to a policy π is given by:

$$V_\pi(s) = R(s,\pi(s)) + \sum_{s' \in S} T(s,\pi(s),s').\gamma V_\pi(s')$$

Where:

- $R(s,\pi(s))$ is the expected value of the reward R_{t+1} when the policy π is followed and such that at time *t*, the state is *s*, and
- A parameter γ $(0 \leq \gamma \leq 1)$ is the discount rate, which is necessary to have a present value of the future rewards. In this case, the return is given by the infinite sum:

$$R_t = \sum_{k=0}^{\infty} \gamma^k r_{t+1+k}$$

The Bellman operator is defined as [5]:

$$\mathrm{T}^* V_\pi(s) = \underset{a \in A}{MAX}[R(s,a) + \sum_{s' \in S} T(s,a,s').\gamma V(s')]$$

We denote V^*, the fixed point of the Bellman operator. $V^*(s)$ is the maximum, according to actions, of $V(s)$ for a particular state s. When considering this maximum for every state s ∈ S, we can therefore define $V^*(s)$, which is called the optimal state-value function.

$$V^*(s) = \underset{a \in A}{MAX}[R(s,a) + \sum_{s' \in S} T(s,a,s').\gamma V^*(s')]$$

In each case, there is one equation per state in S. Therefore, finding the policy π to get the right action to do for every state of the system is now equivalent to solving $/S/$ equations, with $/S/$ unknowns. Solving a MDP consists of looking for an efficient algorithm to solve this system. In the literature, several algorithms are proposed to calculate the optimal policy (value iteration, Q-learning, SARSA etc.) [7]. In our case study, we chose value iteration algorithm.

3 Formalization of the Problem as a Markov Decision Processes

Now, let us characterize the berth control through Markov Decision Processes:
(1) Set of berth states $BS = QC^{(1)}xQC^{(2)}x...xQC^{(n)}$ the Cartesian product of $QC^{(i)}$, where $QC^{(i)}$ is the set of states of the ith quay crane allocated to serve a containership $(i=1, ..., n$; where n is the number of working quay crane). We note that idle quay cranes are not considered in this case. Each $QC^{(i)}$ corresponds to the following variables:

(a) The type of operations (noted $OP^{(i)}$) that is performed in the ith berth. In our case, we have tow values:

- ✓ *'loading operations'* and
- ✓ *'unloading operations'* .

(b) The variable corresponding to the states of the ith quay crane (noted $ST^{(i)}$). To unload a ship, the quay crane picks up containers from the ship and puts them on shuttle trucks that move them to the storage yard in the terminal. To load a ship the quay crane unloads a container from a shuttle and put it on the ship. This operation forms a closed loop that is traveled by shuttles servicing a ship (see fig. 1). If any shuttle truck is available underneath the crane, work ceases until a loaded or unloaded shuttle truck is arrived from the yard or until another is allocated to continue service. We consider three possible states of the quay crane:

- ✓ *'at work'* when the crane is positioned to serve a containership.
- ✓ *'wait'* when the crane waits the arrival of a shuttle truck for loading or unloading a container, and
- ✓ *'Idle'* when the crane breaks down.

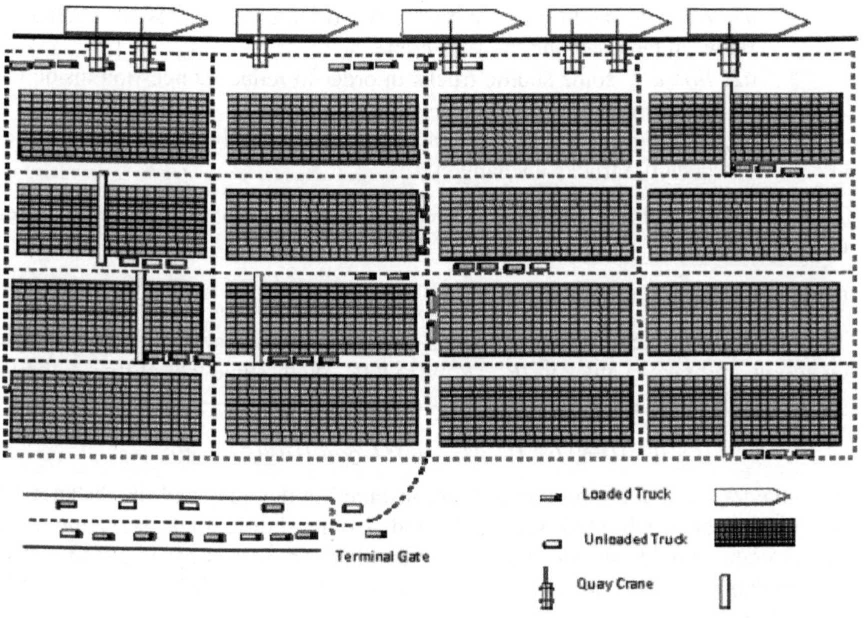

Fig. 1 Representation of a typical container terminal.

(c) The variable describing the state of the queue underneath the quay crane formed by allocated tracks that move containers between the quay and the storage yards during loading or unloading operations. Let us consider the number of 'waiting trucks' on the quay queue:

- ✓ $WT_{QC\,(i)}$ is the number of the shuttle trucks, which wait their turns for loading or unloading containers.

To summarize, an example of a state of the ith quay crane looks like: $qc^{(i)}$=($OP^{(i)}$= *'loading operations'*, $ST^{(i)}$= *'at work'*, $WT_{QC(i)}=4$) (i.e. the quay crane performs the loading operations, the crane is working in order to pick up a container from the ship or the shuttle truck, in the same time four loaded trucks are waiting in the queue).

(1) Set of actions A

The amount of work a container port deals with depends on the quantity of containers in transit. The quay crane plays an important role in the production system. Indeed, the quay crane is the critical element of the container port and is served by all other port operations. Because the quay crane is the only direct link between the storage yard and the ship, it is important to make decisions, regarding labor and equipment assignments that directly affect quay crane productivity. For example, too many trucks in the system cause long files at the cranes and long waiting times for service. Conversely, few trucks in the system will result in idle quay crane or stacking equipment. In our model, the agent controls the flow of containers between the quay and storage yards. So, the possible actions are:

- *a1:'allocate'* additional shuttle trucks to increase the flow of containers between the quay and the yard crane,
- *a2:'liberate'* some shuttle trucks in order to reduce congestion inside the terminal, and
- *a3:'stay the same'*: this action consists of keeping the number of allocated shuttle trucks unchanged.

The frequency of doing an action depends on the values of waiting time at each quay queue: $WT_{QC(i)}$

(2) Set of rewards

The reward is a function which depends on the state of the system and on the taken action, and takes values in *IR*: $R : AxS \to IR$. We assume that at time *t*, the ith quay crane is at the state:

$$qc^{(i)}(t)=[OP^{(i)}(t), ST^{(i)}(t), WT_{QC(i)}(t)],\ i=1,\ldots,n.$$

In container terminal, several cranes are installed in the quay and the distance between tow cranes is limited (see fig. 1), and also the shuttle trucks servicing a particular crane should not wait underneath another. Thus, it is necessary to restrict the number of waiting truck under each crane. In our case $WT_{QC(i)}(t)$ must be lower than 6.

Let's define the reward function as the following (table 1):

Table 1 Definition of the reward function.

States: $qc^{(i)}(t)$			**Reward : $r(qc^{(i)}(t),a)$**
$OP^{(i)}(t)$	**$ST^{(i)}(t)$**	**$WT_{QC(i)}(t)$**	
'loading operation'	*'at work'*	<=6	*1 if (a=a3)*
			-1 otherwise
'loading operation'	*'at work'*	>6	*1 if (a=a2)*
			-1 otherwise
'loading operation'	*'wait'*	*0*	*1 if (a=a1)*
			-1 otherwise
'loading operation'	*'idle'*	*Any*	*1 if (a=a2)*
			-1 otherwise
'unloading operation'	*'at work'*	<=6	*1 if (a=a3)*
			-1 otherwise
'unloading operation'	*'at work'*	>6	*1 if (a=a2)*
			-1 otherwise
'unloading operation'	*'wait'*	*0*	*1 if (a=a1)*
			-1 otherwise
'unloading operation'	*'idle'*	*Any*	*1 if (a=a2)*
			-1 otherwise

(3) Transition probability model

The transition probability model P is the probability of being in the state $s(t+1) = s'$ by selecting a certain action $a(t)$ when the system was at the state $s(t) = s$: $P[s(t+1) = s'/a(t);\ s(t)= s] = P_{s,s',a}$

Let us see more closely how to define the transition probability by considering each possible action and each quay crane state. Since the components: $OP^{(i)}(t)$, $ST^{(i)}(t)$, and $WT_{QC(i)}(t)$ of the quay crane state $qc^{(i)}(t)$ are independents and $WT_{QC(i)}(t)$ is the only variable that can be affected by the applied actions, we have the following equation:

$$P[qc^{(i)}(t+1)/a, qc^{(i)}(t)] = P[OP^{(i)}(t+1)/OP^{(i)}(t)].P[ST^{(i)}(t+1)/ST^{(i)}(t)].$$
$$P[WT_{QC(i)}(t+1)/a, WT\,QC^{(i)}(t)]$$

Since the system changes the type of operations (from loading to unloading) only once throughout the period of servicing a ship, the probability of changing from one type of operation to another between two times t and t +1 is close to zero. So we have:

$$P[OP^{(i)}(t+1) = v / OP((t) = v] \cong 1 \text{ ; and}$$
$$P[OP^{(i)}(t+1) = v' / OP^{(i)}(t) = v] \cong 0$$

Where: v and $v' \in$ {*'loading', 'unloading'*} and $v \neq v'$

Similarly, we consider that the failure rate of each quay crane is too small:

$P[ST^{(i)}(t+1)=\text{'idle'}/ ST^{(i)}(t)=u] \cong 0$; where $u \in$ {*'at work', 'wait'*}.

We note that the action *a1* consists to allocate an additional shuttle truck that is immediately added to the quay queue. Thus, we have the following equation:

$$P[WT_{QC(i)}(t+1)/a = a1, WT_{QC(i)}(t)] = P[WT_{QC(i)}(t+1)/ WT_{QC(i)}(t)+1]$$

Similarly,

$$P[WT_{QC(i)}(t+1)/a = a2, WT_{QC(i)}(t)] = P[WT_{QC(i)}(t+1)/ WT_{QC(i)}(t)-1]$$

and

$$P[WT_{QC(i)}(t+1)/a = a3, WT_{QC(i)}(t)] = P[WT_{QC(i)}(t+1)/ WT_{QC(i)}(t)]$$

We note also that the number of waiting trucks under each quay crane at each time depends of the following parameters:

- $R_{QC(i)}$: the service rate of ith quay crane,
- $R_{YC(i)}$: the service rate of ith yard crane,
- $D_{QC(i)\_YC(i)}$: the distance between the quay crane and the yard crane,
- $N_{QC(i)}$: the number of allocated shuttle trucks, and
- $J_{QC(i)\_YC(i)}$: is a parameter that calculates the jam in the journey between the quay crane and the yard crane.

Thus, it is difficult to calculate the probability:

$$P[WT_{QC(i)}(t+1)/a, WT_{QC(i)}(t)]$$

Our approach consists to simulate the operations of ship loading/unloading, the storage within the yards, and the movement of the shuttle trucks between the quay and the storage yards. Starting from a terminal configuration in term of occupancy level, the expected plan of ship arrivals, expected import/export flows, the available equipments in each work shift (quay cranes, yard cranes, straddle carriers, and shuttle trucks), and a storage policy based on some reservation/allocation criteria for all the different areas inside the terminal, we run the simulation and we retrieve the data that describe the number of waiting shuttle trucks under each quay crane at each minute. The parameters that characterize the terminal system (such as: $R_{QC(i)}$, $R_{YC(i)}$, $D_{QC(i)\_YC(i)}$, $N_{QC(i)}$, $J_{QC(i)\_YC(i)}$, etc.) are also introduced to the simulator. The service time of each quay or yard crane is considered as a random variable that follows the exponential laws. Similarly, travel time of a truck between two points inside the terminal is a random variable following the Gaussian law. Other probability lows are proposed by the simulator, but they are not adapted to this case. The data are used to estimate the probability:

$$P[WT_{QC(i)}(t+1)/a, WT_{QC(i)}(t)], \forall a \in A$$

This has enabled us to calculate the matrix: $[P_{s,s',a}]; s \in S, s' \in S, a \in A$

4 Solution and Interpretation

The value iteration algorithm applied to this problem converges to the following policy:

Table 2 The optimal policy obtained by value iteration algorithm.

$qc^{(i)}(t)$			Action
$OP^{(i)}(t)$	$ST^{(i)}(t)$	$WT_{QC(i)}(t)$	
'loading operation'	*'at work'*	*<=6*	*a3:'Stay the same'*
'loading operation'	*'at work'*	*>6*	*a2:'liberate'*
'loading operation'	*'wait'*	*0*	*a1:'allocate'*
'loading operation'	*'idle'*	*Any*	*a2:'liberate'*
'unloading operation'	*'at work'*	*<=6*	*a3:'Stay the same'*
'unloading operation'	*'at work'*	*>6*	*a2:'liberate'*
'unloading operation'	*'wait'*	*0*	*a1:'allocate'*
'unloading operation'	*'idle'*	*Any*	*a2:'liberate'*

In order to test the obtained solution, we simulated the arrival of a ship with 2500 containers to unload. We save the number of waiting shuttle trucks under the quay crane each five minutes. The fig. 2 reports the queue length without policy and the fig. 3 displays the number of waiting shuttle trucks when we apply the optimal policy obtained by value iteration algorithm. The adoption of this policy allows the quay crane to work incessantly, while reducing the congestion of trucks traveling between the quay and storage areas. But this solution requires the availability of a sufficient number of trucks and stacking equipments. We also note that the number of trucks in the queue becomes increasingly stable over time.

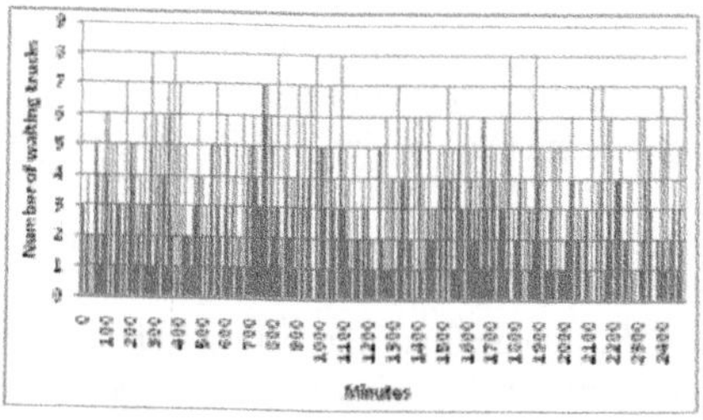

Fig. 2 Queue length without policy

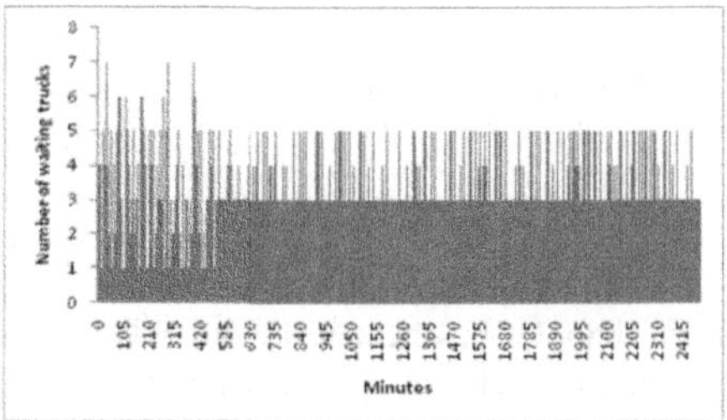

Fig. 3 Queue length with the optimal policy

5 Conclusion

In this paper, Markov Decision Process (MDP) methodology was proposed and applied to the container terminal in order to improve efficiency of loading and unloading operations. MDP provides an interesting framework to model the decision making, and the compact representation of the system provided by a MDP makes the decision strategy easier to implement. Value iteration algorithm, as well as another reinforcement learning methods like Q-learning, SARSA, mainly consist of finding the optimal policy that associates each state of the system to the action to be taken. The simulation is valuable tool that allowed us to calculate the transition matrix required to implement the algorithm, and also allowed us to simulate the obtained solution.

Even the approach using value iteration algorithm has been rarely used in the domain of controlling a container terminal system, this work, by showing its performance, has made it an attractive one to pursue.

Nevertheless, some possible perspectives to this work could be to develop the model more in order to make it more realistic by avoiding some assumptions such as the selected lows of probability and the possibility of vehicles and stack equipments failure.

References

[1] Bielli, M., Boulmakoul, A., Rida, M.: Object oriented model for container terminal distr buted simulation. European Journal of Operational Research 175(3), 1731–1751 (2006)

[2] Cartenì, A., Cantarella, G.E., de Luca, S.: A simulation model for a container terminal. In: Proceedings of the European Transport Annual Meeting: Transportation Planning Methods, PTRC, London (2005)

[3] Cartenì, A., de Luca, S.: Simulation of the Salerno Container Terminal. In: Technical Paper, Dept. Of Civil Engineering, University of Salerno (2009)

[4] Dutech, A.: Solving POMDP's Using Selected Past Events. In: Proceedings of the 14th ECAI 2000 (2000)

[5] Guestrin, C., Koller, D., Parr, R., Venkataraman, S.: Efficient Solution Algorithm for Factored MDPs. Journal of Artificial Intelligence 19, 399–468 (2003)
[6] Stahlbock, R., Voß, S.: Operations research at container terminals: a literature update. OR Spectrum 30, 1–52 (2008)
[7] Sutton, R.S., Barto, A.G.: Reinforcement Learning: An Introduction. The MIT Press, Cambridge (1998)
[8] Zhao, Q.H., Chen, S., Leung, S.C.H., Lai, K.K.: Integration of inventory and transportation decisions in a logistics system. Transportation Research Part E: Logistics and Transportation Review, vol. 46(6), pp. 913–925 (2010)

Calibration of Equilibrium Traffic Assignment Models and O-D Matrix by Network Aggregate Data

Leonardo Caggiani and Michele Ottomanelli

Abstract. In this paper a Generalized Least Square estimator for the simultaneous estimation of O-D matrix and equilibrium traffic assignment model parameters is presented. The problem is formulated as fixed-point model (equilibrium programming) assuming the congested network case. In the optimization step the variability of both O-D demand vector and the matrix of link choice probabilities is considered. We assume as input information a set of observable network data, such as link traffic counts and travel time, as well starting estimates of both O-D matrix and models parameters. Along the paper, the theoretical aspects of the proposed estimator, the solution algorithm as well as the results of numerical applications are discussed.

1 Introduction

Achieving reliable estimation of link flow over road networks is an important issue for transportation planners and engineers. Traffic Assignment Models (TAM) allow to simulate how Origin-Destination (O-D) travel demand loads the links of a given traffic network. The simulation performances of TAM are linked to the quality of O-D matrix as well as to the path-choice and supply models. Another requirements for specifying an effective TAM is the calibration of the model coefficients (e.g. link cost functions parameters or demand model parameters) whose value could strongly affect the estimation performances of the assignment process. Most of the methods presented in literature are addressed to the O-D matrix estimation and link flow simulation neglecting the importance of the traffic model parameters or assume as given their values. In fact, usually, such parameters are assumed "a priori" on the basis of subjective knowledge or transferring calibration results relevant to similar study areas as well as using values from handbooks. Besides, the calibration of such

Leonardo Caggiani · Michele Ottomanelli
Department of Environmental Engineering and Sustainable Development,
Technical University of Bari, viale De Gasperi, 74100 Taranto, Italy
e-mail: {l.caggiani,m.ottomanelli}@poliba.it

A. Gaspar-Cunha et al. (Eds.): Soft Computing in Industrial Applications, AISC 96, pp. 359–367.
springerlink.com

parameters is very difficult for the lack of data and/or because of the high cost of direct surveys, thus it is necessary to define calibration methods based on cheap and easy and immediate to collect data. A lot of effective and theoretically consistent methodologies are proposed in literature to estimate O-D matrix by using aggregate data such as traffic counts (i.e. observed link traffic flows) [4] but very few deal with the calibration of traffic model parameters by aggregate data (e.g. [12, 13, 8, 9, 7, 6]). Particularly in [10] a simultaneous Multinomial Logit calibration and O-D matrix estimation using traffic counts is proposed.

In this paper a Generalized Least Square (GLS) estimator [2] is presented to solve the simultaneous assignment model calibration and O-D matrix estimation based on aggregate data (traffic counts and/or travel time measurements). The calibration problem is specified as an optimization problem using a fixed point formulation assuming the congested network case (equilibrium programming). We assume as available information a set of link traffic counts and/or travel time measurements and starting estimates of the OD travel demand vector and of the unknown parameters.

The GLS estimator determines the O-D matrix and the TAM parameters by minimizing the weighted Euclidean distances between the vectors of the available data and the solution vectors. The solution of the proposed problem can be carried out using heuristic algorithms based on the method of successive averages. Besides, we assume the assignment matrix (link choice probabilities) as variable within the optimization algorithm, taking into account the effects of congestion on user's choice for each iteration. With respect to the methods addressed to adjust only the O-D matrix, we also obtain a calibrated assignment model that can be used to simulate new interventions on the transportation network. In the paper, we will present the statistical performance of the proposed estimator and the results, that will also be compared to the performance of other approaches, such as the traditional O-D matrix estimation with traffic counts. The early obtained results show that the proposed approach is very interesting.

2 Proposed Calibration Model

We assume the case of congested networks, therefore the considered traffic assignment model is based on a Stochastic User Equilibrium (SUE) [1]. User's path choice behaviour is usually modelled on the basis of random utility theory [5] that estimates the probability p_k that a user i chooses the path k belonging to his/her path choice set I_i. The simplest and most applied model to specify the probability p_k that user chooses the path k is the Multinomial Logit model; this model assumes the distribution of random residual term as a Gumbel variate with parameter θ that leads to the following formulation:

$$p_k = \frac{\exp(-C_k \cdot \theta)}{\sum_{h \in I}(-C_k \cdot \theta)}$$

where C_k is a linear function of path cost attributes $x_w \in \mathbf{x}$ and parameters $\beta_w \in \mathbf{b}$:

$$-C_k = -\sum_w \beta_w x_w \quad \forall k \in I_i$$

In general, let $\mathbf{t} = (\mathbf{b}, \theta)$ and assumed $\mathbf{x}$ fixed, the links choice probability is a function of $\mathbf{t}$. The parameters vector $\mathbf{t}$ has to be estimated by using a model calibration procedure. The aim of this paper is to calibrate simultaneously β_w parameters and the O-D matrix using traffic counts assuming the congested network case. The most general form to solve the estimation of O-D matrix using traffic counts can be formulated as an optimization problem [13, 2, 3]:

$$\mathbf{d} = \arg\min_{d \in S_d} \left[\mathrm{F}_1(\mathbf{d}, \bar{\mathbf{d}}) + \mathrm{F}_2(\mathbf{f}, \hat{\mathbf{f}}) \right]$$

where:

- S_d is the feasible solution set to the problem.
- F_1 and F_2 are, respectively, measures of the "distance" between the starting estimate $\bar{\mathbf{d}}$ (target demand vector) and the unknown vector $\mathbf{d}$ and between estimated link flows vector $\mathbf{f}$ and traffic counts vector $\hat{\mathbf{f}}$.
- $\mathbf{f} = \mathbf{Hd}$ with entries $f_l = \Sigma_{od} h_{l,od}\, d_{od}$.
- $\mathbf{H} = \mathbf{Ap}(\Theta) = \{a_{l,k}\ p_k\} = \{h_{l,od}\}$ is the assignment matrix (l = link index, od = OD couple index).
- $\mathbf{A} = \{a_{l,k}\}$ is the link-path incidence matrix.

According to this approach the formulation proposed to calibrate also the distribution parameter is:

$$(\mathbf{d}^*, \mathbf{b}^*) = \arg\min_{\substack{d \in S_d \\ \beta_w \in S_b}} \left[\mathrm{F}_1(\mathbf{d}, \bar{\mathbf{d}}) + \mathrm{F}_2(\mathbf{f}, \hat{\mathbf{f}}) + \mathrm{F}_3(\mathbf{c}(\mathbf{b}), \hat{\mathbf{c}}) \right] \quad (1)$$

where:

- S_b is the feasible solution set for parameters vector $\mathbf{b}$.
- F_3 is a measure of the "distances" between the estimated link costs $\mathbf{c}(\mathbf{b})$ (calculated with the estimated parameters vector $\mathbf{b}^*$) and link costs measured on observed links $\hat{\mathbf{c}}$.

The problem 1 can be represented as a bi-level programming problem or as a fixed point formulation [11]. In this paper the calibration problem is solved using a fixed point formulation (equilibrium programming):

$$(\mathbf{d}^*, \mathbf{b}^*) = \arg\min_{\substack{d \in S_d \\ \beta_w \in S_b}} \left[\mathrm{F}_1(\mathbf{d}, \bar{\mathbf{d}}) + \mathrm{F}_2(\mathbf{H}((\mathbf{c}(\mathbf{f}(\mathbf{d}))), \mathbf{b})\mathbf{d}, \hat{\mathbf{f}}) + \mathrm{F}_3(\mathbf{c}((\mathbf{f}(\mathbf{d})), \mathbf{b}), \hat{\mathbf{c}}) \right] \quad (2)$$

The distances F_1, F_2 and F_3 can be defined following different statistical approaches and assumptions. In this work the Generalized Least Square (GLS) estimator has been assumed because it is robust and it is the most efficient linear unbiased estimator [2]. For the GLS estimator the measure of distance assumes the form of a weighted Euclidean metric; then the F_1, F_2 and F_3 of optimization 2 becomes:

$$F_1^{GLS} = (\mathbf{d}, \bar{\mathbf{d}})^T \mathbf{W}^{-1}(\mathbf{d}, \bar{\mathbf{d}}) \tag{3}$$

$$F_2^{GLS} = (\mathbf{H}((\mathbf{c}(\mathbf{f}(\mathbf{d}))), \mathbf{b})\mathbf{d} - \hat{\mathbf{f}})^T \mathbf{V}^{-1}(\mathbf{H}((\mathbf{c}(\mathbf{f}(\mathbf{d}))), \mathbf{b})\mathbf{d} - \hat{\mathbf{f}}) \tag{4}$$

$$F_3^{GLS} = (\mathbf{c}((\mathbf{f}(\mathbf{d})), \mathbf{b}) - \hat{\mathbf{c}})^T \mathbf{Y}^{-1}(\mathbf{c}((\mathbf{f}(\mathbf{d})), \mathbf{b}) - \hat{\mathbf{c}}) \tag{5}$$

where **W**, **V** and **Y** are the variance-covariance matrices of the error for the target demand vector, for the link flows and for the link choice probability respectively. **W**, **V** and **Y** represent the weights of the available information that can be interpreted as the level of confidence (or the reliability) in the available starting data.

In general, these optimization problems are solved with algorithms assuming the assignment matrix **H** as fixed into the whole procedure [4, 9, 3, 11]. This procedure is suitable when the link costs are known or the network is not congested. If we assume a congested network, the user's choice is affected by the congestion since link flows and link costs are mutually dependent. For this reason it is necessary to consider, in TAM and O-D calibration problem, the effects due to congestion on link flow estimation too. For instance, this problem is solved fixing a new assignment matrix for each iteration [3] but in the optimization step matrix **H** is a fixed constant. As it will be highlighted, in the proposed method, assignment matrix is assumed varying also within each optimization step.

3 Algorithm for Problem Solution

In this section we propose an algorithm resolution for the simultaneous calibration algorithm using GLS estimator and stochastic assignment model (SUE). We consider, as variables, O-D demand vector and **b** parameters. So, assignment matrix will be variable too.

- *Step 0:* inizialize counter ($z = 0$), initialize demand vector ($\mathbf{d}_0 = \bar{\mathbf{d}}$), initialize parameter values ($\mathbf{b}_0 = \bar{\mathbf{b}}$);
- *Step 1:* update $z = z + 1$;
- *Step 2:* solve the optimization problem 2 (distances specified in 3, 4 and 5) with $\mathbf{H}_z^{SUE} = \mathbf{H}(\mathbf{c}(\mathbf{f}(\mathbf{d}_k))$ (SUE Logit assignment problem is solved with MSA-FA algorithm: Method of Successive Averages Flow Averaging). Optimization is solved by using the SQP algorithm (Sequential Quadratic programming); the starting point of the optimization algorithm (SQP) is $\mathbf{d}_z = \mathbf{d}_z^s$ and $\mathbf{b}_z = \mathbf{b}_z^s$;
- *Step 3:* stop test on maximum relative difference between demand vector elements obtained in Step 2 and demand vector elements of previous iteration: $\max \left| d_z^{od} - d_{z-1}^{od} \right| / d_z^{od} < \varepsilon$;
- *Step 4:* new starting point of the optimization is $\mathbf{d}_{z+1}^s = \mathbf{d}_z^*$ and $\mathbf{b}_{z+1}^s = \mathbf{b}_z^*$.

4 Numerical Application

To test numerically the performances of the algorithm proposed, we have carried out the following application. The network and data considered by [13] are used as our test network and data, since the problem formulation is similar to the one they propose. The considered network has 9 nodes (3 origin centroids and 3 destination centroids), and 14 links as depicted in Figure 1.

The traffic assignment model used is a SUE Logit and the relevant Logit path choice model is given by the well known equation 1. The paths choice set I_{od} is constituted of all the possible paths of the considered network. We assume as link cost function the BPR function 6 with α and γ model parameters, where free flow travel time (tr) and capacity (ca) vary link by link.

$$c_l(f_l) = tr_l \left[1.0 + \alpha \left(\frac{f_l}{ca_l} \right)^{\gamma} \right]$$

Because of the capacity of the links we considered the linear approximation of the BPR cost function starting from the 95% of the link capacity value. The value of Logit parameter is fixed to 1.5, then the parameters of the path choice model (eq. 1) are only the parameters of the BPR function ($\mathbf{b}$ = (α, γ)). In this framework the proposed calibration model is formulated as the following fixed-point problem:

$$(\mathbf{d}^*, \alpha^*, \gamma^*) = \underset{\substack{d \in S_d \\ \alpha \in S_a, \gamma \in S_g}}{\arg\min} \left[\mathrm{F}_1(\mathbf{d}, \bar{\mathbf{d}}) + \mathrm{F}_2(\mathbf{H}(\mathbf{d}, \alpha, \gamma)\mathbf{d}, \hat{\mathbf{f}}) + \mathrm{F}_3(\mathbf{c}(\mathbf{d}, \alpha, \gamma), \hat{\mathbf{c}}) \right]$$

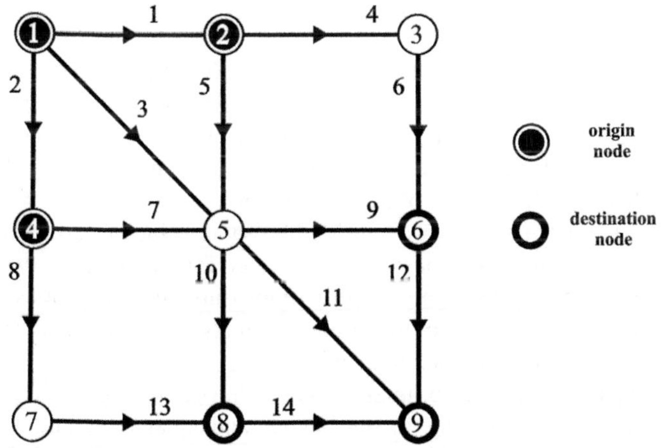

Fig. 1 Test network [13]

Table 1 True O-D matrix

O-D	1-6	1-8	1-9	2-6	2-8	2-9	4-6	4-8	4-9
$\mathbf{d}^t$	120	150	100	130	200	90	80	180	110

where:

- S_a and S_g are the feasible solution sets respectively for α and γ parameters.
- F_3 is a measure of the "distances" between the estimated link costs (calculated with the estimated α^* and γ^* parameters) and link costs measured on observed links.

The "true" values of BPR parameters are $\alpha^t = 0.15$ and $\gamma^t = 4$. The "true" O-D vector is shown in Table 1. The "true" value of link flows and link costs has been generated by SUE-Logit traffic network assignment of the assumed true O-D matrix and with the true value of θ, α and γ. Target vector of demand $\bar{\mathbf{d}}$, measured link flows vector $\hat{\mathbf{f}}$ and measured link costs vector $\hat{\mathbf{c}}$ were obtained, starting from true values, through random extractions from a normal distribution.

As many authors suggest, [4, 12] the variance-covariance matrices $\mathbf{W}$ (for demand), $\mathbf{V}$ (for flows) and $\mathbf{Y}$ (for costs) can be assumed diagonal. Let $\mathbf{d}^t$, $\mathbf{f}^t$ and $\mathbf{c}^t$ be the true values of the demand vector, flows vector and costs vector; the relevant values of variances have been computed through the following expressions:

$$\sigma^2_{d,od} = (cv_d \cdot d^t_{od})$$
$$\sigma^2_{f,l} = (cv_f \cdot f^t_l)$$
$$\sigma^2_{c,l} = (cv_c \cdot c^t_l)$$

The algorithm was applied on test network with $\varepsilon = 0.001$ as stop test for calibration algorithm. To evaluate the effect of variation coefficients (or reliability of starting data) on estimated values, numeric application has been carried out considering all the combinations between the values of demand variation coefficients ($cv_d = \{0.2, 0.4, 0.6, 0.8, 1\}$) an the values of flows variation coefficients ($cv_f = \{0.02, 0.04, 0.06, 0.08, 0.1\}$). For the costs we consider the same value of variation coefficient employed for the flows. We consider a low value for flows (and costs) variation coefficient because we give high reliability to measurements as proposed by [3]. In order to evaluate the influence of the number of links observed we have considered five different sets of links defined in Table 2.

The starting point for SQP optimization algorithm (Step 2) for the first iteration ($z = 1$) is fixed a priori considering the starting demand equal to the target demand ($\mathbf{d}_1 = \bar{\mathbf{d}}$) and all the sixteen combinations among the parameter values ($\alpha^s_1 = \{0.0375, 0.075, 0.3, 0.6\}$, $\gamma^s_1 = \{1, 2, 8, 16\}$). Similarly to [3], the numeric analysis has been made considering 30 demand target vectors, 30 measured flows vectors and 30 measured link costs for each combination of variation coefficients, observed link sets and starting points. To evaluate the statistical performances of the proposed method

Table 2 Observed link sets

sets	Set 1 (2/14)	Set 2 (5/14)	Set 3 (7/14)	Set 4 (11/14)	Set 5 (14/14)
links	6-9	6-9-10-11-13	6-9-10-11-12-13-14	1-3-4-6-8-9-10-11-12-13-14	all

the reduction of MSE (Mean Square Error) of median estimated values in comparison to the true one have been used. For example the reduction of MSE of estimated demand is:

$$red_{\mathrm{MSE}(d^*)} = \frac{\mathrm{MSE}(\overline{\mathbf{d}}, \mathbf{d}^t) - \mathrm{MSE}(\mathbf{d}^*, \mathbf{d}^t)}{\mathrm{MSE}(\overline{\mathbf{d}}, \mathbf{d}^t)}$$

where:

$$\mathrm{MSE}(\overline{\mathbf{d}}, \mathbf{d}^t) = \frac{1}{N_{OD}} \cdot \sum_{od} (\overline{d}_{od} - d^t_{od})$$

$$\mathrm{MSE}(\mathbf{d}^*, \mathbf{d}^t) = \frac{1}{N_{OD}} \cdot \sum_{od} (d^*_{od} - d^t_{od})$$

with N_{OD} equal to the number of O-D couples.

5 Model Results

For each combination proposed we obtain 30 different estimates. For this analysis (different combination of variation coefficients and starting points) the considerations and results related are relative to the obtained median values. In general, we observe that the worst values of MSE reduction are obtained for higher values of the variation coefficients. Conversely, the best results are obtained for lower values of the variation coefficients. The demand estimation performances depend on the quality of the target demand vector and traffic costs counts. As shown in Figure 2, for Set 1 of observed links, the value of MSE reduction increases as cv_d increases and decreases as cv_f decreases.

The more the number of observed links increases, the more the MSE reduction increases (from Set 1 to Set 5). The estimation performances relevant to the validation link set (hold-out sample) are better than the demand one since MSE reduction is greater than 0.15 in all cases and for the smallest observed set (Figure 2). As far as the estimation of the BPR parameter (i.e. α^* and γ^*) is concerned, we obtained median values that tend to the true one (Figure 3) with α^* between 0.150 and 0.173 and γ^* between 2.74 and 4.08.

The cv_d seems to affect slightly the estimated parameters. Instead, as reported in Figure 3 for Set 1, the more cv_f is high the more the estimated parameter is far from the true one. Considering all the results relevant to all the considered combination (not the median values) the more the starting error on demand, flows and costs is

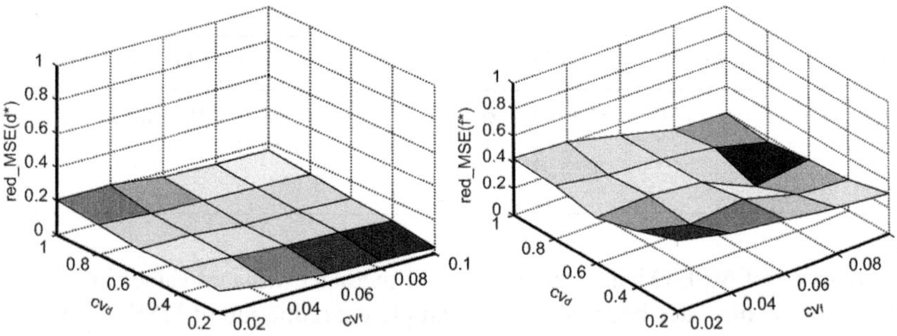

Fig. 2 Median reduction of MSE for estimated demand and hold out sample link flows (Set 1)

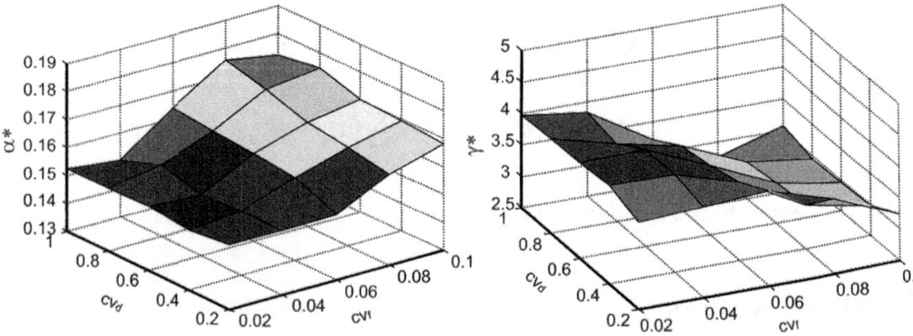

Fig. 3 Estimated median values of BPR parameters (Set 1)

low, the more the estimated values are reliable. In fact, the higher the cv_d and cv_f, the more the number of estimated values worse than the starting values increases. In particular, whereas all the values obtained of α^* are near to the median ones, the estimated values γ^* are very scattered.

6 Conclusion and Further Research

In many practical applications, O-D matrix estimation and link flow simulation are carried out without giving the proper importance to the traffic model parameters. In fact, usually, such parameters are assumed "a priori" on the basis of subjective knowledge or assuming calibration results obtained in different study areas. In this work, a GLS estimator with fixed-point formulation is proposed to estimate simultaneously and consistently the SUE-Logit assignment model parameters and the O-D matrix by using a set of available traffic counts. The application of the proposed estimator to a test network has shown the effectiveness of the approach that is able to reduce the bias induced on the considered traffic variables. Such results have been obtained without using expensive and time consuming data and computing effort, providing a method that should be helpful for practical application when limited resources for data collecting are available. It also allows to explicitly take into account the analyst's level of confidence (say data reliability) in the assumed starting data by weighting the distances between starting and final estimates.

Application to more extended networks will be a further step. In addition, the effect of the selection of traffic counts to be used in calibration is under investigation, as well as the improvement of the proposed algorithms.

References

1. Cantarella, G.E.: A General Fixed-Point Approach to Multi-Mode Multi-User Equilibrium Assignment with Elastic Demand. Transportation Science 31, 305–329 (1997)
2. Cascetta, E., Nguyen, S.: A unified framework for estimating or updating origin/destination matrices from traffic counts. Transportation Research Part B 22, 437–455 (1988)
3. Cascetta, E., Postorino, M.N.: Fixed Point Approaches to the Estimation of O/D Matrices Using Traffic Counts on Congested Networks. Transportation Science 35, 134–147 (2001)
4. Cascetta, E., Russo, F.: Calibrating Aggregate Travel Demand Model with Traffic Counts: Estimators and Statistical Performance. Transportation 24, 271–293 (1997)
5. Domencich, F.A., McFadden, D.: Urban Travel Demand: a Behavioural Analysis. North Holland, Amsterdam (1975)
6. García-Ródenas, R., Marìn, A.: Simultaneous estimation of the origin-destination matrices and the parameters of a nested logit model in a combined network equilibrium model. European Journal of Operational Research 197, 320–331 (2009)
7. Liu, S., Fricker, J.D.: Estimation of a trip table and the θ parameter in a stochastic network. Transportation Research Part A 30, 287–305 (1996)
8. Lo, H.P., Chan, C.P.: Simultaneous estimation of an origin-destination matrix and link choice proportions using traffic counts. Transportation Research Part A 37, 771–788 (2003)
9. Ottomanelli, M.: Effect of data accuracy in aggregate travel demand models calibration with traffic counts. In: Pursula, M., Niittymaki, J. (eds.) Mathematical methods on optimization in transportation systems. Kluwer Academic Publisher, Dordrecht (2001)
10. Ottomanelli, M., Caggiani, L., Sassanelli, D.: Simultaneous path choice model calibration and O-D matrix estimation using traffic counts: a GLS estimator for congested network. In: Proceedings of WCTR 2007, Berkeley, California (2007)
11. Ottomanelli, M., Di Gangi, M.: Aggregate Calibration of Path Choice Models for Congested Networks. In: Proceedings of the 8th EURO-Working Group Meeting on Transportation, Rome, Italy (2000)
12. Russo, F., Vitetta, A.: Updating O/D matrices and calibrating link cost functions jointly from traffic counts and time measurements. In: Proceedings of TRISTAN V, Le Goisier, Guadalupe, French West Indies (2004)
13. Yang, H., Meng, Q., Bell, M.G.H.: Simultaneous estimation of the origin-destination matrices and travel-cost coefficient for congested networks in a stochastic user equilibrium. Transportation Science 35, 107–123 (2001)

compared with [illegible]

References

1. [illegible] with [illegible] (1997)
2. [illegible]
3. [illegible]
4. [illegible]
5. [illegible] (2007)
6. [illegible]
7. [illegible]
8. [illegible]

[illegible]

[illegible]

[illegible]

A Fuzzy Logic-Based Methodology for Ranking Transport Infrastructures

Giuseppe Iannucci, Michele Ottomanelli, and Domenico Sassanelli

Abstract. Transport companies in many cases have to evaluate their competitiveness, comparing it with that of their competitors. Usually this assessment is performed through one or more indices representing facility performances, derived from a set of indicators relevant to problem representation. If the aim is to estimate the user evaluation for the service offered by a facility, the development of a synthetic index can be difficult since user's choice is often characterized by significant uncertainties and it is not always governed by certain rules and rational behaviour, so that it could not be easily and explicitly represented by traditional mathematical techniques and models. Such uncertainties in the relationship between indicator values and facility attractiveness can be properly defined by explicitly specifying them in an approximate way using fuzzy sets theory. In this paper an innovative approach for the classification of Transport Facilities is proposed. The method is based on a Fuzzy Inference System and may be employed both as a benchmarking/ranking procedure and as a decision support tool to evaluate future scenarios as a result of facilities remodelling.

1 Introduction and Background

In a world context characterized by the lack of resources due to economic crisis and, at the same time, by a high competition on global scale it is increasingly important for company managers to assess reliably their corporate performance.

Giuseppe Iannucci · Domenico Sassanelli
Dept. of Roads and Transportation Technical University of Bari - Via Orabona, 4 - 70125 Bari, Italy
e-mail: g.iannucci@poliba.it, d.sassanelli@poliba.it

Michele Ottomanelli
Dept. of Environmental Engineering and Sustainable Development
Technical University of Bari - Viale De Gasperi - 74100 Taranto, Italy
e-mail: m.ottomanelli@poliba.it

A. Gaspar-Cunha et al. (Eds.): Soft Computing in Industrial Applications, AISC 96, pp. 369–377.
springerlink.com

Also transport companies, whether in public, private or mixed management, in many cases have to evaluate their competitiveness in comparison to their competitors. There are many possible examples: airport managers, logistics platform managers, container terminal operators, ferries or cruise ships ports managers, etc., have to face frequently with concurrent facilities.

The ability to expand, or at least maintain their current level of business is therefore subject to continuous evaluation and improvement of business competitiveness.

Many methodological approaches to this problem have been proposed in the field of Transport Facilities (TF). For example, in the field of road infrastructures a well-established methodology for classifying the quality of travel conditions of vehicles on a given road section, is based on the concept of Level of Service [1]. A similar methodology has been introduced by Ballis [2] to evaluate the performances of intermodal freight terminals. Several studies [3, 4, 5, 6, 7] propose methodologies for the benchmarking of port container terminals. From the methodological point of view they are usually based on the statistical analysis of data resulting from observation of a set of facilities, or on operational research methods such as Data Envelopment Analysis, Stochastic Frontier Analysis, multicriteria classification. These methods have the disadvantage that, to provide consistent results, require significant amount of data to calibrate model parameters. Moreover, to obtain an adequate representation of the problem is often indispensable to include in the model a considerable number of TF features: this fact leads inevitably to an increase in parameters to be calibrated. As often in TF field the amount of calibration data needed for classic operational research methods are not available, and, when available, these data are affected by imprecision and vagueness of information, in our opinion a soft computing approach allows to balance these deficiencies with the need to get useful information, with a degree of accuracy not very high, but consistent with analysis purposes.

2 Problem Statement

All methods described in the previous section are based on the use of a set of indicators relevant to the problem representation, leading to a quantification of facility performances based on one or more synthetic indices.

The choice of these indices, and the criterion for the classification of the facility, depends on the purposes of the analysis. If the objective of the analysis is to evaluate company efficiency, it could be made through the evaluation of appropriate productivity indices, possibly combined with appropriate weights to get a synthetic efficiency index. If the aim is to evaluate the user rating of the facility for the offered service, therefore the ability of a given TF in attracting shares of transport demand, then the development of a synthetic index becomes more problematic. In fact, in this case it is difficult to combine the indicators, because they are often characterized by heterogeneity and vagueness. This is due mainly to the fact that user's choice is characterised by significant uncertainties and it is not always governed by certain rules and rational behaviour, so that it can't be easily and explicitly represented by traditional mathematical techniques and models. In our opinion, such uncertainties in the relationship between indicator parameters and facility attractiveness can be properly defined by explicitly specifying them in an approximate way by using the theoretical framework of fuzzy sets.

For these reasons, in this paper it is proposed an approach to the classification of a certain class of Transport Facilities, based on a Fuzzy Inference System (FIS).

The method may be employed both as a benchmarking/ranking procedure and as a decision support tool to evaluate future scenarios as a result of facilities remodelling.

3 Benchmarking through a Fuzzy Inference System

Starting from a set of input indicators (facility features), the proposed methodology has as output one indicator that describes an overall attractiveness measure of a Transport Facility.

Assuming that the attraction of a given TF, for a generic user, is related to its features, the behaviour of a human decision-maker that has to choose the best TF, or has to rank a group of TF, is simulated through a Fuzzy Inference System (FIS) [8, 9].

3.1 FIS Input Parameters

The inputs of the FIS algorithm are the features c_i with $i \in [1, 2, \ldots n]$ of a TF, the output is a Level of Attractiveness (LA) p of a TF. The possible values of c_i and p are defined into respective bounded definition sets ($c_i \in S_{ci}$, $p \in S_p$).

In classical benchmarking procedures the definition sets are divided into a finite number of subsets and a value of a variable may belong or not to a defined subset. Given a number m of subsets (levels), in the proposed method each feature definition set S_{ci} is divided into m fuzzy subsets $I_{i,v}$ with $v \in [1, 2, \ldots m]$. In the same way, S_p is split into m fuzzy subsets O_v with $v \in [1, 2, \ldots m]$.

In fuzzy logic a value belongs to a set with a certain degree of membership defined in the *interval* [0, 1], rather than to the *set* {0, 1}. Usually, each fuzzy subset is defined by a linguistic value that could be "low", "high", "short", "long", etc. In the current case, given m LA, the linguistic judgment corresponding to each fuzzy subset is just the corresponding LA.

The degree of membership to a set is defined by a *membership function (MF)*. In this framework a value c_i^* belongs to a subset $I_{i,v}$ depending on the membership function $\mu_{i,v}(c_i) \in [0, 1]$ and, in the same way, a value p^* belongs to a subset O_v depending on the membership function $\mu_v(p) \in [0, 1]$. The more a value belongs to a LA, the more the degree of membership is near to one. Given a shape for each MF, they may be identified by their typical parameters. For example a triangular or a trapezoidal shape can be defined by the position of the vertexes.

If we consider, as an example, a set of airport facilities and we want to classify them according to their LA for airline companies, we may assume, among others, as one of feature inputs to the classification process, the number of boarding gates in the air terminal. If c_1 is the number of boarding gates, assuming to classify airports according to 3 LA ($m = 3$), $I_{1,1}$ would represent the degree of membership to "Level C", $I_{1,2}$ to "Level B" and $I_{1,3}$ to "Level A", where level A means the level of facilities with higher power of attractiveness, and level C being the lower. For triangular membership functions the fuzzy subsets can be defined as depicted in Figure 1. In this way an airport with 8 gates ($c_1 = 8$) is "level C" with a degree of membership equal to $\mu_{1,1}(8) = 0.79$ and is "Level B" with a degree of membership equal to $\mu_{1,2}(8) = 0.2$; in other words, 8 boarding gates belong more to the subset "Level C" than to the subset "Level B" and do not belong to the subset "Level A".

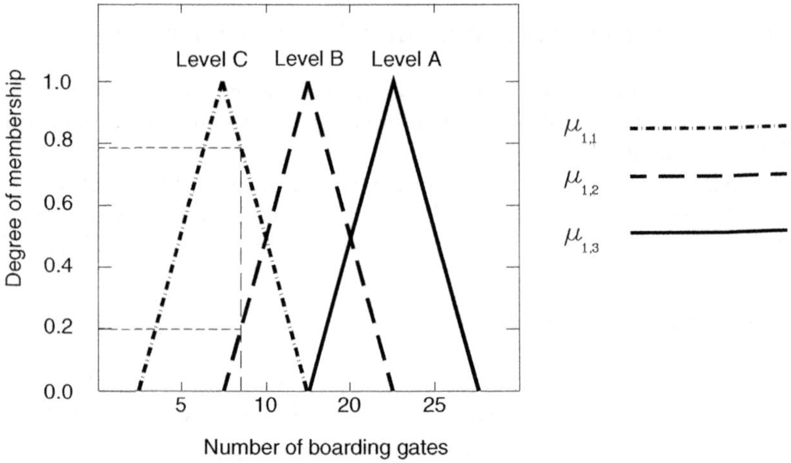

Fig. 1 Membership functions of the feature "Number of boarding gates"

3.2 *FIS Logical Rules and Output*

Generally the number of TF relevant features n is greater than one. Then it is necessary to combine different input values with fuzzy logical operators like fuzzy intersection or conjunction (AND), fuzzy union or disjunction (OR) and fuzzy complement (NOT). In most cases the most suitable for feature aggregation is the AND operator and the general form of FIS rules can be summarized into m rules as follows:

$$\textbf{\textit{IF}}\ c_1 \text{ is } I_{1,1} \textbf{ AND } c_2 \text{ is } I_{2,1} \textbf{ AND } c_3 \text{ is } I_{3,1} \ldots \textbf{ AND } c_n \text{ is } I_{n,1}\ \textbf{\textit{THEN}}\ p \text{ is } O_1$$

....

$$\textbf{\textit{IF}}\ c_m \text{ is } I_{1,m} \textbf{ AND } c_2 \text{ is } I_{2,m} \textbf{ AND } c_3 \text{ is } I_{3,m} \ldots \textbf{ AND } c_n \text{ is } I_{n,m}\ \textbf{\textit{THEN}}\ p \text{ is } O_m$$

The proposed model is therefore a FIS with different rules; each rule may consider all or a portion of n features combined with each other in the if-then statements with one type or different logical fuzzy operators. In this FIS structure, given the input values for a TF, after the application of an aggregation method [8] of the results of each rule and after defuzzification, it is possible to obtain the output value for LA.

3.3 Algorithm Tuning

To obtain accurate and realistic results, the algorithm has to be calibrated. First of all, it is important a preliminary selection of c_i relevant for the particular class of TF to be examined. Then the process involves MF shape choice, definition of MF feature parameters, and identification of appropriate logical rule of combination of membership degrees to different TF features.

Depending on the quantity of data available fortuning or on the complexity of the phenomenon to analyze, three different methods of calibration can be implemented.

3.3.1 Tuning by Expert Assessments

This approach is based on performing a series of market surveys/interviews to TF members and users, to link their assessment on the ranking of a TF sample, to the corresponding c_i^* values.

3.3.2 Data Correlation Analysis

The approach is based on the correlation of c_i^* values for a sample of TF, with corresponding p^* values. To this purpose, a measurable parameter has to be taken as indicator of membership to a certain LA for that type of TF. In order to minimize the size of calibration sample, compared to a large number of parameters to be calibrated, this procedure may be carried out by splitting the definition subsets into x equispaced intervals and adopting for $I_{i,v}$ and O_v shapes with few definition parameters (triangular or trapezoidal). MF vertex exposition will undergo adjustments to suit FIS output data with calibration data.

3.3.3 ANFIS Learning Procedure

In this case a series of c_i^*, relevant to the considered TFs, is correlated to the corresponding p^* values by an Adaptive Neural Fuzzy Inference system [9]. However, this procedure requires a large amount of calibration data to be performed.

4 Application of the Methodology

The methodology described has been implemented for the classification of Mediterranean Sea container terminals, on the basis of their attitude to attract container shipping companies, and thus on their container throughput per year (TEUs /year).

The classification is based on 4 levels (A, B, C, D) representative, in descending order, of terminal importance and thus related to the number of TEUs/year handled in the terminal. The methodology for classification in this case involves the identification of features relevant to port classification (Table 1).

The definition sets of each of the 8 features and of the LA indicator have been split into 4 fuzzy equispaced subsystems, represented by 4 MF with trapezoidal shape. An example of MF for feature1 is shown in Figure 2. Then 4 fuzzy logic combination rules have been implemented (Table2). It may be noted that, as all the features considered in Table 1 contribute to increase performances and attractiveness of a terminal container, they have been combined by taking only AND logic operator.

Table 1 Features relevant to maritime terminal container classification.

I	Feature (c_i)	Units of measure
1	Distance of port position from Gibraltar-Suez course	nautical miles
2	Rates of TEUs sorted by road or rail	%
3	Quay length	Meters
4	Stacking area	square meters
5	Quay cranes	number of cranes
6	Maximum water draft	Meters
7	Connected ports	number of ports
8	Connected HUB ports	number of ports

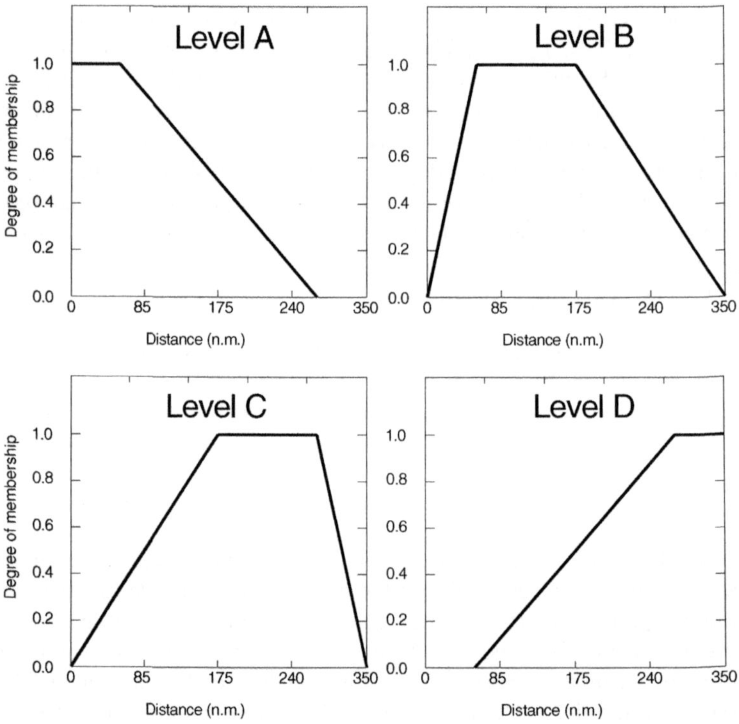

Fig. 2 Membership functions for "Distance of port position from Gibraltar-Suez course" feature.

For algorithm calibration and validation, 18 ports have been considered (Table 3). For each of them c_i^* values have been collected. Given the limited number of available data, the methodology of data correlation analysis showed in previous sections has been applied. Thus c_i^* values were related, for each port, to the corresponding TEUs/year handled. In this case p^* values are the number of TEUs handled in each port in the year 2008. The sample has been divided into two groups each consisting of 9 ports. Data on the ports of the first group were used for model calibration, while data from the second group were used to validate the model. In calibration process, MF have been tuned to obtain as output from the FIS model, p values comparable with those of the sample. In Table 4 are shown the results of process validation.

Table 2 FIS logical rules

N.	Rule
1	***IF*** c_1 is*Level A* **AND** c_2 is*Level A* **AND** c_3 is *Level A* **AND** c_4 is *Level A* **AND** c_5 is*Level A* **AND** c_6 is *Level A* **AND** c_7 is *Level A* **AND** c_8 is *Level A* ***THEN*** *LA is "A"*
2	***IF*** c_1 is *Level B* **AND** c_2 is *Level B* **AND** c_3 is *Level B* **AND** c_4 is *Level B* **AND** c_5 is *Level B* **AND** c_6 is *Level B* **AND** c_7 is *Level B* **AND** c_8 is *Level B* ***THEN*** *LA is "B"*
3	***IF*** c_1 is *Level C* **AND** c_2 is *Level C* **AND** c_3 is *Level C* **AND** c_4 is *Level C* **AND** c_5 is *Level C* **AND** c_6 is *Level C* **AND** c_7 is *Level C* **AND** c_8 is *Level C* ***THEN*** *LA is "C"*
4	***IF*** c_1 is *Level D* **AND** c_2 is *Level D* **AND** c_3 is *Level D* **AND** c_4 is *Level D* **AND** c_5 is *Level D* **AND** c_6 is *Level D* **AND** c_7 is *Level D* **AND** c_8 is *Level D* ***THEN*** *LA is "D"*

Table 3 Data base for FIS calibration and validation

Port	c_1	c_2	c_3	c_4	c_5	c_6	c_7	c_8	p
Valencia	140	0.56	4162	1365420	30	16	27	10	3602112
GioiaTauro	66	0.07	3395	1700000	30	18	44	14	3467772
Algeciras	0	0.05	9823	866132	37	17	19	10	3324310
Port said est/west	0	0.13	2400	1242000	19	16.5	20	10	3257984
Barcelona	209	0.61	4065	908000	17	16	40	12	2569549
MaltaFreeport	6	0.00	2426	683000	23	17	57	14	2330000
Genoa	350	0.92	4141	1619355	18	15	31	12	1766605
Piraeus	178	0.39	2774	626000	14	18	39	13	1403408
Haifa	169	0.59	1360	500000	12	14	28	11	1395900
Alexandria/El Dekh	32	0.92	2045	571304	8	14	38	14	1259000
Damietta	0	0.26	1050	254231	10	14.5	20	10	1236502
Izmir	345	0.87	1050	295000	5	14.5	43	14	884000
Mersin	339	0.90	1470	1100000	5	14	35	14	868000
Marseille	275	1.00	2127	560000	13	14.5	41	14	847651
Ashdod	125	1.00	2850	500000	11	15.5	26	11	827900
Taranto	172	0.10	1500	650000	10	15	20	7	786655
Lattakia	310	0.83	4280	500000	4	13.3	29	10	570000
Cagliari	70	0.60	1580	435000	7	16	43	13	252837

The value of LA calculated with the proposed procedure match the expected ones in 9 out of 9 cases, while the mean percentage error on the number of TEUs/year handled is of 38.9%.

As expected, the small number of data used for calibration, compared with the number of features considered, leads to results characterized by a relevant, though not excessive, mean percentage error. However the proposed methodology allows to obtain the required information, namely to assign a given port to the right class, according to its relevant characteristics, with an high level of accuracy. In conclusion, given the complexity of the problem, and the lack of available data, the proposed approach can still provide useful information for the analyst, with an appropriate degree of accuracy.

Table 4 Results of algorithm validation.

Port	TEUs/year (p)	Expected Level	TEus/year from FIS	Level from FIS
Valencia	3602112	A	3120000	A
Algeciras	3324310	A	4000000	A
Barcelona	2569549	B	1960000	B
MaltaFreeport	2330000	B	2490000	B
Genoa	1766605	C	1370000	C
Haifa	1395900	C	1280000	C
Izmir	884000	C	1370000	C
Taranto	786655	D	406000	D
Lattakia	570000	D	406000	D

5 Conclusions and Further Development of the Research

The proposed classification procedure is able to provide useful results for evaluating comparatively the performance of different TF. The approach based on fuzzy Level of Attractiveness, makes it possible to get information with a degree of approximation sufficient and useful for analysis purposes, even in presence of uncertainty due to the high number of calibration parameters and, at the same time, to a low number of calibration data.

Moreover, the described procedure is suitable for strategic evaluation of the effectiveness of any action aimed at increasing the competitiveness of a TF. In fact, for a given TF, implementing change scenarios of the numerical values of one or more features, it is possible to evaluate the consequent potential attractiveness.

Future development of the research will concern application of the methodology proposed to a TF field characterized by the availability of enough data for model calibration with ANFIS. In this context, in order to raise as much as possible the size of the calibration database, time series of c_i^* and p^* values for each TF should be considered.

The final goal of the research is to propose, for TF fields in which this approach would result feasible and useful, a benchmarking system able to produce a unique and universally accepted classification of TFs, easily updatable during time.

Acknowledgements. The work was partially supported by the Italian Ministry for Education, University and Research (MIUR) in the Program PRIN2007, project 2007LPAM25_003.

References

1. Transportation Research Board. Highway Capacity Manual (HCM), National Academies, Washington D.C (2000)
2. Ballis, A.: Introducing Level-of-Service Standards for Intermodal Freight Terminal. J. of the Transportation Research Board 1873, 79–88 (2004)
3. Quaresma Dias, J.C., Garrido Azevedo, S., Ferreira, J., et al.: A comparative benchmarking analysis of main Iberian container terminals: a DEA approach. Int. J. of Shipping and Transport Logistics 1, 260–275 (2009)
4. Organization for Economic Co-operation and Development. Benchmarking Intermodal freight transport IRTD No E112021, OECD, Paris (2002)
5. Cullinane, K., Wang, T.F., Song, D.W., et al.: The technical efficiency of container sea ports: Comparing data envelopment analysis and stochastic frontier analysis. Transp. Res. Part A 40, 354–374 (2006)
6. Roll, Y., Hayuth, Y.: Port performance comparison applying data envelopment analysis (DEA). Maritime Policy and Management 20, 161–163 (1993)
7. Sharma, M.J., Yu, S.J.: Performance based stratification and clustering for benchmarking of container terminals. Expert Systems with Applications 36, 5016–5022 (2009)
8. Zimmermann, H.J.: Fuzzy Set Theory and Its Applications. Kluwer, Dordrecht (1996)
9. Jang, J.S.R.: ANFIS: Adaptive-network-based fuzzy inference systems. IEEE Trans. on Systems, Man, and Cybernetics 23, 665–685 (1993)

Transferability of Fuzzy Models of Gap-Acceptance Behavior

Rossi Riccardo, Gastaldi Massimiliano, Gecchele Gregorio, and Meneguzzer Claudio

Abstract. The transferability of fuzzy models of gap-acceptance behavior between different intersections is evaluated in this paper using a method known as ROC curve analysis. The results of an application to four unsignalized intersections indicate that, even if transferred models generally perform adequately, intersection geometric and traffic characteristics may be very important in determining the capability of a model developed in a given context to reproduce gap-acceptance behavior observed in other contexts.

Keywords: Gap-acceptance, Fuzzy Models, Transferability.

1 Introduction

In studies of vehicular gap-acceptance behavior, the choice to accept or reject a gap of a certain size is generally considered the result of a driver decision process which includes, as inputs, subjective estimates of a set of explanatory variables, given specific objective factors. These subjective evaluations are usually affected by a high degree of uncertainty, which can be properly treated both by classical probabilistic models ([4],[18],[19],[12],[7],[16]) and by fuzzy system theory ([15],[17]); calibration and validation of these models is usually based on gap-acceptance data collected at real intersections using, for instance, observations based on video surveys. The primary objective of the present work is to evaluate, with specific reference to gap-acceptance fuzzy models, the possibility of performing a successful transfer of such models from an original context to a different context. This objective is directly

Rossi Riccardo · Gastaldi Massimiliano · Gecchele Gregorio · Meneguzzer Claudio
University of Padova - Department of Structural and Transportation Engineering
Via Marzolo 9 35131, Padova, Italy
e-mail: riccardo.rossi@unipd.it, massimiliano.gastaldi@unipd.it, gecchele@dic.unipd.it, claudio.meneguzzer@unipd.it

A. Gaspar-Cunha et al. (Eds.): Soft Computing in Industrial Applications, AISC 96, pp. 379–390.
springerlink.com

related to the practical requirement to include in micro-simulation models or, more generally, in intersection operational analysis, behavioural models (for instance, car-following, lane merging, and gap-acceptance) easily adaptable to different contexts.

We consider the effectiveness of the full model transfer (direct transfer), which does not involve any updating of model knowledge base. Using video survey system gap-acceptance data were collected at four intersections and then used for building four corresponding fuzzy gap-acceptance models. Each of these models (called "original") has been directly transferred to the other three intersections (called "application contexts") and the capability of these transferred models to represent the real behavior (drivers gap-acceptance decisions) has been validated using the corresponding gap-acceptance data collected in the application contexts. A comparative analysis among locally built and transferred models has been performed. The method used to carry out this comparison is the so-called ROC (*Receiver Operating Characteristic*) curve analysis ([5],[17]); the objective of the analysis is to evaluate the performance of the transferred models and the locally built models in terms of their ability to predict the real behavior of drivers in gap-acceptance situations.

The paper is organized as follows. Section 2 briefly summarizes literature on the analysis of gap-acceptance behavior and on model transferability. In Section 3 a description of the experimental data used in this study is provided. Section 4 describes the main characteristics of the fuzzy gap-acceptance models under analysis. Section 5 presents the application of ROC curve analysis to the evaluation of the transferability of fuzzy gap-acceptance models. Concluding remarks are presented in Section 6.

2 Related Works

The gap-acceptance problem considered in this paper refers to the situation in which a driver, starting from the secondary approach of a priority intersection, wants to perform a crossing or merging maneuver into a primary road. Essentially, this requires the choice between two mutually exclusive alternative actions: to accept or reject a gap (or lag[1]) of a given time size in the primary traffic stream. Evidently, such a choice is the result of a decision process affected both by driver characteristics (for example, driving experience, gender and age; see [21],[18],[19]) and characteristics of the gap/lag and of the choice situation (for example, gap/lag size, waiting time and speed of vehicles on the primary road; see [1],[14]). Thus, as shown by several previous studies, gap-acceptance behavior varies among drivers and, for the same driver, over time.

Fuzzy set theory-based and probabilistic discrete choice models (e.g. Logit models) have been considered to be appropriate for modeling the choice behavior under

[1] *gap* is the time interval between two successive vehicles passing a section of the main street, measured from the rear bumper to the front bumper; *lag* is the time interval between the arrival of a vehicle at the stopline of the secondary road and the passage at the conflict point of the first vehicle on the main road.

examination ([15],[16],[4],[18],[19],[12],[13],[20]). In general terms model transferability (spatial and/or temporal) refers to situations in which a model specified and estimated in a given original (estimation) context is subsequently transferred and applied to another (application) context.

This operation should have two primary advantages:

- to reduce the efforts in model development (using the same structure of the model previously identified);
- to reduce or eliminate the need for a large data collection in the application context.

Model transferability has been widely studied in the past with reference to the transferability of trip generation models ([2]), mode choice models ([3],[10],[11]), and four-step models [8]. These authors studied the effectiveness of both full model transfer (direct transfer) and of updating the original model using a small dataset from the application context. In all these studies the transferability of probabilistic models has been considered; in a previous work [16] the authors have analyzed the transferability of a gap-acceptance discrete choice model (Logit model).

3 Experimental Data

The case studies refer to four three-leg priority intersection. The observations relate to the right turn movement from a minor street controlled by "Stop" or "Yield" sign. The intersections have a different geometric layout in terms of the angle between the two intersecting roads and are located both in rural and in urban areas; in Figure 1 the intersection layouts are shown and in Table 1 some details about the intersections are reported.

The experimental observations have been collected during peak-hour periods through video camera recorder. The videos have been processed using an application software that allows the user to record the secondary vehicle arrival and departure at the stop line (SL in Figure 1), the primary vehicle arrival at the conflict point (C9-2) together with the vehicle category (car, van, truck, etc.).

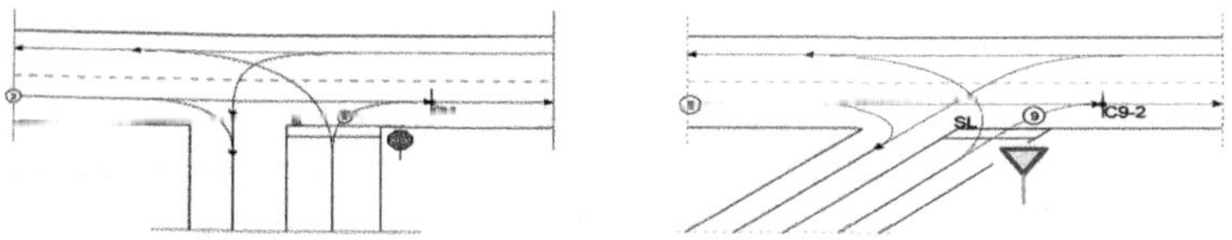

(a) Type "A90": 90° approaching angle

(b) Type "A45": 45° approaching angle

Fig. 1 Layout of the analyzed intersections.

Table 1 Analyzed intersections. Layout type and location.

Intersection	Name	Location	Type	Sign
I1	Vicenza	Sub-urban	A90	Stop
I2	Padova	Urban	A90	Stop
I3	Salzano	Sub-urban	A90	Stop
I4	Limena	Rural	A45	Yield

The data have been organized in a database and then processed through a software procedure that allows to extract the following information associated to each drivers decision:

- Type of time interval (lag or gap)
- Interval time size
- Secondary street vehicle waiting time at the stop line
- Category of secondary street vehicle
- Category of primary street vehicle closing the interval
- Driver decision (interval acceptance or rejection)

In Table 2 some details about intersection flows during the survey are shown and in Table 3 a summary of the data used in the transferability analysis is shown.

Table 2 Analyzed intersections. Characteristics of the observed major and minor flows.

			Primary street flow (maneuver 2)					Right turn from minor (maneuver 9)			
I.	O.P.L.	S.L.	A.H.V.	A.H.	F.C.(%)			A.H.V.	F.C.(%)		
	(hours)	(km/h)	(veh./hour)	(sec)	Car	Van	Truck	(veh./hour)	Car	Van	Truck
I1	1,5	50	540	6,7	93,4	5,3	1,1	222	95,5	3,6	0,9
I2	10	50	655	5,5	84	13	3	206	96,1	3,4	0,5
I3	6	50	780	4,6	86,5	11,7	1,8	130	93,8	6,0	0,2
I4	6	70	710	5,1	70,4		29,6	149	84,5		15,5

I stands for "Intersection", O.P.L. for "Observation Period Length", S.L. for "Speed Limit", A.H.V. for "Average Hourly Volume", A.H. for "Average Headway", F.C. for "Flow Composition".

Table 3 Analyzed intersections. Sample sizes of observed decisions.

Maneuver	Intersection	Total number of decisions (acceptances and rejections)			Average number of decisions per vehicle
		Gap	Lag	Total	
	I1	341	334	675	2,02
Right turn	I2	2955	2055	5001	2,43
from minor	I3	1502	777	2279	2,93
	I4	1448	892	2340	2,62

4 Identification of Fuzzy Models

Starting from the consideration that the time headway between vehicles on the primary street is the most important factor affecting gap-acceptance behavior (as widely reported in literature), and considering that drivers evaluate this variable in subjective terms, in this work we consider time interval as a fuzzy variable (it is realistic to think that a driver decides to accept or reject a gap based on simple rules like "if the time interval is *large* then I will accept it") and interval type (lag or gap) as an objective factor.

With reference to the original models locally built (named OM), for the identification of the membership functions of the premise and consequence fuzzy sets and rules of inferences the so-called FPA (Fast Prototype Algorithm, [6]) has been used. A satisfactory value of goodness-of-fit has been obtained (see Table 4). The fuzzy systems knowledge base obtained was characterized by five triangular fuzzy sets in the domain of the time interval size and by two "singletons" in the domain of the crisp variable representing the type of interval.

A simple Mamdani-type method of inference has been adopted (see [9]).

The identification/calibration processes did not produce the same rules for all the cases. In Table 4 the knowledge base characteristics and the goodness of fit values of the four fuzzy models locally built are shown (OM-Ii stands for gap-acceptance original model locally built with reference to intersection Ii gap-acceptance data).

Table 4 Fuzzy gap-acceptance original models. Knowledge base characteristics after identification process.

Model	Premise				Consequence		Number of Rules			Goodness
	Time Interval		Interval Type		Acceptance		Non	Comp.	Tot.	of Fit
	shape	number	shape	number	shape	number	Comp.			R^2
OM-I1	triangular	5	singleton	2	triangular	2	5	0	5	0,78
OM-I2	triangular	5	singleton	2	triangular	2	5	0	5	0,78
OM-I3	triangular	5	singleton	2	triangular	2	4	2	6	0,74
OM-I4	triangular	5	singleton	2	triangular	2	3	4	7	0,74

In Figure 2, for each of the original models, a graphical representation of the fuzzy sets both of the premises and of the consequence is shown. The fuzzy variable "Time interval size" is characterized by five triangular fuzzy sets over the variable domain. The resulting membership functions (their position and shape) are very similar regardless of the intersection. This result seems to indicate that, regardless of the intersection characteristics, drivers classified the time interval size in the same way, i.e. a "large" or a "medium" or a "small" interval are approximately the same in the sense that correspond to "around 5,5 seconds", "around 4,5 seconds" and "around 3,5 seconds".

The rules obtained are shown in compact form in Table 5.

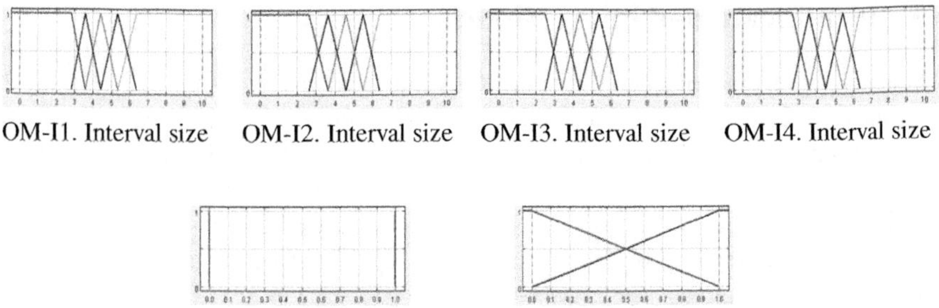

OM-I1. Interval size OM-I2. Interval size OM-I3. Interval size OM-I4. Interval size

Interval type Acceptance

Fig. 2 Fuzzy gap-acceptance original models. Premise fuzzy variable "Time interval size", crisp variable "Interval type" and fuzzy consequence variable "Acceptance".

First, note that models OM-I1 and OM-I2 are characterized by the same set of 5 non-compensatory rules, which do not include the "Interval Type" variable. This means that in the first two intersections driver decisions depend only on the size of the time interval between main street vehicles. The fact that model rules are identical for the two intersections is explained by the similarity of their characteristics (see Tables 1 and 2). Models OM-I3 and OM-I4, instead, have different sets of rules; in particular, only 3 of the 5 non-compensatory rules are maintained in both models, and compensatory rules taking into account both interval size and type are introduced. These results indicate that the four cases analyzed have some similarities, determining two ranges of interval size in which decisions are independent of interval type (Acceptance for IS $>$ 6,5 s and Refusal for IS $<$ 2,5 s). Nevertheless, there are differences between the intersections, which explain the modifications of the inference rules appearing for intersections OM-I3 and OM-I4. In particular, the layout, location and traffic characteristics of intersection I4 are significantly different from those of I1 and I2. The characteristics of intersection I3 appear to be intermediate, being similar to I1-I2 for some aspects, and to I4 for other aspects (see Tables 1 and 2).

Using the so called "acceptance index" obtained from the fuzzy inference system through a simple defuzzification (centroid method, see [9]) of the fuzzy output variable "acceptance", it is possible to build "acceptance curves" that allow to use the models as predictive tools (and to validate them over the calibration sample). When a gap/lag of a certain size has an acceptance index greater than or equal to the 0.5 threshold, it is considered "acceptable", otherwise it is considered "unacceptable". In Figure 3 the acceptance curves for the original models are shown. The analysis of this figure confirms the previous considerations regarding the inference rules, in particular the invariance of the curves with respect to interval type for intersections I1 and I2, and the gradual modification of the position and shape of the curves for the "Gap" interval type from intersections I1-I2 to I3 and I4.

We observe that the oscillations of the acceptance index that are visible in all the curves may be explained by the specific shape of the membership functions (triangular) and by the structure of the adopted inference rules.

Table 5 Fuzzy gap-acceptance original models. Rules of inference.

Kind of Rule	Model rules			
	OM-I1	OM-I2	OM-I3	OM-I4
Non Compensatory	If IS is VS then Refusal	If IS is VS then Refusal	If IS is VS then Refusal	If IS is VS then Refusal
Non Compensatory	If IS is S then Refusal	If IS is S then Refusal	If IS is S then Refusal	If IS is S then Refusal
Non Compensatory	If IS is VL then Acceptance	If IS is VL then Acceptance	If IS is VL then Acceptance	If IS is VL then Acceptance
Non Compensatory	If IS is L then Acceptance	If IS is L then Acceptance	If IS is L then Acceptance	
Non Compensatory	If IS is M then Acceptance	If IS is M then Acceptance		
Compensatory				If IS is L and IT is L then Acceptance
Compensatory				If IS is L and IT is G then Refusal
Compensatory			If IS is M and IT is L then Acceptance	If IS is M and IT is L then Acceptance
Compensatory			If IS is M and IT is G then Refusal	If IS is M and IT is G then Refusal

IS stands for Interval Size: VS "very small", S stands for "small", M "medium", L "large", VL "very large". IT stands for "Interval Type": G for "gap", L stands for "lag".

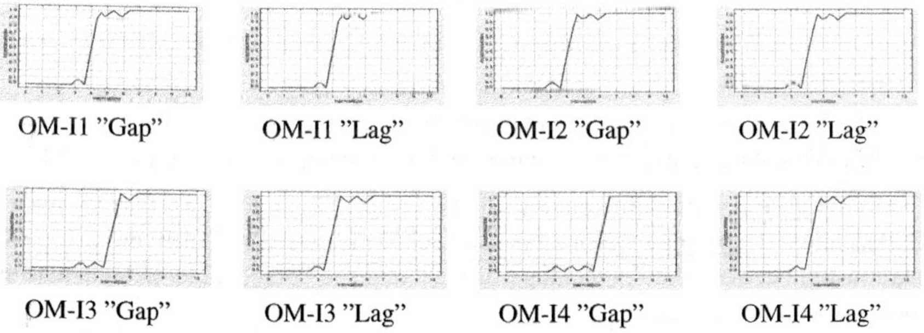

Fig. 3 Fuzzy gap-acceptance original models. Acceptance curves as a function of "Time interval size" and "Interval type".

5 Transferability Analysis

Each of the four original models (OM-Ii) has been directly transferred to the other (application) contexts; for each intersection, four couples of datasets were defined associating, in each case, the real drivers decisions (acceptance or rejections) R-Ii to the corresponding modelled decisions OM-Ii (see Table 6).

Table 6 Transferability analysis. Couples of drivers decision dataset (real and modelled) compared using ROC curve analysis.

	Application context			
Original Context	I1	I2	I3	I4
I1	OM-I1/R-I1	OM-I1/R-I2	OM-I1/R-I3	OM-I1/R-I4
I2	OM-I2/R-I1	OM-I2/R-I2	OM-I2/R-I3	OM-I2/R-I4
I3	OM-I3/R-I1	OM-I3/R-I2	OM-I3/R-I3	OM-I3/R-I4
I4	OM-I4/R-I1	OM-I4/R-I2	OM-I4/R-I3	OM-I4/R-I4

The transferability effectiveness of the models has been evaluated by means of the ROC curve analysis [5], a method used in various research fields for evaluating and comparing the discriminatory power of models having binary outputs. The basic idea of ROC curve analysis may be explained by considering an experiment with only two possible outcomes, 1 and 0, that are denoted as *positive* and *negative* outcomes. Suppose we are using a model that predicts the outcome of the experiment based on a classification threshold (cutoff), and denote by:

- True Positive (TP): the model predicts 1 and the actual outcome is 1
- False Positive (FP): the model predicts 1 but the actual outcome is 0
- True Negative (TN): the model predicts 0 and the actual outcome is 0
- False Negative (FN): the model predicts 0 but the actual outcome is 1

the probability of correctly identifying positive and negative outcomes may be defined, respectively, as:

- TPR (True Positive Rate) = number of TP/(number of TP + number of FN)
- TNR (True Negative Rate) = number of TN/(number of TN + number of FP)

The ROC curve describes the relationship between TPR, called "sensitivity", and (1-TNR), called "1-specificity", for all possible classification thresholds, describing the relationship between the "percentage of hits" and the "percentage of false alarms" obtained with the model.

The Area Under the ROC Curve (AUC) is related to the accuracy of the model predictions, and increases with it; in particular, when this area is equal to $1{,}0$ the model produces perfect forecasts, and when it is equal to $0{,}5$ the model produces random forecasts (no discriminatory power). The AUC is equivalent to the Gini coefficient = 2AUC-1, and also to the MannWhitneyWilcoxon two-independent sample non-parametric test statistic, which tests the null hypothesis that two samples of ordinal measurements are drawn from a single distribution.

Additional performance metrics adopted are also [5]:

- F-measure = 2/{[(number of TP + number of FP)/number of TP] + 1/TPR}
- Percent right = [(number of TP + number of TN)/(number of outcomes)]*100
- Youden Index = TPR + TNR - 1

Clearly a model with high discriminatory power should have high values of metrics AUC, TPR, TNR, F-measure, Percent right and Youden Index.

In the case under analysis the "acceptance" domain was divided in two parts with a threshold equal to 0,5: the left side was defined as a negative outcome (gap/lag rejection) and the right side (gap/lag acceptance) as a positive one. The outcomes obtained from model rules, and represented by fuzzy sets, have been determined using the centroid defuzzification method. This method allows to identify a synthetic measure representative of the output fuzzy set, giving a compact information about the "acceptance" of a gap/lag of a certain size.

Under these assumptions the basic metrics of ROC analysis become:

- TP: the model predicts an acceptance and the driver accepted the gap/lag;
- FP: the model predicts an acceptance and the driver rejected the gap/lag;
- TN: the model predicts a rejection and the driver rejected the gap/lag;
- FN: the model predicts a rejection and the driver accepted the gap/lag;

Consequently all the other metrics can be derived.

The results of the transferability ROC analysis are reported in Tables 7, 8, and 9, each for a different metric. For each model built in the original context (row in the tables) the dark grey cells correspond to situations where the model (original or transferred) has the best value of the measure considered, while the light grey cells correspond to situations where the model has the worst value of the measure. The cells corresponding to the matrix diagonal contain the value of the measure computed with reference to the original model. We observe that the original model is generally the best, and that there are a few situations where the computed metrics have the same value for the directly transferred models and for the original one. Looking at the absolute values of the measures, we note that, while transferability between the contexts having similar geometric characteristics (I1, I2 and I3) appears to be satisfactory, model transfer from and to intersection I4 suffers from the specificity of this intersection, that is located in a rural area and has an approach angle very different from the other intersections; this result confirms the previous considerations regarding the model OM-I4 rules of inference (see Section 4).

Examples of ROC curves can be seen in Figures 4. In the first case (Figure 4(a)), models OM-I1/R-I1 e OM-I2/R-I1 show the best values of AUC ($0,95 \pm 0,008$), followed by models OM-I3/R-I1 ($0,94 \pm 0,010$) and OM-I4/R-I1 ($0,91 \pm 0,012$). This result is in agreement with the already noted specificity of model OM-I4, which has a low capability of reproducing decisions observed at the other intersections. On the other hand (Figure 4(b), the transferability analysis for intersection I4 shows an acceptable value of AUC for all the specifications, even if model OM-I4 appears to be superior according to the other metrics.

Table 7 Transferability analysis. Youden Index.

Original Context	Application context I1	I2	I3	I4
I1	0,86	0,87	0,85	0,82
I2	0,86	0,87	0,85	0,82
I3	0,85	0,84	0,85	0,83
I4	0,80	0,78	0,76	0,84

Table 8 Transferability analysis. F-measure.

Original Context	Application context I1	I2	I3	I4
I1	0,93	0,92	0,89	0,89
I2	0,93	0,92	0,89	0,89
I3	0,92	0,91	0,91	0,90
I4	0,89	0,88	0,86	0,91

Table 9 Transferability analysis. Percent right.

Original Context	Application context I1	I2	I3	I4
I1	93,19%	93,57%	92,67%	90,94%
I2	93,19%	93,59%	92,67%	90,94%
I3	92,59%	92,46%	93,77%	91,80%
I4	89,93%	90,44%	91,14%	92,70%

This result seems to indicate an asymmetric transferability of different model specifications ([11]): while the model calibrated on intersections with similar geometric characteristics (A90) may perform well in other situations, the same is not true for the inverse transfer.

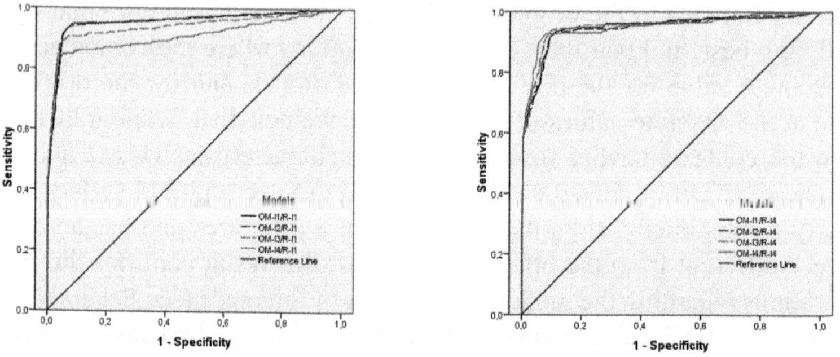

(a) Intersection I1 (b) Intersection I4

Fig. 4 Transferability ROC analysis at intersections I1 and I4.

6 Conclusions

In this paper an analysis of the transferability of fuzzy gap-acceptance models between different intersections has been performed using metrics of ROC curve analysis. The results indicate that the effectiveness of the model transfer is significantly influenced by the geometric, location and traffic characteristics (in particular the approach angle of the minor road) of the intersections in which the model is developed and to which the model is transferred. The analysis of results conducted with the use of ROC curves also indicates the presence of an asymmetric transferability of different model specifications. As expected, a model developed in a given context generally performs better in that context as compared to models transferred from other contexts. Future research should analyze in more depth the effect of intersection characteristics on model transferability, in order to support the possibility of integrating fuzzy models of gap-acceptance behavior into traffic microsimulation models. In fact a deeper knowledge of model transferability may be very useful in reducing the efforts in model development, particularly in the context of intersection planning and design.

References

1. Adebisi, O., Sama, G.N.: Influence of stopped delay on driver gap acceptance behavior. J. Transp. Engrg. 115, 305–315 (1989)
2. Agyemang-Duah, K., Hall, F.L.: Spatial transferability of an ordered response model of trip generation. Transport Res. A–Pol. 31, 389–402 (1997)
3. Badoe, D.A., Miller, E.J.: Comparison of Alternative Methods for Updating Disaggregate Logit Mode Choice Models. Transp. Res. Record 1493, 90–100 (1995)
4. Cassidy, M., Madanat, S.M., Wang, M.H., Yang, F.: Unsignalized intersection capacity and level of service: revisiting critical gap. Transp. Res. Record 1484, 16–22 (1995)
5. Fawcett, T.: An introduction to ROC analysis. Pattern Recognit. Lett. 27, 861–874 (2006)
6. Glorennec, P.Y.: Algorithmes d'apprentissage pour systemes d'inference floue. Editions Hermes, Paris (1999)
7. Hwang, S.Y., Park, C.H.: Modeling of the gap acceptance behavior at a merging section of urban freeway. In: Proceedings of the Eastern Asia Society for Transportation Studies, pp. 1641–1656 (2005)
8. Karasmaa, N.: Evaluation of transfer methods for spatial travel demand models. Transport Res. A–Pol. 41, 411–427 (2007)
9. Klir, G.J., Yuan, B.: Fuzzy sets and fuzzy logic. Theory and applications. Prentice Hall PTR, Upper Saddle River (1995)
10. Koppelman, F.S., Kuah, G., Wilmot, C.G.: Transfer model updating with disaggregate data. Transp. Res. Record 1037, 102–107 (1985)
11. Koppelman, F.S., Wilmot, C.G.: Transferability analysis of disaggregate choice models. Transp. Res. Record 895, 18–24 (1982)
12. Maze, T.H.: A probabilistic model of gap acceptance behavior. Transp. Res. Record 795, 8–13 (1981)
13. Pollatschek, M.A., Polus, A., Livneh, M.: A decision model for gap acceptance and capacity at intersections. Transport Res. B–Meth. 36, 649–663 (2002)
14. Polus, A., Kraus, J., Reshetnik, I.: Non-stationary gap acceptance assuming drivers learning and impatience. Traffic Eng. Control 37, 395–402 (1996)

15. Rossi, R., Meneguzzer, C.: The effect of crisp variables on fuzzy models of gap-acceptance behaviour. In: Proceedings of the 13th Mini-EURO Conference Handling Uncertainty in the Analysis of Traffic and Transportation Systems, Bari, Italy, pp. 240–246 (2002)
16. Rossi, R., Meneguzzer, C., Gastaldi, M.: A comparison of methods for transferring Logit models of gap-acceptance behaviour. In: Proceedings of the 12th World Conference on Transport Research Society, Lisbon, Portugal, (2010) ISBN/ISSN:978-989-96986-1-1
17. Rossi, R., Meneguzzer, C., Gastaldi, M., Gecchele, G.: Comparative evaluation of Logit and Fuzzy Logic models of gap-acceptance behavior. In: Proceedings of the TRISTAN VII, Seventh Triennial Symposium on Transportation Analysis, Tromsoe, Norway, pp. 649–652 (2010)
18. Teply, S., Abou-Henaidy, M.I., Hunt, J.D.: Gap acceptance behaviour aggregate and Logit perspectives: Part 1. Traffic Eng. Control 38, 474–482 (1997)
19. Teply, S., Abou-Henaidy, M.I., Hunt, J.D.: Gap acceptance behaviour aggregate and Logit perspectives: Part 2. Traffic Eng. Control 38, 540–544 (1997)
20. Toledo, T.: Driving behavior: models and challenges. Transp. Rev. 27, 65–84 (2007)
21. Wennel, J., Cooper, D.F.: Vehicle and driver effects on junction gap acceptance. Traffic Eng. Control 22, 628–632 (1981)

Part VIII: Optimization Techniques

Logic Minimization of QCA Circuits Using Genetic Algorithms

Mahboobeh Houshmand, Razieh Rezaee Saleh, and Monireh Houshmand

Abstract. Quantum-dot cellular automata (QCA) are proposed as one of the foremost candidates to replace the complementary metal-oxide semiconductor (CMOS) technology. The majority gate and the inverter gate together make a universal set of Boolean primitives in QCA technology. Reducing the number of required primitives to implement a given Boolean function is an important step in designing QCA logic circuits. Previous research has shown how to use genetic algorithms to minimize the number of majority gates implementing a given Boolean function with one output. In this paper we show how to minimize Boolean functions with an arbitrary number of outputs. Simulation results for the circuits with three, four and five outputs show our method on the average results in 25.41, 28.82, 30.89 percentage decrease in the number of required gates in comparison with optimizing each output independently.

1 Introduction

Scaling of the predominant silicon complementary metal-oxide semiconductor (CMOS) technology is ultimately approaching its limit after decades of exponential growth[1]. This has prompted the development of many nano-scale molecular devices. Quantum-dot Cellular Automata(QCA)[2-5], Single Electron Tunneling (SET)[6,7], and Tunneling Phase Logic (TPL)[8], have been proposed as possible

Mahboobeh Houshmand
Dept. of Computer Engineering and Information Technology, Amirkabir University of Technology, Tehran, Iran
e-mail: houshmand@aut.ac.ir

Razieh Rezaee Saleh
Dept. of Computer Engineering, Ferdowsi University of Mashhad, Mashhad, Iran
e-mail: razie.rezaei@stu-mail.um.ac.ir

Monireh Houshmand
Dept. of Electrical Engineering, Imam Reza University, Mashhad, Iran
e-mail: monirehhoushmand@gmail.com

A. Gaspar-Cunha et al. (Eds.): Soft Computing in Industrial Applications, AISC 96, pp. 393–403.
springerlink.com

replacement candidates for CMOS. QCA was first suggested by Lent, et al. in 1993 [9, 10]. Promising size density of several orders of magnitude smaller than CMOS, fast switching time and extremely low power, has caused QCA to become a topic of intense research. QCA performs computation not on electron flow but on Columbic interactions of electrons trapped in quantum dots. The fundamental QCA logic primitives are wire, inverter and majority gate. Traditional logic reduction methods, such as Karnaugh maps (K-maps), always produce simplified expressions in the two standard forms: sum of products (SOP) or product of sums (POS) and logic CMOS circuits are implemented using AND and OR gates based on these SOP or POS formulae. However, due to the complexity of multi-level majority circuits, it is difficult to convert these two forms into majority expressions. This fact initiated a number of studies[11-14]aimed to find an effective method for synthesis of QCA based logic structures, most of them, with a drawback that they are suitable only for synthesizing small networks by hand or they can process only three-variable Boolean functions. Zhang et al.[12]develop a Boolean algebra based on a geometrical interpretation of only three-variable Boolean functions to facilitate the conversion of sum-of-products expressions into reduced majority logic. Observing the three-cube representation of such functions, there are 8 minterms that may be included, providing a total of 256 possible Boolean functions. They identify thirteen standard functions which can be used with rotation and translation to represent any of these arbitrary 256 Boolean functions. Walus et al. [15]modify their previous work in [12] to improve three of the standard function implementations using Karnaugh maps minimizations. Improvement is made when the total number of majority gates is reduced or when the number of levels is reduced. Boolean functions of more than three inputs must be factored in order to decompose the function into three-input functions that use this and the previous work. Huo et al. in[16] employ a one-to-one mapping table consisting of twenty standard functions and their corresponding majority expressions. They first simplify a given Boolean function to one form of these twenty standard functions. Then the majority expression corresponding to this standard function is chosen as the result. Zhang et al. in [17] develop the MALS tool that produces optimized majority logic circuits. They use factorization methods in a pre-synthesis step in order to reduce the size of the circuit. Their factorization technique is applied to the SOP representation of the desired logic. Bonyadi et al. in [18]present a method based on genetic algorithms with tree representation chromosomes for optimizing the number of majority and inverter gates to compute a given single output Boolean function, which we call Single Output-Genetic Algorithms, SO-GA. This method can efficiently synthesize arbitrary multi-variable Boolean functions. In[19], we extended the algorithm in[18] for the circuits with two outputs. We showed that for the two output circuits (e.g. a full adder with the sum and carry outputs) independently optimizing each output using SO-GA does not necessarily produce an overall optimized solution and our approach can result in synthesis of the circuit with less number of gates. In this paper we extend the approach for circuits with more than two outputs, Multi Output-Genetic Algorithms, MO-GA. Our proposed algorithm uses an inductive approach to implement each new output noticing the circuit implementing previous ones to benefit the possible common parts.

We applied our approach for the circuits with three, four and five outputs and three and four inputs. Simulation results showed our method can efficiently produce less number of majority and inverter gates than optimizing each output independently using the SO-GA method. The rest of this paper is organized as follows. In the next section some related background materials are presented. In section 3, the optimization of a single-output QCA circuit using SO-GA is described. Our proposed method (MO-GA) is explained in detail in Section 4. In Section 5 simulation results are presented and finally Section 6 concludes the paper.

2 Background Material

A. QCA Basic

A QCA (Fig.1) cell contains four quantum-dots positioned at the corners of a square and two free electrons that can move to any quantum-dot within the cell through electron tunneling[20]. The electrons are forced to the corner positions by Coulombic interactions. As shown in Fig. 2, only two stable configurations of an electron pair exist. These configurations are denoted as cell polarization P = +1and P = -1, respectively. Binary information is encoded using cell polarization P = +1 to represent logic '1'and P = -1to represent logic'0'.

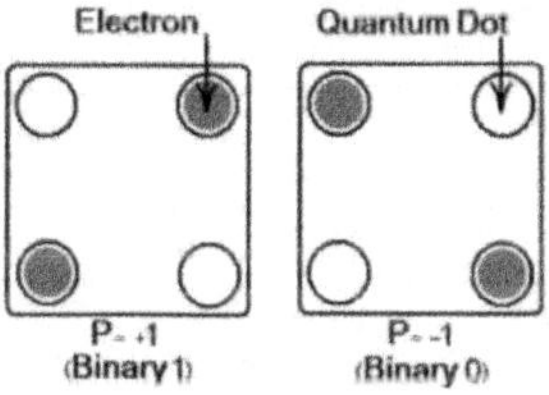

Fig. 1 Two polarizations in a QCA cell

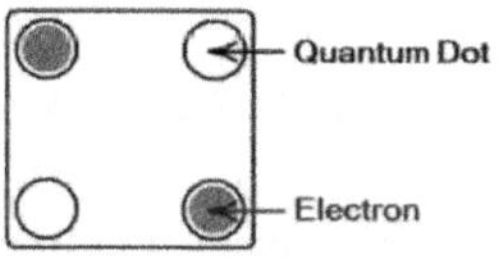

Fig. 2 A QCA cell

B. QCA Logic Device

The basic QCA logic devices are the QCA wire, majority gate and inverter. Fig.3 shows QCA wire, majority and inverter gate.

- QCA wire: A QCA wire is just a line of QCA cells. In a QCA wire a binary signal propagates from input to output because of the Coulombic interactions between cells.
- QCA Inverter: In the QCA inverter cells oriented at 45° to each other take on opposing polarization. This characteristic is exploited to implement an inverter, such as the one shown in Fig. 3(c).

- QCA majority gate: The QCA majority gate implements a three input majority function.

The tendency of the majority device cell to move to a ground state ensures that it takes on the polarization of the majority of its neighbors. It is in this polarization state that the Coulombic repulsion between electrons in the input cells is minimized. Assuming A, B, and C as inputs, the logic function of the majority gate is:

$$M(A,B,C) = AB + AC + BC$$

By fixing the polarization of one input as logic '1'or '0'we can obtain an OR gate and an AND gate, respectively (reduction).

$$M(A,B,1) = A + B, M(A,B,0) = AB$$

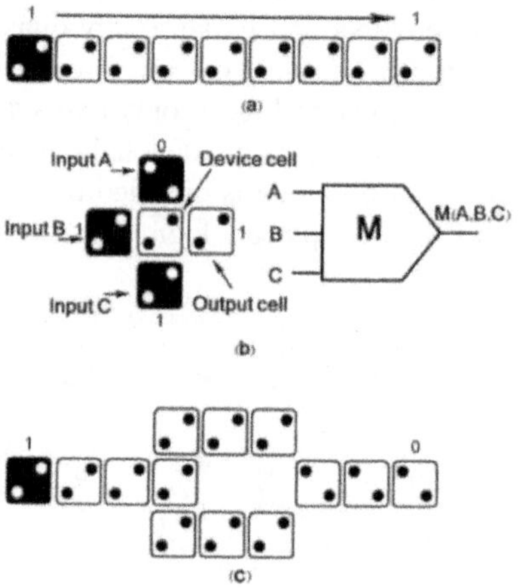

Fig. 3 a) QCA wire b) QCA majority gate c) inverter gate.

3 QCA Circuit Optimization Using SO-GA

In [18], an efficient method for optimizing the number of majority and inverter gates required to compute a given Boolean function is presented. In the following we will review the method.

A. Chromosome Representation

In this method, a tree structure is used for chromosome representation. The internal nodes are the majority and inverter gates and the leaves can be either logical '1's or Boolean variables. The preorder traversal of the tree produces a majority and inverter expression. As an example, Fig. 4 shows M (A,C',M(A,B,1)').

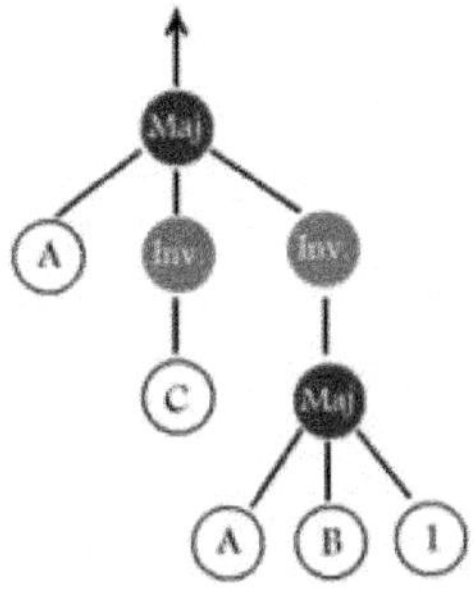

Fig. 4 A chromosome for *M* (*A*,*C*',*M*(*A*,*B*,1)')

B. Fitness Function

There are two main parameters to calculate the fitness of a given chromosome. First is the similarity of the function it represents to the given Boolean function. Therefore the fitness of a chromosome C is defined as Eq. (1).

$$Fit(C) = \frac{N(F,C)}{2^n} \tag{1}$$

Where n is the number of variables, F is the Boolean function, and $N(F,C)$ is the number of identical minterms between the function that chromosome C represents and the function F. The second important parameter is the inverse number of gates (internal nodes) in the chromosome. If the fitness of the chromosome is equal to 1, fitness is defined as Eq. (2).

$$Fit(C) = 1 + \frac{1}{Gates(C)} \tag{2}$$

Where Gates(C) returns the number of internal nodes in chromosome C. These two equations together rank the chromosomes according to both their similarity to the given Boolean function F and the inverse number of internal nodes in the chromosome C. The selection can be performed using any algorithm explained in [23].

C. Mutation and Crossover

Mutation cannot be performed by replacing a random gate by another one because it may lead to an invalid chromosome (e.g. consider the case in which a three input majority node is altered by a one input inverter node). So in this proposed method the mutation is performed in a new way in which the chromosome that has the worst fitness in the population is replaced by a new randomly produced chromosome. This method performs the mutation function, i.e. restoring the diversity that may be lost from the repeated application of selection and crossover.

A double point crossover is carried out by exchanging two random nodes and their subtrees in a chromosome with two other random node and their subtrees in some other chromosome. Fig. 5 shows a double point crossover. Mutation and crossover take place with their corresponding probabilities, i.e. *Pm* and *Pc*, respectively.

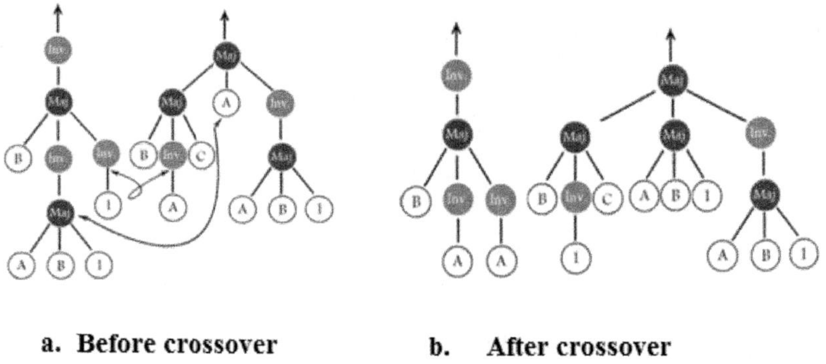

Fig. 5 A double point crossover. a. Before crossover, b. After crossover

4 QCA Circuit Optimization Using MO-GA

The algorithm proposed in [16], reduces the number of required QCA gates in the circuits with a single output. In the approach presented in[19], we extended the SO-GA algorithm for optimizing the circuits with two outputs. Using SO-GA for optimizing the two outputs separately is not necessarily efficient, because it cannot benefit the common parts of the circuit implementing different outputs. We applied SO-GA for one of the two outputs and produced the optimized circuits to compute that function. Then, we optimized the circuit implementing the second output noticing the produced circuits for the first output, in a way to benefit the common parts of the circuits. In the paper we mentioned that the main idea of algorithm is inductively extensible for more than two outputs, but we did not explain and implement it. In this paper we extend the presented method in [19]and propose an algorithm for circuits with an arbitrary number of outputs. Fig. 6 shows the pseudo code of the algorithm. This algorithm uses an inductive approach to implement each new output noticing the circuits implementing previous outputs. For convenience we initially number the outputs in some arbitrary order. First SO-GA is run for the first output. For implementing a circuit for the *i*th output, the circuits for *i-1* previous outputs are considered o obtain a chromosome with *i* outputs. This process is repeated for each output until achieving a circuit synthesizing all of them. Crossover and mutation are performed in the same way as explained in section 3.C. The fitness of a chromosome with *i* outputs is calculated as Eq. (3).

$$Fit(C_i) = \frac{N(F_i, C_i)}{2^n} \tag{3}$$

Where n is the number of variables, *Fi* is the Boolean function for the *ith* output, *Ci* is a chromosome with i outputs, and *N(Fi,Ci)* is the number of identical minterms between the function that chromosome *Ci* represents and the function *Fi.* Fitness equal to 1implies that the function which chromosome *Ci* represents has the same minterms as the function *Fi*. If the fitness of the chromosome is equal to 1, the number of gates in the chromosome is observed and the fitness is defined as Eq. (4).

$$Fit(C_i) = 1 + \frac{1}{Gates(C_i)} \tag{4}$$

Where $Gates(C_i)$ returns the number of gates in chromosome *Ci* ignoring repeated gates. These two equations together rank the chromosomes according to both their similarity to the given Boolean function and the inverse number of gates in the chromosome. As seen in the algorithm (Fig. 6) we consider the first output a single output circuit and obtain chromosomes implementing it using SO-GA. SO-GA is run for a sufficient number of generations to get some chromosomes with fitness greater than 1. These chromosomes are stored in an array named *Correct_Chromosomes[1]*. The index of the outer loop, n, varies from 2 to the number of outputs. For producing circuit for the *n*th output, we use the previous n-1 output chromosomes whose fitness is greater than 1 i.e. *Correct_Chromosomes[n-1]*. In the second loop, i varies to point to the chromosomes in *Correct_Chromosomes[n-1]*. GA is run to generate chromosomes for the *n*th output based on each chromosome in *Correct_Chromosomes[n-1]*. The third loop with index *j* denotes the generations of GA. Initial population of this GA, *Temp_Chromosomes* are produced randomly. The index of the inner loop, k points to the chromosomes in *Temp_Chromosomes*. *M* and *N*, two random numbers are generated as cut off points for *Temp_Chromosomes[k]* and *Correct_Chromosomes[n-1,i]* respectively. *M*th node and its subtree in *Temp_Chromosmes[k]* are replaced with *N*th node and its subtree in *Correct_Chromosomes[n-1,i]*. In this process *Temp_Chromosmes[k]* obtains a common subtree with *Correct_Chromosomes[n-1,i]* without altering it. The combine function gets two chromosomes - one with *n*-1outputs and the other with one output- and combines them together into a chromosome with n outputs. Fig. 7 shows an example of the combine function. *Correct_Chromosomes[2,i]* is a two output chromosome generated in the previous iterations for the first and second outputs. *Temp_Chromosomes[k]* is a single output chromosome for the third output. The subtree of the third node of *Temp_Chromosomes[k]* is replaced with the subtree of second node of *Correct_Chromosomes[2,i]*, then these two chromosomes are combined to produce a chromosome with three outputs. In the combined chromosome the common subtrees between two initial chromosomes are circled in the figure. After performing combine function for each chromosome in the initial population, fitness is calculated. The total number of gates used for calculating the fitness (Eq. (4)) of the combined chromosome, is defined as follows:

$$\begin{aligned} &Gates(Final\_Temp\_Chromosoms\ [k]) = \\ &\qquad Gates(Correct\_Chromosome\ [n-1, i]) \\ &\qquad + Gates(Temp\_Chromosoms\ [k]) - Num\_of\_Common\_Gates \end{aligned}$$

In this equation number of common gates is subtracted from the sum of number of gates of the two chromosomes. Then selection is performed. Crossover and mutation with the probabilities *Pc* and *Pm*, respectively are carried out for *Temp_Chromosomes* in order to produce the next generation of *Temp_Chromosomes*. The whole procedure is repeated for a predetermined number of generations. In the end, the chromosome with the highest fitness among the chromosomes in *Correct_Chromosomes[Num_of_Outputs]* is selected as the final result.

5 Simulation Results

We used Roulette Wheel algorithm for selection and set the parameters *Pm* and *Pc*, to 0.06 and 0.75, respectively. We applied our proposed algorithm on 10000 four-input/three-output, 10000 four-input/four-output, and 10000 four-input/five-output randomly generated circuits. Fig. 8 shows the different amount of decrease percentage in the number of gates for four-input circuits, with three, four and five outputs. For example, the figure implies that for 3712 four-input circuits with four outputs, we had between 30-40% decreases in the number of required gates. Our approach on the average results in 25.41, 28.82, 30.89 decrease percentages in the number of the required gates for three, four and five output circuits, respectively. Table 1 shows a sample of comparison between SO-GA and MO-GA for circuits, with for outputs.

```
Run the SO-GA for the first output for sufficient number of generation to obtain some chromosomes with the fitness greater
than 1, and store them in the first row of Correct_Chromosomes array, i.e. Correct_Chromosomes[1]. Then do this algorithm:
   for n•2 to Num_of_Outputs
     do for i•1 to length(Correct_Chromosomes[n-1])
       do Generate Temp_Chromosomes, a random initial population for nth output
          for j•1 to Num_of_Generations
            do for k•1 to Num_of_Chromosomes in current generation
              do Generate a random number M between 1 and Gates(Temp_Chromosomes[k])
                 Generate a random number N between 1 and Gates (Correct_Chromosomes[n-1,i]).
                 Replace Mth node and its subtree in Temp_Chromosmes[k] with Nth nodean its subtree in the
                 Correct_Chromosomes[n-1,i]//notice that Correct_Chromosomes[n-1,i] does not change.
                 Combined_Chromosmes[k]= Combine(Correct_Chromosomes[n1,i],Temp_Chromosmes[k])
                 Do the selection, crossover with probability P_c, and mutation with probability P_m, for
                 Temp_Chromosomes in order to generate the next generation of Temp_Chromosomes.
                 The number of gates in theCombined_Chromosomes[k] which is used in the calculation of fitness
                 is obtained as follows:
                 Gates (Combined_Chromosomes[k]) = Gates(Correct_Chromosomes[n-1,i])+
                 Gates(Temp_Chromosomes[k] )- Num_of_Common_Gates
     Save all chromosomes in Combined_Chromosomes whose fitness is greater than 1 in Correct_Chromosomes [n]
Return the best chromosome in Correct_Chromosome [Num_of_Outputs]
```

Fig. 6 The pseudo code of MO-GA

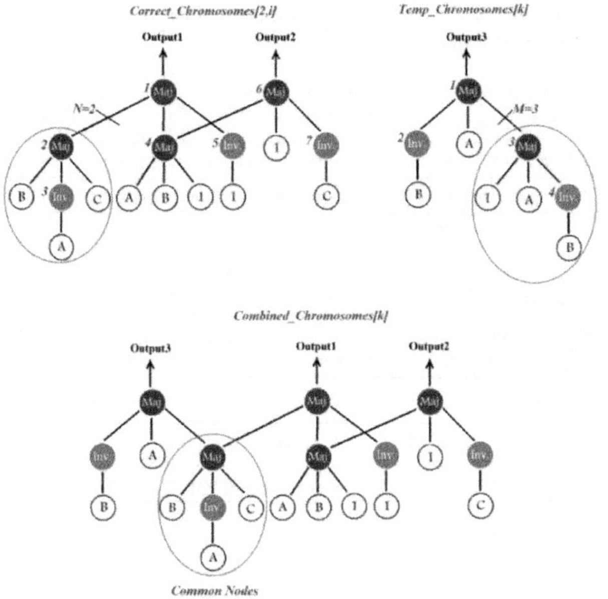

Fig. 7 An example of combine function. The chromosome on top left has two outputs, Output1 and Output2. The chromosome on top right is generated to implement Output3. The combined chromosome is a chromosome implementing three outputs with some common nodes.

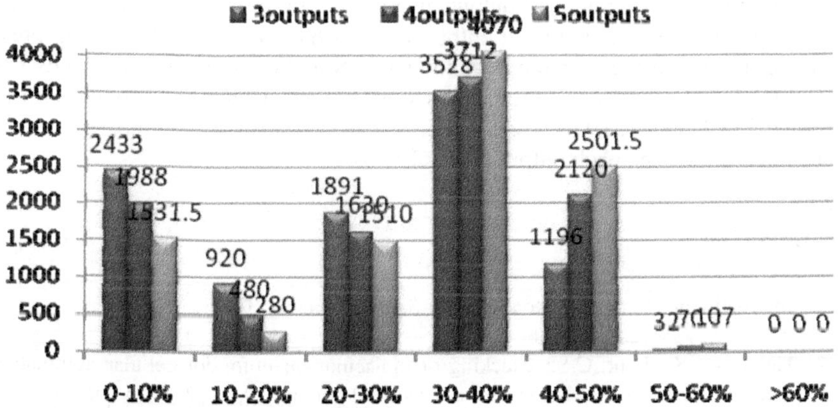

Fig. 8 The number of four-input circuits and their corresponding decrease percentage in the number of gates, for different number of outputs.

Table 1 Simulation results for a four input circuit with four outputs. In the case of the MO-GA, the common parts of the circuit are underlined. In this case, the proposed MO-GA algorithm uses 9 gates less than SO-GA.

Approach	Minterms	Circuit	NOI	NOM	NTG
SO-GA	$F_1 = \sum(m_3, m_4, m_7, m_{15})$	M(M(B,A',D),M(C,D,1)',M(C,D,1'))	3	4	7
	$F_2 = \sum(m_1, m_3, m_4, m_9, m_{13}, m_{15})$	M(M(A,M(D,B,C),M(D',1,B))',B,M(A,B',D))	3	5	8
	$F_3 = \sum(m_3, m_6, m_7, m_{11}, m_{13}, m_{14}, m_{15})$	M(C,D,M(B,A,M(1',B,C)))	1	3	4
	$F_4 = \sum(m_2, m_6, m_{10}, m_{11}, m_{14})$	M(D',M(M(A,B,D')',A,1'),C)	4	3	7
	Total number of gates		**11**	**15**	**26**
MO-GA	$F_1 = \sum(m_3, m_4, m_7, m_{15})$	M(M(C,1',D),M(D,M(C,B,A)',B),D')	3	4	7
	$F_2 = \sum(m_1, m_3, m_4, m_9, m_{13}, m_{15})$	M(M(1',M(D',A,1),B),M(D,M(C,B,A)',B),B')	4	5	9
	$F_3 = \sum(m_3, m_6, m_7, m_{11}, m_{13}, m_{14}, m_{15})$	M(C,D, M(1',M(D',A,1),B))	2	3	5
	$F_4 = \sum(m_2, m_6, m_{10}, m_{11}, m_{14})$	M(M(M(C,1',D),M(D,M(C,B,A)',B),D')',1',C)	5	5	10
	Total number of gates		**14-6=8**	**17-8=9**	**31-14=17**

6 Conclusion

In this paper we presented an approach for reducing the number of gates for QCA circuits with an arbitrary number of inputs and outputs based on genetic algorithms. In this paper, we presented an approach to optimize the circuits with an arbitrary number of inputs and outputs using GA. In an inductive manner it optimizes the circuit producing each output noticing the circuits implementing previous ones to benefit the possible common parts. Simulation results for circuits with three, four, five outputs and four inputs showed that our method produces on the average 25.41, 28.82, 30.89 less number of gates, respectively compared to independently optimizing each output using GA.

References

[1] International Technology Roadmap for Semiconductors (ITRS), `http://www.itrs.net`

[2] Hennessy, K., Lent, C.S.: Clocking of molecular quantum-dot cellular automata. J. Vac. Sci. Technol. B, Microelectron. Process. Phenom. 19, 1752–1755 (2001)

[3] Lent, C.S., Isaksen, B., Lieberman, M.: Molecular quantum-dot cellular automata. J. Am. Chem. Soc. 125, 1056–1063 (2003)

[4] Lent, C.S., Tougaw, P.D.: A device architecture for computing with quantum dots. Proceedings of the IEEE 85, 541–557 (1997)

[5] Tougaw, P.D., Lent, C.S.: Logical devices implemented using quantum cellular automata. Journal of Applied Physics 75, 1818–1825 (1994)

[6] Oya, T., Asai, T., Fukui, T., Amemiya, Y.: A majority-logic nanodevice using a balanced pair of single-electron boxes. J. Nanosci. Nanotech. 2, 333–342 (2002)
[7] Oya, T., Asai, T., Fukui, T., Amemiya, Y.: A majority-logic device using an irreversible single-electron box. IEEE Trans. Nanotechnol. 2, 15–22 (2003)
[8] Fahmy, H.A., Kiehl, R.A.: Complete logic family using tunnelingphase-logic devices. In:Proc. Int. Conf. Microelectron, pp.22–24 (1999)
[9] Lent, C.S., Tougaw, P.D., Porod, W.: Bistable saturation of in coupled quantum dot for quantum cellular automata. Appl. Phys. Letters 62, 714–716 (1993)
[10] Tougaw, P.D., Lent, C.S., Porod, W.: Bistable saturation in coupled quantum dot cells. J. Appl. Phys. 74, 3558–3566 (1993)
[11] Karnaugh, M.: The map method for synthesis of combinational logic circuits. Transactions of the American Institute of Electrical Engineers 72, 593–599
[12] Zhang, R., Walus, K., Wang, W., Jullien, G.: A method of majority logic reduction for quantum cellular automata. IEEE Trans. Nainventerechnology 3 (December 2004)
[13] Akers, S.B.: Synthesis of combinational logic using three-input majority gates. In: Proc. 3rd Annu. Symp. Switching Circuit Theory and Logical Des., pp. 149–157 (October 1962)
[14] Miller, H.S., Winder, R.O.: Majority logic synthesis by geometric methods. IRE Trans. Electron. Comput. EC-11, 89–90 (1962)
[15] Walus, K., Schulhof, G., Jullien, G.A., Zhang, R., Wang, W.: Circuit design based on majority gates for applications with quantum-dot cellular automata. In: 38th Asilomar Conference on Signals, Systems and Computers, vol. 2, pp. 1354–1357 (2004)
[16] Huo, Z., Zhang, Q., Huo, Z., Zhang, Q.: Logic optimization for Majority Gate-Based Nanoelectronic Circuits. In: Proc. Int. Symp. Circuits and Systems (ISCAS), May 2006 pp. 1307–1310 (2006)
[17] Zhang, R., Jha, N.K.: Threshold/majority logic synthesis and concurrent error detection targeting nanoelectronic implementations. In: Proc. of the 16th ACM Great Lakes symposium on VLSI (April 2007)
[18] Bonyadi, M.R., Azghadi, S.M.R., Rad, N.M., Navi, K., Afjei, E.: Logic optimization for Majority Gate-Based Nanoelectronic Circuits Based on Genetic Algorithm. In: Proc. International Conference on Electrical Engineering, ICEE 2007, pp. 1–5 (2007)
[19] Houshmand, M., Khayyat, S.H., Rezayee, R.: Genetic algorithm based logic optimization for multi-output majority gate-based nano-electronic circuits. In: Proc. of 2009 IEEE International Conference on Intelligent Computing and Intelligent Systems, ICIS 2009 (2009)
[20] Amlani, I.: Experimental demonstration of a leadless quantum-dot cellular automata cell. Appl. Phys. Lett. 77, 738–740 (2000)

Optimization of Combinational Logic Circuits Using NAND Gates and Genetic Programming

Arezoo Rajaei, Mahboobeh Houshmand, and Modjtaba Rouhani

Abstract. The design of an optimized logic circuit that implements a desired Boolean function is of interest. Optimization can be performed in terms of different objectives. They include optimizing the number of gates, the number of levels, the number of transistors of the circuit, etc. In this paper, we describe an approach using genetic programming to optimize a given Boolean function concerning the above mentioned objectives. Instead of commonly used set of gates, i.e. {AND, OR, NOT, XOR}, we use the universal NAND gates which lead to a faster and more compact circuit. The traditional gate minimization techniques produce simplified expressions in the two standard forms: sum of products (SOP) or product of sums (POS). The SOP form can be transformed to a NAND expression by a routine, but the transformation does not lead to optimized circuit; neither in terms of the number of gates, nor the number of levels. Experimental results show our approach produces better results compared to transforming the SOP form to the NAND expression, with respect to the number of gates, levels and transistors of the circuit.

1 Introduction

A combinational circuit is one for which the output value is determined solely by the values of the inputs. Such a circuit can be represented by a truth table and computes a Boolean function. The design of an optimized logic circuit that implements a desired Boolean function is of interest.

Arezoo Rajaei
Dept. of Computer Engineering, Sheikhbahaee University, Isfahan, Iran
e-mail: arezoo_rajaee@yahoo.com

Mahboobeh Houshmand
Dept. of Computer Engineering and Information Technology, Amirkabir University of Technology, Tehran, Iran
e-mail: houshmand@aut.ac.ir

Modjtaba Rouhani
Dept. of Electrical Engineering, Islamic Azad University, Gonabad Branch, Iran
e-mail: m.rouhani@ieee.org

A. Gaspar-Cunha et al. (Eds.): Soft Computing in Industrial Applications, AISC 96, pp. 405–414.
springerlink.com

Karnaugh map [1] and Quine-Mc Cluskey method [2, 3] are the popular traditional gate minimization techniques. Karnaugh maps are useful in minimization of circuits only up to six variables and according to [4] Qunie-Mc Cluskey is not very time efficient.

Karnaugh map and Quine-Mc Cluskey method always produce simplified expressions in the two standard forms: sum of products (SOP) or product of sums (POS) and logic CMOS circuits are implemented using AND, OR and NOT gates based on these SOP or POS formulae.

The NAND gates are said to be universal gates since combinations of them can be used to accomplish any of the basic operations and can thus produce a NOT, an OR gate or an AND gate. Writing the SOP and POS forms in terms of NAND gates saves on cost, because implementing the circuits using NAND gate yields a more compact result than the alternatives. SOP formula can be transformed to the NAND expressions. But this transformation does not lead to an optimized circuit, neither in terms of number of gates, nor the number of levels.

In [5-8] also genetic algorithms are used to solve the optimization of combinational logic circuits by a routine. The process of evolutionary circuit design is fundamentally different form traditional design process, because it is not based on designer knowledge and experience, but on the evolution process [9].

Genetic ïprogrammingï(GP) is an extension of the genetic algorithm. Koza [10] introduced GP which synthesizes programs or functions that reproduce a desired behavior. He has designed, for example a two bit adder, using a set of AND, OR, NOT gates, but his emphasis was on generating functional circuits rather than optimizing them. In [6, 11] GP is used to gate level synthesis of Boolean functions. In [12] Cartesian Genetic Programming (CGP) is used to circuit optimization/resynthesis of difficult circuits.

Instead of the commonly used set of gates, i.e. {AND, OR, NOT, XOR}, we propose an approach which fully describes how GP can be used in combinational logic optimization using the universal NAND gates. Our approach consists of two major parts. In the first one we only use two-input NAND gates and optimize the circuit with respect to number of gates. Then we use our approach in a way to optimize the levels of the circuit. And finally we apply the weighted sum of objective functions as a simple and efficient technique to consider both objectives.

In the second part of our approach, we concentrate on transistor-level optimization. In this part, we use multi-input NAND gates and show how our proposed approach can be used to optimize the number of transistors of the implementing circuit.

Experimental results demonstrated our approach efficiently optimizes the Boolean functions and always led to better results compared to the ones obtained by a human designer, with respect to the number of gates, levels and transistors of implementing circuit.

The organization of this paper is as follows. In Section 2, we propose our approach. In Section 3, our experimental results are expressed. Finally in Section 4, we conclude the paper.

2 Proposed Approach

2.1 Chromosome Representation

The program functions (i.e. internal nodes in the tree) are two-input NAND gates and the terminal nodes are {Boolean variables, true, false}. The preorder traversal [13] of the tree produces an expression. As an example, Fig. 1 shows NAND (NAND (False, A), NAND (A, B)).

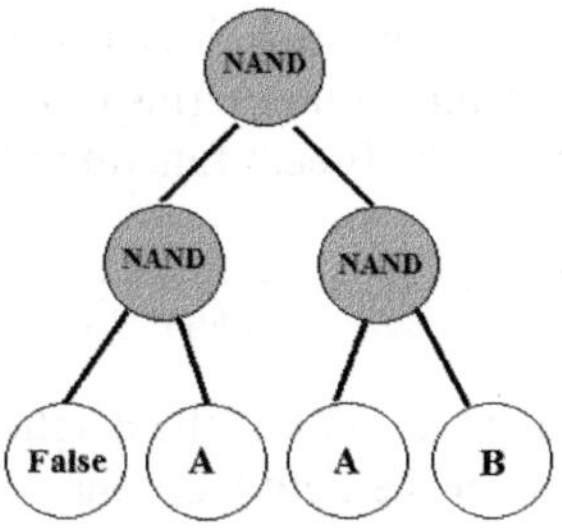

Fig. 1 A chromosome to represent NAND (NAND (False, A), NAND (A, B))

2.2 Fitness Function

- **Optimization based on the number of gates**

We use a two stage fitness function. In the first stage, the similarity of the function the chromosome represents to the given Boolean function is calculated. Therefore the fitness of a chromosome C is defined as (1).

$$Fit_1(C) = \frac{N(F,C)}{2^n} \tag{1}$$

Where n is the number of variables, F is the Boolean function, and $N(F,C)$ is the number of identical minterms between the function that chromosome C represents and the function F. If fitness in (1) is equal to 1, it implies that the chromosome is a feasible solution and implements the corresponding function. In this situation, because in this section we aim to optimize the number of gates, the fitness in (1) is replaced with (2).

$$Fit_1(C) = \alpha_1 + \frac{1}{Gates(C)^{\beta_1}} \tag{2}$$

Where Gates(C) returns the number of internal nodes (gates) in chromosome C and α_1, β_1 are two user defined parameters greater equal than 1. A chromosome whose fitness is more than 1 is a feasible solution.

The equations (1) and (2) together rank the chromosomes according to both their similarity to the given Boolean function *F* and the inverse number of gates in the chromosomes.

- **Optimization based on levels**

Optimizing the number of levels optimizes the propagation delay of signals in the circuit and makes the circuit faster.

Optimizing the overall number of gates does not necessarily lead to an optimized implementing circuit with respect to the number of levels.

Due to the flexibility of our approach, it can be easily modified in a way that we optimize a given Boolean function in terms of the number of levels of the implementing circuit. In this way, the fitness function (2) is replaced with (3):

$$Fit_2(C) = \alpha_2 + \frac{1}{Height(C)^{\beta_2}} \tag{3}$$

Where *Height(C)* is a function that returns the height of the chromosome tree. The height of a tree is the length of the path from the root to the deepest node in the tree and we assume a tree with only one node has a height of zero. For example in Fig 1, the height of tree is two and the numbers of levels of circuit is 2.

- **Muti-objective GA**

To meet the multiple targets in the design of a combinational digital circuit, the problem is turned to a Multi-Objective Optimization Problem (MOOP) that can be expressed as (4).

$$Maximize \quad f(x) = (f_1(x), f_2(x), \dots, f_n(x)), x \in \Omega \tag{4}$$

Where Ω is the feasible solution space, $f_1(x), f_2(x), \dots, f_n(x)$ are *n* objective functions.

A well-known fitness function widely used with GAs is the weighted sum of objective functions, which converts a MOOP to its single-objective equivalent as (5).

$$Maximize \quad Fit(X) = \sum_{i=1}^{n} w_i Fit_i(X) \tag{5}$$

Where $Fit_i(x)$ denotes the normalized objective function that corresponds to $f_i(x)$; w_i denotes the corresponding nonnegative weight for i=1…N, which satisfies

$$\sum_{i=1}^{n} w_i = 1$$

We considered the two objectives, i.e. to optimize the number of levels and optimize the number of gates of the circuit. w_i expresses the user preferences which we currently set to $\frac{1}{2}$. So the fitness of a chromosome is defined as (6).

$$Fit(C)=\frac{1}{2}Fit_1(C)+\frac{1}{2}Fit_2(C) \tag{6}$$

- **Optimization based on the number of transistors**

When we aim to optimize the number of gates, we actually ignore the real circuit cost on the chip, because the circuits are ultimately implemented using transistors. Gate-level optimization does not necessarily lead to transistor-level optimization. The lower number of transistors leads to lower circuit size on chip and production cost will be cheaper too.

When considering the number of transistors in the circuit, we modify our chromosome representation presented in section 2.1 in a way to use multi-input NAND gates. We currently used two, three, four and five- input NAND gates. In CMOS technology, a two-input NAND gate is implemented using four transistors [14], and each extra input requires two extra transistors; i.e. to implement a three, four and five-input NAND gate we need 6, 8 and 10 transistors respectively. In the other words, 2n transistors are required to implement n-input NAND gates. To optimize a circuit based on the number of its transistors the fitness function is defined as (7).

$$Fit_3(C)=\alpha_3+\frac{1}{(4*NAND2(C)+6*NAND3(C)+8*NAND4(C)+10*NAND5(C))^{\beta_3}} \tag{7}$$

Where NAND2(C), NAND3(C), NAND4(C) and NAND5(C) respectively return the number of two, three, four and five input NANDs in the circuit. Selection can be performed using any algorithm explained in [15].

2.3 *Mutation and crossover*

To perform mutation, we use the method introduced in [16]. In this paper, it is proposed to replace the chromosome with the worst fitness in the population with a new randomly produced chromosome. This method performs the mutation function, i.e. restoring the diversity that may be lost from the repeated application of selection and crossover.

In order to perform the crossover, a random node and its sub-tree in one chromosome are exchanged by another random node and its sub-tree in another chromosome. Fig. 2 shows the crossover operation.

Mutation and crossover take place with their corresponding probabilities, i.e. Pm and Pc, respectively.

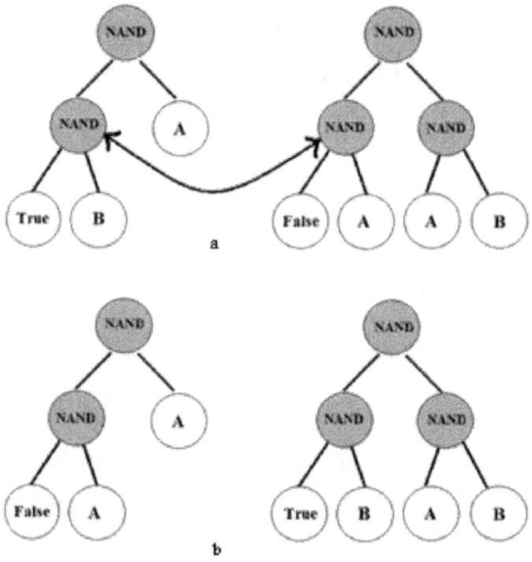

Fig. 2 Crossover. a. Before crossover. b. After crossover

3 Experimental Results

We implemented our approach with C++. We used a population of 100 chromosomes and ran GP for 500 generations. We used Roulette Wheel algorithm for selection and set the parameters P_m and P_c, to 0.06 and 0.75, respectively.

We tentatively set α_1, α_2, α_3 to 100, and β_1, β_2 and β_3 to 2 (in (2), (3) and (4)).

To evaluate the first part of our approach, we randomly generated some three and four-input Boolean functions and compared the results obtained by our approach with the ones obtained by a human designer. A human designer first should simplify the function using the Karnaugh map and then transfer the SOP form to the NAND2 expression. Table 1 and Table 2 show the comparison of our approach with human designer, for three-input and four-input circuits, respectively. NG and NL in these two tables denote the number of gates and the number of levels of implementing circuits respectively.

To evaluate the second part of our approach, we randomly generated some other three and four-input Boolean functions. We compared the results our approach produced with the ones obtained by a human designer with respect to the number of transistors. The human designer1 simplifies the circuit using Karnaugh map and obtains the number of transistors required in the SOP form. It should be mentioned that AND2, AND3 require 6 and 8 transistors respectively. OR2, OR3 require 6 and 8 transistors respectively and NOT requires 2 transistors.

Human designer2 first transforms the SOP expression to NAND expression and then counts the required number of transistors as mentioned in Section 2.2.

Table 3 and Table 4 show a sample comparison of our approach with two human designers, for three-input and four-input circuits, respectively. NT in these two tables denotes the number of transistors of implementing circuits.

Table 1 Comparison of our approach (Genetic Programming) with a circuit obtained by a human designer for two three-input Boolean functions, in terms of Number of Gates (NG) and number of levels (NL).

Approach	Minterms	Circuit	NG	NL
Genetic programming	$F_1 = \Sigma(0,1,3,5,7)$	NAND (NAND (1 , C) , NAND (NAND (1 , B) , NAND (A , 1)))	5	3
Human designer		NAND(NAND(1, C), NAND(1, NAND(1, NAND(NAND(A, 1), NAND(B, 1)))))	7	5
Genetic programming	$F_2 = \Sigma(0,3,5,6)$	NAND (NAND (NAND (NAND (B , C) , NAND (NAND (1 , C) , NAND (1 , B))) , NAND (1 , A)) , NAND (NAND (NAND (C , A) , NAND (A , B)) , NAND (B , C)))	13	5
Human designer		NAND(NAND(A, NAND(1, NAND(B, NAND(C, 1)))), NAND(NAND(A, NAND(1, NAND(C, NAND(B, 1)))), NAND(1, NAND(NAND(NAND(1,A), NAND(1, NAND(B, C))), NAND(NAND(1, A), NAND(1, NAND(NAND(1, B), NAND(1, C)))))))	22	8

Table 2 Comparison of our approach (Genetic Programming) with a circuit obtained by a human designer for two four-input Boolean function, in terms of Number of Gates (NG) and Number of Levels (NL).

Approach	Minterms	Circuit	NG	NL
Genetic programming	$F_1 = \Sigma(1,3,5,6,9,11,14)$	NAND (NAND (NAND (B , 1) , D) , NAND (NAND (D , C) , NAND (NAND (B , C) , NAND (NAND (A , A) , D))))	9	5
Human designer		NAND(NAND(1, NAND(NAND(NAND(1, B), D), NAND(B, C))), NAND(D, NAND(NAND(1, A), NAND(1, C))))	10	5
Genetic programming	$F_2 = \Sigma(0,2,4,6,7,9,13)$	NAND (NAND (NAND (C , 1) , NAND (NAND (A , D) , S)) , NAND (NAND (D , NAND (B , C)) , NAND (S , A)))	9	4
Human designer		NAND(NAND(1, NAND(NAND(NAND(1, A), NAND(1, D)),NAND(C, NAND(1, NAND(NAND(1,A), NAND(1, B)))))), NAND(D, NAND(1, NAND(A, NAND(1, C)))))	15	7

Table 3 Comparison of our approach (Genetic Programming) with a circuit obtained by two human designers for two three-input Boolean functions, in terms of the Number of Transistors (NT).

Approach	Minterms	Circuit	NT
Genetic programming	$F_1 = \Sigma(0,2,4,7)$	NAND (1 , NAND4 (A , B , C , 1) , NAND (NAND (A , B) , NAND (C , 1)))	26
Human designer1 (SOP form)		OR(AND(NOT(B),NOT(C)), AND(NOT(A),NOT(C)), AND(A ,B,C)))	36
Human designer2		NAND(NAND(NAND(B, 1), NAND(C, 1)), NAND(NAND(A , 1) , NAND(C, 1)), NAND(A,B,C))	32
Genetic programming	$F_2 = \Sigma(3,5)$	NAND (NAND (NAND (A , 1) , C , B) , NAND (NAND (B , 1) , C , A))	24
Human designer1 (SOP form)		OR(AND(NOT(A),B,C),AND(A,NOT(B),C))	26
Human designer2		NAND(1, NAND(C, NAND(NAND(A, NAND(1, B)), NAND(B, NAND(1, A)))))	28

Table 4 Comparison of our approach (Genetic Programming) with a circuit obtained by two human designers for two four-input Boolean functions, in terms of Number of Transistors (NT).

Approach	Minterms	Circuit	NT
Genetic programming	$F_1 = \Sigma(0,1,3,5,9,11,13)$	NAND (NAND (NAND (B , C) , D) , NAND (NAND (1 , B) , NAND (1 , A , 1) , NAND (1 , C)) , 1)	34
Human designer1 (SOP form)		OR(AND(NOT(C),D),AND(NOT(B),D),AND(NOT(A), NOT(B), NOT(C)))	36
Human designer2		NAND(NADN(D, NAND(1, B)), NAND(NAND(1, A), NAND(1,B), NAND(1,C)), NAND(NAND(1,C), D))	40
Genetic programming	$F_2 = \Sigma(1,4,6,7,9,13)$	NAND (NAND (NAND (NAND (1, C), D ,1), 1,NAND (A , A) , B) , NAND (1 , NAND (1 , C) , D , 1 , NAND (NAND (A , D) , 1 , B)))	50
Human designer1 (SOP form)		OR(AND(NOT(B),NOT(C),D),AND(A,NOT(C),D),AND(NOT(A),B,C),AND(NOT(A),B,NOT(D)))	54
Human designer2		NAND(NAND(NAND(1, B), NAND(1, C) , D), NAND(A, NAND(1, C) , D), NAND(NAND(1, A), B, C), NAND(NAND(1, A), B, NAND(1, D)))	56

Tables 1 through 4 demonstrate our approach always produced better results than the ones obtained by human designers, with respect to the number of gates, levels and the transistors of the implementing circuit.

4 Conclusion

We have shown a genetic programming approach for the synthesis of combinational logic circuits. Instead of commonly used set of gates, i.e. {AND, OR, NOT, XOR} we used the universal NAND gates. First we used 2-input NAND gates and aimed to optimize the circuit based on the number of gates and levels. Then we used multi-input NAND gates to achieve a circuit with an optimized number of transistors. The experimental results demonstrated that our approach always produced better results than the results obtained by a human designer.

References

[1] Karnaugh, M.: A map method for synthesis odf combinational logic circuits. Transactions of the AIEE, communications and electronics 72, 593–599 (1953)

[2] Quine, W.V.: A way to simplify truth functions. American mathematical monthly 62, 627–631 (1955)

[3] McCluskey, E.J.: Minimization of boolean functions. Bell systems technical journals 35, 627–631 (1956)

[4] Katz, R.H., Borriello, G.: Contemporary logic design. Prentice-Hall, Englewood Cliffs (2004)

[5] Coello, C.A., Christiansen, A.D., Aguirre, A.H.: Automated design of combinational logic circuits using genetic algorithms. In: Int. Conf. on Artificial Neural Nets and Genetic Algorithms, pp. 335–338 (1997)

[6] Hernandez-Aguirre, A., Buckles, B.P., Coello-Coello, C.A.: Gate-level synthesis of Boolean functions using binary multiplexers and genetic programming. In: Proceedings of the 2000 Congress on Evolutionary Computation, vol. 1, pp. 675–682 (2000)

[7] Zhao, S., Jiao, L.: Multi-objective evolutionary design and knowledge discovery of logic circuits based on an adaptive genetic algorithm. Genetic Programming and Evolvable Machines 7, 195–210 (2006)

[8] Chong, K.H., Aris, I.B., Bashi, S.M.: Application of evolutionary algorithm in optimization digital structure design. AIML Journal 6 (2006)

[9] Greene, J.: Simulated evolution and adaptive search in engineering design. In: 2nd Online Workshop on Soft Computing (1997)

[10] Koza, J.R.: Genetic programming: on the programming of computers by means of natural selection. MIT Press, Cambridge (1992)

[11] Dill, K.M., Herzog, J.H., Perkowski, M.A.: Genetic programming and its applications to the synthesis of digital logic. In: 1997 IEEE Pacific Rim Conference on Communications, Computers and Signal Processing, 1997. '10 Years PACRIM 1987-1997 - Networking the Pacific Rim, vol. 2, pp. 823–826 (1997)

[12] Fišer, P., Schmidt, J., Vašíček, Z., Sekanina, L.: On logic synthesis of conventionally hard to synthesize circuits using genetic programming. In: 2010 IEEE 13th International Symposium on Design and Diagnostics of Electronic Circuits and Systems (DDECS), pp. 346–351 (2010)

[13] Horowitz, E., Sahni, S.: Fundamentals of data structures. W H Freeman & Co (Sd), New York (1983)

[14] Weste, N.H.E., Eshraghian, K.: Principles of CMOS VLSI Design: Addison Wesley (October 1994)
[15] Melanie, M.: An Introduction to genetic algorithms. The MIT Press, Cambridge (1999)
[16] Bonyadi, M.R., Azghadi, S.M.R., Rad, N.M., Navi, K., Afjei, E.: Logic optimization for majority gate-Based nanoelectronic circuits based on genetic Algorithm. In: International Conference on Electrical Engineering, ICEE (2007)

Electromagnetism-Like Augmented Lagrangian Algorithm for Global Optimization

Ana Maria A.C. Rocha and Edite M.G.P. Fernandes

Abstract. This paper presents an augmented Lagrangian algorithm to solve continuous constrained global optimization problems. The algorithm approximately solves a sequence of bound constrained subproblems whose objective function penalizes equality and inequality constraints violation and depends on the Lagrange multiplier vectors and a penalty parameter. Each subproblem is solved by a population-based method that uses an electromagnetism-like mechanism to move points towards optimality. Benchmark problems are solved in a performance evaluation of the proposed augmented Lagrangian methodology. A comparison with a well-known technique is also reported.

1 Introduction

This paper presents a numerical study of an augmented Lagrangian methodology, where the subproblems are solved by a stochastic population based algorithm, for continuous constrained global optimization. We aim to address the problem in the form:

$$\min f(x) \text{ subject to } g(x) \leq 0,\, h(x) = 0,\, x \in \Omega, \tag{1}$$

where $f : \mathbb{R}^n \to \mathbb{R}$, $g : \mathbb{R}^n \to \mathbb{R}^p$ and $h : \mathbb{R}^n \to \mathbb{R}^m$ are nonlinear continuous functions and $\Omega = \{x \in \mathbb{R}^n : lb \leq x \leq ub\}$. We do not assume that the objective function f is convex. There may be many local minima in the feasible region. This class of global optimization problems arises frequently in engineering applications.

Ana Maria A.C. Rocha
Department of Production and Systems, University of Minho,
4710-057 Braga, Portugal
e-mail: arocha@dps.uminho.pt

Edite M.G.P. Fernandes
Department of Production and Systems, University of Minho,
4710-057 Braga, Portugal
e-mail: emgpf@dps.uminho.pt

A. Gaspar-Cunha et al. (Eds.): Soft Computing in Industrial Applications, AISC 96, pp. 415–425.
springerlink.com

Specially for large scale problems, derivative-free and stochastic methods are the most well-known and used methods. The two main categories of methods to handle constraints in these algorithms are listed below.

1. Methods based on penalty functions. The constraints violation is combined with the objective function to define a penalty function. This function aims to penalize infeasible solutions by increasing their fitness values proportionally to their level of constraints violation. Static, dynamic, annealing and adaptive penalties are the most popular [2, 4, 8, 14, 16, 19]. Methods based on augmented Lagrangians are common in deterministic type methods for global optimization, for example in [6, 7, 15], but rare when combined with heuristics that rely on a population of points to converge to the solution [1, 22, 23, 24].
2. Methods based on biasing feasible over infeasible solutions. They seem to be nowadays interesting alternatives to penalty methods for handling constraints. In this type of methods, constraints violation and the objective function are used separately and optimized by some sort of order, being the constraints violation the most important. See, for example, [9, 17, 18, 20, 21].

Here, we aim to show the functionality of an augmented Lagrangian methodology to handle the equality and inequality constraints of the problem (1), where the subproblems are approximately solved by a stochastic global population based algorithm. Due to its simplicity, the electromagnetism-like (EM) algorithm proposed in [4, 5] is used to obtain the solution of each subproblem. The EM algorithm simulates the electromagnetism theory of physics by considering each point in the population as an electrical charge. The method uses an attraction-repulsion mechanism to move a population of points towards optimality.

Since the EM algorithm has been designed to find a minimizer which satisfies $x \in \Omega$, our subproblem has bound constraints. Although other constraint-handling techniques have been implemented by the authors with the EM algorithm, namely, the feasibility and dominance rules [17, 18], the separate feasibility and optimality measures based on sufficient reduction conditions [20], and the adaptive penalty technique [19], the new proposed augmented Lagrangian strategy has been given the best results so far.

Our implementation of an augmented Lagrangian methodology is as follows: we reformulate problem (1) converting each equality constraint into an inequality as herein shown: $|h_j(x)| \leq \beta$, where β is a positive relaxation parameter. This is an usual procedure in stochastic based methods [9, 18, 19]. In general, the relaxation parameter is fixed over the entire iterative process. Typically, 10^{-3}, 10^{-4} and 10^{-5} are common values in the literature. Our proposal defines a sequence $\{\beta^k\}$ of decreasing nonnegative numbers such that $\lim_{k\to\infty} \beta^k = \beta^* > 0$. The idea is to tighten the equality constraints relaxation scheme as iterations proceed. Further, a different updating scheme for the penalty parameter is also proposed. When the level of constraints violation is under a specified tolerance, even if the infeasibility did not improve, the penalty is allowed to decrease instead of increasing (see Algorithm 2.1).

The remainder of this paper is organized as follows. Sect. 2 describes the proposed stochastic augmented Lagrangian paradigm and Sect. 3 briefly introduces the

EM algorithm. Sect. 4 displays the results of the numerical experiments, including a comparison with a well-known penalty method, and Sect. 5 contains some concluding remarks.

2 An Augmented Lagrangian Method

Most stochastic methods for global optimization are developed primarily for unconstrained or simple bound constrained problems. Then they are extended to constrained optimization problems using, for example, a penalty technique. This type of technique transforms the constrained problem into a sequence of unconstrained subproblems by penalizing the objective function when constraints are violated. The objective penalty function, in the unconstrained subproblem, depends on a positive penalty parameter that must be updated throughout the iterative process. With most penalty functions, the solution of the constrained problem is reached for an infinite value of the penalty parameter. An augmented Lagrangian is a more sophisticated penalty function for which a finite penalty parameter value is sufficient to yield convergence to the solution of the constrained problem [3].

We now show the functionality of the proposed augmented Lagrangian function when solving constrained global optimization problems. Practical and theoretical issues from the augmented Lagrangian methodology are used with this population based algorithm, the EM algorithm as proposed in [5], to compute approximate solutions of the sequence of bound constrained subproblems.

Since equality constraints are the most difficult to be satisfied, our augmented Lagrangian methodology considers problems only with inequality constraints, using a common procedure in stochastic methods for global optimization to convert the equality constraints of the problem into inequality constraints, as follows: $|h_j| \leq \beta$, $j = 1,\ldots,m$ for a fixed $\beta > 0$. For simplicity, the problem (1) is rewritten as

$$\min f(x) \text{ subject to } G(x) \leq 0,\ x \in \Omega, \tag{2}$$

where the vector of the inequality constraints is now defined by

$$G(x) = (g_1(x),\ldots,g_p(x),|h_1(x)| - \beta,\ldots,|h_m(x)| - \beta).$$

Our proposal concerning the relaxed equality constraints aims to tighten the relaxation scheme as iterations proceed, using variable relaxation parameter values. Thus, a sequence of decreasing nonnegative values bounded by $\beta^* > 0$ is defined as:

$$\beta^{k+1} = \max\left\{\beta^*, \frac{1}{\sigma}\beta^k\right\},\ \sigma > 1. \tag{3}$$

The formula of the augmented Lagrangian that corresponds to the inequality constraints in the converted problem (2) is:

$$\mathscr{L}_\rho(x,\mu) = f(x) + \frac{\rho}{2}\sum_{i=1}^{p+m}\left[\max\left(0, G_i(x) + \frac{\mu_i}{\rho}\right)\right]^2 \tag{4}$$

where ρ is a positive penalty parameter and $\mu \in \mathbb{R}^{p+m}$ represents the Lagrange multiplier vector associated with the $p+m$ constraints. Our proposed stochastic augmented Lagrangian algorithm adapted to the reformulation (2), of the original problem (1), and based on the Lagrangian (4) is presented in Algorithm 2.1:

Algorithm 2.1 (Augmented Lagrangian algorithm)

Given: $\mu^+ > 0$, $\varepsilon^* > 0$, $0 < \alpha < 1$, $\gamma > 1$, $\sigma > 1$, $k_{\max}$, $l_{\max}$, $\beta^* > 0$, $0 < \rho^- < \rho^+$, $\mu^1 \in [0, \mu^+]$

randomly generate x^0 in Ω; compute ρ^1; set $k = 1$

while $\|v^{k-1}\| > \varepsilon^*$ and $k \le k_{\max}$ **do**

  $\varepsilon^k = \max\left\{\varepsilon^*, 10^{-k}\right\}$; update β^k using (3); set $l = 1$

  while $\left(\mathscr{L}_{\text{avg}} - \mathscr{L}_{\rho^k}(x(\text{best}), \mu^k)\right) > \varepsilon^k$ and $l \le l_{\max}$ **do**

    use x^{k-1} and randomly initialize a population of $p_{\text{size}} - 1$ points in Ω

    run EM to compute a population of solutions to $\min_x \mathscr{L}^l_{\rho^k}(x, \mu^k)$ subject to $x \in \Omega$

    $l = l + 1$

  end while

  $x^k = x(\text{best})$

  compute $v_i^k = \max\left\{G_i(x^k), -\dfrac{\mu_i^k}{\rho^k}\right\}, i = 1, \ldots, m+p$

  if $k = 1$ or $\|v^k\| \le \alpha\|v^{k-1}\|$ **then**

    $\rho^{k+1} = \rho^k$

  else

    if $\|v^k\| \le \varepsilon^k$ **then**

      $\rho^{k+1} = \max\{\rho^-, \dfrac{1}{\gamma}\rho^k\}$

    else

      $\rho^{k+1} = \min\{\rho^+, \gamma\rho^k\}$

    end if

  end if

  update $\mu_i^{k+1} = \min\left\{\max\left\{0, \mu_i^k + \rho^k G_i(x^k)\right\}, \mu^+\right\}, i = 1, \ldots, m+p$

  $k = k + 1$

end while

This algorithm extends recent work with the Powell-Hestenes-Rockafellar augmented Lagrangian function presented in [6, 7], where equalities and inequalities are treated separately. In Algorithm 2.1 the penalty parameter ρ, besides being increased, when infeasibility is not reduced, it is also reduced whenever the constraints violation is under a specified tolerance ε^k; otherwise it is not changed. Further, a safeguarded scheme is also included in the process. This is motivated by the need to keep the penalty parameter bounded and the subproblems well conditioned. This procedure is reported in the lines 12–20 of Algorithm 2.1.The initial value for the penalty parameter is defined by

$$\rho^1 = \max\left\{10^{-6}, \min\left\{10, 2|f(x^0)|/(\|\max(0, G(x^0))\|^2)\right\}\right\}$$

for an arbitrary initial approximation x^0 [6, 7]. The algorithm also updates the Lagrange multipliers using first order estimates and safeguarded schemes. This is a crucial issue to maintain the sequence $\{\mu^k\}$ bounded.

Further, lines 5-9 of the algorithm show details of the inner iterative process to compute an approximation to the solution of subproblem (5), at each iteration k :

$$\min_x \mathscr{L}_{\rho^k}(x, \mu^k) \text{ subject to } x \in \Omega \tag{5}$$

for fixed values of the parameters ρ^k and μ^k. For each set of fixed values of penalties and Lagrange multipliers, a reasonable approximate solution of subproblem (5) is required so that convergence could be promoted [3]. Since the EM algorithm is based on a population of points, with size p_{size}, the point which yields the least objective function value, denoted by the best point of the population, $x(\text{best})$, after stopping, is taken as the next approximation to the problem (1). We also note that our stochastic EM algorithm uses the approximation x^{k-1} as one of the points of the population to initialize the EM algorithm. The remaining $p_{\text{size}} - 1$ points are randomly generated. The inner iteration counter is represented by l. This process terminates when the difference between the function value at the best point, $\mathscr{L}_{\rho^k}(x(\text{best}), \mu^k)$, and the average of the function values of the population, $\mathscr{L}_{\text{avg}}$, is under a specified tolerance ε^k. This tolerance decreases as outer iterations proceed. A limit of $l_{\max}$ iterations is also imposed.

3 The Electromagnetism-Like Mechanism

In this section, we briefly present the EM mechanism, proposed in [5], for solving the subproblems in the Algorithm 2.1. Here, the objective is to compute an approximate minimizer of $\mathscr{L}_{\rho^k}(x, \mu^k)$, for fixed values of the parameters ρ^k and μ^k. For simplicity we use the notation $\mathscr{L}^k(x) = \mathscr{L}_{\rho^k}(x, \mu^k)$. Because EM is a population based algorithm, the inner iterative process begins with a population of p_{size} solutions (line 6 in Algorithm 2.1). The best found solution, denoted by $x(\text{best})$, and the average of function values, are defined by

$$x(\text{best}) = \arg\min\left\{\mathscr{L}^k(x(s)) : s = 1, \ldots, p_{\text{size}}\right\} \text{ and } \mathscr{L}^k_{\text{avg}} = \sum_{s=1}^{p_{\text{size}}} \mathscr{L}^k(x(s))/p_{\text{size}}, \tag{6}$$

respectively, where $x(s), s = 1, \ldots, p_{\text{size}}$ represent the points of the population. The main steps of the EM mechanism are shown in Algorithm 3.1. Details of each step follow.

Algorithm 3.1 (EM algorithm)

1: **Given:** $x(s), s = 1, \ldots, p_{\text{size}}$
2: evaluate the population and select $x(\text{best})$
3: compute the charges $c(s)$, $s = 1, \ldots, p_{\text{size}}$
4: compute the total forces $F(s)$, $s = 1, \ldots, p_{\text{size}}$
5: move the points except $x(\text{best})$
6: evaluate the new population and select $x(\text{best})$
7: apply local search to $x(\text{best})$
8: compute $\mathscr{L}^k(x(\text{best}))$ and $\mathscr{L}^k_{\text{avg}}$.

The EM mechanism starts by identifying the best point, $x(\text{best})$, of the population using the augmented Lagrangian $\mathscr{L}^k$ for point assessment, see (6). According to the electromagnetism theory, the total force exerted on each point $x(s)$ by the other $p_{\text{size}} - 1$ points is inversely proportional to the square of the distance between the points and directly proportional to the product of their charges:

$$F(s) = \sum_{r \neq s}^{p_{\text{size}}} F_r^s \equiv \begin{cases} (x(r) - x(s)) \dfrac{c(s)c(r)}{\|x(r) - x(s)\|^2}, & \text{if } \mathscr{L}^k(x(r)) < \mathscr{L}^k(x(s)) \\ (x(s) - x(r)) \dfrac{c(s)c(r)}{\|x(r) - x(s)\|^2}, & \text{otherwise} \end{cases},$$

for $s = 1, \ldots, p_{\text{size}}$, where the charge $c(s)$ of point $x(s)$ determines the magnitude of attraction of that point over the others through

$$c(s) = \exp\left(\frac{-n\left(\mathscr{L}^k(x(s)) - \mathscr{L}^k(x(\text{best}))\right)}{\sum_{r=1}^{p_{\text{size}}} \left(\mathscr{L}^k(x(r)) - \mathscr{L}^k(x(\text{best}))\right)} \right).$$

Then, the normalized total force vector exerted on each point $x(s)$ is used to move the point in the direction of the force by a random step size $\iota \sim U[0,1]$, maintaining the point inside the set Ω. Thus for $s = 1, \ldots, p_{\text{size}}$ ($s \neq \text{best}$) and for each component $i = 1, \ldots, n$

$$x_i(s) = \begin{cases} x_i(s) + \iota \dfrac{F_i(s)}{\|F(s)\|}(ub_i - x_i(s)), & \text{if } F_i(s) > 0 \\ x_i(s) + \iota \dfrac{F_i(s)}{\|F(s)\|}(x_i(s) - lb_i), & \text{otherwise} \end{cases}.$$

Finally, a local search is performed around the best point of the population in order to refine the solution. In [5], a simple random local search is proposed. This is described in Subsect. 3.1. In this paper, we also aim to integrate in the EM algorithm a more sophisticated local search (see Subsect. 3.2). Our numerical experiments show that our proposal significantly improves the performance of the overall augmented Lagrangian algorithm.

3.1 Random Local Search

Here, we briefly describe the local search proposed in [5] for the population-based EM algorithm. This is a simple random line search applied component by component to $x(\text{best})$. For each component i, $x(\text{best})$ is assigned to a temporary point y. Then a random movement of maximum length $\delta_{EM} \max_j(ub_j - lb_j)$, $\delta_{EM} > 0$ is carried out and if a better position is obtained within $\max_{local}$ iterations, $x(\text{best})$ is replaced by y, the search ends for that component and proceeds to another one. Although this local search is based on a simple random procedure, it has been shown to greatly improve accuracy of the EM algorithm.

3.2 Hooke and Jeeves Local Search

In this section, we describe our modification to the original EM algorithm. In our proposal, the local procedure is based on the Hooke and Jeeves (HJ) pattern search algorithm. This is a derivative-free method that searches in the neighborhood of a point x for a better approximation *via* exploratory and pattern moves [11, 12]. To reduce the number of function evaluations, the HJ pattern search algorithm is applied to the current best point only, $x(\text{best})$. This algorithm is a variant of the coordinate search, in the sense that incorporates a pattern move to accelerate the progress of the algorithm, by exploiting information obtained from the search in previous successful iterations.

The exploratory move carries out a coordinate search (a search along the coordinate axis) about the best point, with a step length δ_{HJ}. If a new trial point, y, with a better fitness value than $x(\text{best})$ is encountered, in the sense that the augmented Lagrangian $\mathscr{L}^k$ value is better, the iteration is successful. Otherwise, the iteration is unsuccessful and δ_{HJ} is reduced by a factor $0 < \Delta_{HJ} < 1$. If the previous iteration was successful, the vector $y - x(\text{best})$ defines a promising direction and a pattern move is then implemented, meaning that the exploratory move is carried out about the trial point $y + (y - x(\text{best}))$, rather than about the current point y. Then, if the coordinate search is successful, the returned point is accepted as the new point; otherwise, the pattern move is rejected and the method reduces to coordinate search about y. Please see [11] for details. To ensure that only points inside Ω are tested in the HJ pattern search algorithm, a projection scheme is used, i.e., for each component j, if $y_j < lb_j$ or $y_j > ub_j$ then y_j is set to lb_j or ub_j respectively.

4 Numerical Experiments

In this section, we report the results of our numerical study, after running a set of 24 benchmark constrained global problems, described in full detail in [13]. The problems are known as g01-g24 (the 'g' suit, where six problems only have equality constraints, thirteen have inequality constraints, five have both equalities and inequalities and all have simple bounds). Not all problems have multi-modal objective functions, although some are difficult to solve. The best known solution for problem g20 is slightly infeasible. We remark that g02, g03, g08 and g12 are maximization problems that were transformed and solved as minimization ones.

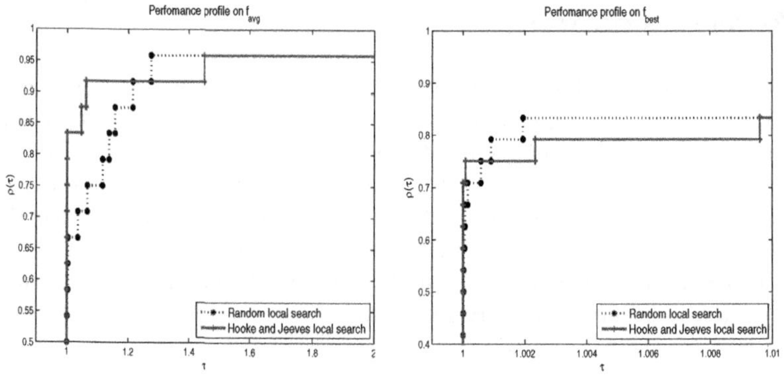

Fig. 1 Comparison of Hooke and Jeeves and Random searches based on f_{avg} (left) and f_{best} (right).

Since the algorithm relies on some random parameters and variables, we solve each problem 30 times and take average of the obtained solutions, herein denoted by f_{avg}. The best of the solutions found after all runs is denoted by f_{best}. The size of the population depends on n, and since some problems have large dimension, $n > 20$, we choose $p_{\text{size}} = \min\{200, 10n\}$. The fixed parameters are set in this study as follows: $\mu^+ = \rho^+ = 10^{12}$, $\varepsilon^* = 10^{-6}$, $\alpha = 0.5$, $\gamma = 2$, $\sigma = 2$, $\beta^* = \rho^- = 10^{-12}$, $\beta^1 = 10^{-3}$. We define $k_{\max} = 50$ and $l_{\max} = 30$ so that a maximum of 1500 iterations are allowed. We remark that the other conditions in the stopping criteria of the Algorithm 2.1 (in the outer and inner iterative processes) may cause the termination of the algorithm before reaching the 1500 iterations. The initial multiplier vector is set to the null vector. The values for the two parameters of the random local search are set as proposed in [5]: $\max_{local} = 10$, $\delta_{EM} = 0.001$. In our implementation of the HJ pattern search, we set the initial δ_{HJ} to 1 and the factor $\Delta_{HJ} = 0.1$.

To compare the performance of the HJ local search procedure, with the random local search of the EM algorithm [5], when incorporated in the proposed augmented Lagrangian context, we use Dolan and Moré's performance profiles [10]. The two profiles are based on the metrics: f_{avg}, the average of the solutions obtained after the 30 runs, and f_{best}, the best solution over the 30 runs. For each solver in comparison, the plot shows the proportion of problems in the set, denoted by $\rho(\tau)$, that has the best value of the metric, for each value of $\tau \in \mathbb{R}$. Thus, to see which solver has the least value of the metric mostly, then its value of $\rho(1)$ should be compared with those of the other solvers in comparison. The higher the ρ the better the solver is. On the other hand, $\rho_s(\tau)$ for large values of τ measures the solver robustness.

Hooke and Jeeves local search performs significantly better than Random local search when comparing the average performance (left plot in Fig. 1) and it gives the best results for about 71% of the problems against 57% of the problems with Random search (plot on the right of Fig. 1).

Table 1 reports f_{best} and f_{avg} obtained by our study and those of [2] for the eleven problems therein registered (g01-g11) and f^* is the best known solution as reported in [13]. In [2], a genetic algorithm combined with an adaptive penalty function is implemented. These are mainly concerned with the frequency of penalty parameters updating and constraints violation computation. For a fair comparison, we use the same conditions described in [2]: $p_{size} = 100$, runs = 25, maximum number of generations = 1000, leading to 100000 fitness function evaluations. We have better performance (both in f_{best} and f_{avg}) than the adaptive penalty algorithm of [2] (best results are in boldface) in all but g02, g03, g06 and g10 problems.

Table 1 Comparison of our results with the best of 5 variants in [2]

Prob.	f^*	our study		[2]	
		f_{best}	f_{avg}	f_{best}	f_{avg}
g01	-15.0000	**-15.0000**	**-15.0000**	-14.9998	-14.9989
g02	-0.80362	-0.76250	-0.63387	**-0.79252**	**-0.72555**
g03	-1.00050	-0.99634	**-0.92664**	**-0.99725**	-0.77797
g04	-30665.54	**-30665.55**	**-30665.55**	-30665.32	-30578.55
g05	5126.497	**5126.497**	**5126.498**	5126.779	5323.866
g06	-6961.814	-6945.097	-6761.717	**-6961.448**	**-6805.229**
g07	24.3062	**24.3062**	**24.3062**	24.5450	27.8486
g08	-0.09583	**-0.09583**	**-0.08992**	**-0.09583**	-0.08769
g09	680.630	**680.630**	**680.691**	680.681	681.470
g10	7049.25	7074.93	**7217.08**	**7070.56**	8063.29
g11	0.74990	**0.75000**	**0.75000**	0.75217	0.88793

5 Final Remarks

From our preliminary numerical tests, we may conclude that the proposed augmented Lagrangian algorithm is able to effectively solve constrained problems till optimality and seems to be competitive with a well-known penalty based algorithm. Practical engineering problems, for example, those reported in [2], will be solved in the near future. We also aim to test our algorithm with a point-to-point search yet stochastic method, when solving the bound constrained subproblems.

References

1. Barbosa, H.J.C.: A coevolutionary genetic algorithm for constrained optimization. In: Proceedings of the, Congress on Evolutionary Computation, vol. 3, pp. 1605–1611 (1999), doi:10.1109/CEC1999, 785466
2. Barbosa, H.J.C., Lemonge, A.C.C.: An adaptive penalty method for genetic algorithms in constrained optimization problems. In: Frontiers in Evolutionary Robotics, 34.p I-Tech Education Publ., Austria (2008) ISBN: 978-3-902613-19-6
3. Bertsekas, D.P.: Nonlinear Programming, 2nd edn. Athena Scientific, Belmont (1999)
4. Birbil, S.I.: Stochastic Global Optimization Techniques. Ph.D. diss., North Carolina State University (2002)

5. Birbil, S.I., Fang, S.-C.: An electromagnetism-like mechanism for global optimization. Journal of Global Optimization 25, 263–282 (2003)
6. Birgin, E.G., Castillo, R., Martinez, J.M.: Numerical comparison of Augmented Lagrangian algorithms for nonconvex problems. Computational Optimization and Applications 31, 31–56 (2004)
7. Birgin, E.G., Floudas, C.A., Martinez, J.M.: Global minimization using an Augmented Lagrangian method with variable lower-level constraints. Mathematical Programming A 125(1), 139–162 (2010)
8. Coello Coello, C.A.: Use of a self-adaptive penalty approach for engineering optimization problems. Computers in Industry 41, 113–127 (2000)
9. Deb, K.: An efficient constraint handling method for genetic algorithms. Computer Methods in Applied Mechanics and Engineering 186, 311–338 (1998)
10. Dolan, E.D., Moré, J.J.: Benchmarking optimization software with performance profiles. Mathematical Programming 91, 201–213 (2001)
11. Hooke, R., Jeeves, T.A.: Direct search solution of numerical and statistical problems. Journal of Association and Computing Machinery 8, 212–229 (1961)
12. Lewis, R.M., Torczon, V.: Pattern search algorithms for bound constrained minimization. SIAM Journal on Optimization 9, 1082–1099 (1999)
13. Liang, J.J., Runarsson, T.P., Mezura-Montes, E., Clerc, M., Suganthan, P.N., Coello, C.A.C., Deb, K.: Problem Definitions and Evaluation Criteria for the CEC, Special Session on Constrained Real-Parameter Optimization. Technical Report, Nanyang Technological University, Singapore (2005)
14. Liu, J.-L., Lin, J.-H.: Evolutionary computation of unconstrained and constrained problems using a novel momentum-type particle swarm optimization. Engineering Optimization 39, 287–305 (2007)
15. Luo, H., Sun, X., Wu, H.: Convergence properties of augmented Lagrangian methods for constrained global optimization. Optimization Methods and Software 23, 763–778 (2008)
16. Petalas, Y.G., Parsopoulos, K.E., Vrahatis, M.N.: Memetic particle swarm optimization. Annals of Operations Research 156, 99–127 (2007)
17. Rocha, A.M.A.C., Fernandes, E.M.G.P.: Feasibility and dominance rules in the electromagnetism-like algorithm for constrained global optimization. In: Gervasi, O., Murgante, B., Laganà, A., Taniar, D., Mun, Y., Gavrilova, M.L. (eds.) ICCSA 2008, Part II. LNCS, vol. 5073, pp. 768–783. Springer, Heidelberg (2008)
18. Rocha, A.M.A.C., Fernandes, E.M.G.P.: Implementation of the electromagnetism-like algorithm with a constraint-handling technique for engineering optimization problems. In: Eighth International Conference on Hybrid Intelligent Systems, pp. 690–695. IEEE Computer Society, Los Alamitos (2008)
19. Rocha, A.M.A.C., Fernandes, E.M.G.P.: Self-adaptive penalties in the electromagnestism-like algorithm for constrained global optimization problems. In: Proceedings of the 8th World Congress on Structural and Multidisciplinary Optimization, Lisbon, 10.p (2009)
20. Rocha, A.M.A.C., Fernandes, E.M.G.P.: Hybridizing the electromagnetism-like algorithm with descent search for solving engineering design problems. International Journal of Computer Mathematics 86, 1932–1946 (2009)

21. Runarsson, T.P., Yao, X.: Stochastic ranking for constrained evolutionary optimization. IEEE Transactions on Evolutionary Computation 4, 284–294 (2000)
22. Sedlaczek, K., Eberhard, P.: Augmented Lagrangian particle swarm optimization in mechanism design. Journal of System Design and Dynamics 1, 410–421 (2007)
23. Wah, B.W., Wang, T.: Efficient and Adaptive Lagrange-Multipliers. Journal of Global Optimization 14, 1–25 (1999)
24. Wang, T., Wah, B.W.: Handling inequality constraints in continuous nonlinear global optimization. In: Integrated Design and Process Science, pp. 267–274 (1996)

Multiobjective Optimization of a Quadruped Robot Locomotion Using a Genetic Algorithm

Miguel Oliveira, Lino Costa, Ana Rocha, Cristina Santos, and Manuel Ferreira

Abstract. In this work, it is described a gait multiobjective optimization system that allows to obtain fast but stable robot quadruped crawl gaits. We combine bio-inspired Central Patterns Generators (CPGs) and Genetic Algorithms (GA). A motion architecture based on CPGs oscillators is used to model the locomotion of the robot dog and a GA is used to search parameterizations of the CPGs parameters which minimize the body vibration, maximize the velocity and maximize the wide stability margin. In this problem, there are several conflicting objectives that leads to a multiobjective formulation that is solved using the Weighted Tchebycheff scalarization method. Several experimental results show the effectiveness of this proposed approach.

1 Introduction

Robot locomotion is a challenging task that involves the control of a large number of degrees of freedom (DOF's). Several previous works, [9, 12] proposed biologic approaches to generate and modulate gait locomotion of quadruped robots, combining biometric sensory information with motion oscillators such as Central Pattern Generators (CPGs).

There are still many open questions in the quadruped locomotion, considering learning gaits or gait optimization. The problem of finding the best possible locomotion is a problem currently addressed in the literature. In [2] it is presented a Genetic Algorithm (GA) robust to the noise in the parameters evolution and that

Miguel Oliveira · Cristina Santos · Manuel Ferreira
Industrial Electronics Department, School of Engineering,
University of Minho, 4800-058 Guimaraes, Portugal
e-mail: {mcampos,cristina,mjf}@dei.uminho.pt

Lino Costa · Ana Rocha
Department of Production and Systems , School of Engineering,
University of Minho, 4800-058 Guimaraes, Portugal
e-mail: {lac,arocha}@dps.uminho.pt

A. Gaspar-Cunha et al. (Eds.): Soft Computing in Industrial Applications, AISC 96, pp. 427–436.
springerlink.com

also avoids premature local optima. The evaluation criterion is to maximize robot velocity. A comparison between several gait learning algorithms, including Genetic and Policy Gradient algorithms, is presented in [11]. The optimization goal is to determine the best 12 parameters of an elliptical locus scheme of locomotion, such that the robot takes less time to walk a certain distance.

In [6] it is presented an evolutionary algorithm based on a GA, in which genetic operators are chosen by an adaptation mechanism. Locomotion is implemented in real time and is evaluated by analysing the forward-backward, side-walk and rotation motion as well as the vibration. In [7] it is presented an evolutionary algorithm to optimize a vector of parameters for locomotion of an ERS110 robot. In [10] it is presented an optimization system for the locomotion of an ERS110 based on the Powell's method. It online optimizes 12 parameters of a locus locomotion scheme.

In this work, we propose a multiobjective approach to optimize a quadruped slow crawl gait, using Central Pattern Generators (CPGs) and a GA. CPGs are neural networks located in the spine of vertebrates, that generate coordinated rhythmic movements, namely locomotion [8]. In this work, a locomotion controller, based on CPGs, generates trajectories for hip robot joints [12]. Speed, vibration and stability are the evaluated criterions used to explore the parameter space of the network of CPGs to identify the best crawl pattern.

In order to achieve the desired crawl gait, it is necessary to appropriately tune these parameters by means of a optimization procedure. The resulting optimization problem has multiobjective nature since there are several conflicting objectives. This multiobjective problem was solved by a GA [5] applied to the corresponding Weighted Tchebycheff scalarized formulation [3]. Optimization is done online in a simulated ers-7 AIBO robot using Webots [13].

We have already addressed a slightly different but related problem in a preliminary experience using a genetic algorithm [16] and the electromagnetism-like algorithm [17]. In these works, we noticed that solving this problem requires a considerable computational effort. Notably because several constraints are imposed in this optimization problem. Thus, alternative techniques for handling constraints can make the search more efficient.

This article is structured as follows. In Sect. 2, we introduce several multiobjective optimization concepts. In Sect. 3 the optimization problem is formulated. Sect. 4 presents the optimization system details. Simulated results are presented in Sect. 5. The paper ends with a discussion and conclusions in Sect. 6.

2 Multiobjective Optimization

Mathematically, a multiobjective optimization problem with s objectives and n decision variables can be formulated as, without loss of generality:

$$\begin{array}{l} \min \mathbf{f}(\mathbf{x}) = (f_1(\mathbf{x}), \ldots, f_s(\mathbf{x})) \\ \text{subject to} \quad \mathbf{g}(\mathbf{x}) \geq 0 \text{ and } \mathbf{h}(\mathbf{x}) = 0 \end{array}$$

where $\mathbf{x} \in \mathbb{X}^n$, $\mathbf{g}(\mathbf{x})$ and $\mathbf{h}(\mathbf{x})$ are the inequality and equality constraints, respectively.

Solving multiobjective problems is a very difficult task due to, in general, for this class of problems, the objectives conflict across a high-dimensional problem space and the computational complexity of the problem (NP-hardness). Thus, the interaction between the multiple objectives gives rise to a set of efficient solutions, known as Pareto-optimal solutions.

For a multiobjective minimization problem, a solution **a** is said to dominate a solution **b**, if and only if, $\forall i \in \{1,...,s\} : f_i(\mathbf{a}) \leq f_i(\mathbf{b})$ and $\exists j \in \{1,...,s\} : f_j(\mathbf{a}) < f_j(\mathbf{b})$. A solution **a** is said to be non-dominated regarding a set $\mathbb{Y}^n \subseteq \mathbb{X}^n$ if and only if, there is no solution in $\mathbb{Y}^n$ which dominates **a**. The solution **a** is Pareto-optimal if and only if **a** is non-dominated regarding $\mathbb{X}^n$.

The main goal of a multiobjective algorithm is to find a good and balanced approximation to the Pareto-optimal set. Multiobjective problems can be addressed by scalarization methods such as the Weighted Tchebycheff function [3]:

$$\min f(\mathbf{x}) = \max\{W_i \sum_{i=1}^{s} |F_i - f_i(\mathbf{x})|\} \tag{1}$$

where $W_i \geq 0$ and $\sum_{i=1}^{s} W_i = 1$. The weighted distance is measured to an utopian objective vector **F** with components F_i. Different combinations of the weights W_i can produce different (weakly) Pareto optimal solutions. The problem defined in (1) is non differentiable and must be solved by a derivative free algorithm such as genetic algorithms [3].

3 Problem Formulation

The proposed network of CPGs generates trajectories for the robot limbs. Different combinations of these trajectories for each joint in terms of amplitude, offset and frequency, result in different gait patterns.

The proposed CPGs are based on Hopf oscillators and one intrinsic property is the possibility to smoothly modulate the generated trajectories according to explicit changes in the CPG parameters: amplitude, offset and the stance knee value. Therefore, in order to tune the CPG parameters, we use a GA to search for an optimal combination of these parameters. Speed, vibration and stability are the objectives to optimize in order to define different walking pattern solutions.

Robot trajectories are generated and modulated by the proposed network of CPGs, by explicitly changing the CPG parameters: amplitude (A), Offset(O), and the stance knee value (K), for each limb. Further, the parameter swing frequency (ω_{sw}) is common for the overall network. This means a total of 13 parameters we need to tune to modulate trajectories. However, left and right fore and left and right hind limb trajectories have the same amplitude, offset and frequency but a relative phase of π among them. These considerations enable us to reduce the number of CPG parameters required to optimize, as follows: amplitude of the fore and hind limbs (A_{FL}, A_{HL}); fore and hind limbs knee stance angle (K_{FL}, K_{HL}); fore and hind limbs offset (O_{FL}, O_{HL}) and swing frequency (ω_{sw}). Thus, the problem has 7 decision variables corresponding to the 7 CPG free parameters.

The goal is the minimization of the body vibration and maximization the velocity and wide stability margin. We have used a scalarization method based on the minimization of the Weighted Tchebycheff function:

$$\min f(\mathbf{x}) = \max\{W_a|F_a - f_a(\mathbf{x})|, W_v|F_v - f_v(\mathbf{x})|, W_{\mathrm{WSM}}|F_{\mathrm{WSM}} - f_{\mathrm{WSM}}(\mathbf{x})|\} \quad (2)$$

where $W_a, W_v, W_{\mathrm{WSM}}$ are weights satisfying $W_a, W_v, W_{\mathrm{WSM}} \geq 0$ and $W_a + W_v + W_{\mathrm{WSM}} = 1$, and $\mathbf{x} = (A_{\mathrm{FL}}, A_{\mathrm{HL}}, K_{\mathrm{FL}}, K_{\mathrm{HL}}, O_{\mathrm{FL}}, O_{\mathrm{HL}}, \omega_{\mathrm{sw}})$ is the vector of decision variables.

The weighted distance is measured to a reference point, e.g., an utopian objective vector with components F_a, F_v and F_{WSM}. Different combinations of the weights W_a, W_v and W_{WSM} can produce different (weakly) Pareto optimal solutions that represent different locomotion compromises.

In (2), $f_a(\mathbf{x})$, $f_v(\mathbf{x})$ and $f_{\mathrm{WSM}}(\mathbf{x})$ are, respectively, the robot body vibration, robot forward velocity and wide stability margin (WSM) computed for CPG parameterization given by $\mathbf{x}$. We consider that a good gait should have less vibration, because the robot is subjected to less strain.

In order to calculate the total vibration we sum the standard deviation of the measures of the (a_x, a_y, a_z) accelerometers built-in onto the robot, *i.e.*, $std(a_x) + std(a_y) + std(a_z)$, similarly to [14, 6, 15].

We calculate forward velocity using the traveled distance of the robot during 12 seconds. A gait is considered better if it achieves higher velocities.

For stability, we calculate the wide stability margin [18]. This is a measure of the locomotion stability that provides the shortest distance between the projection of the center of mass in the ground and the polygon formed by the vertical projection in the ground of robot feet contact points. A gait is considered better for higher WSM values.

The search range of the CPG network parameters directly depend on the Aibo Ers-7 robot. The values of A_{FL} and A_{HL} are limited by the maximum range that the AIBO Hip joints may have. Offset values O_{FL} and O_{HL} for the hips are limited by the same ranges and the calculated amplitude values, A_{FL} and A_{HL}, respectively.

Maximum and minimum values for each knee stance angle are calculated in order to avoid leg collision during locomotion. Thus, the problem has several simple boundary constraints (for A_{FL}, A_{HL}, O_{FL}, O_{HL} and ω_{sw}), as follows

$$\begin{aligned} 0.01 \leq A_{\mathrm{HL}}, A_{\mathrm{FL}} &\leq 60 \\ -40 \leq O_{\mathrm{FL}} &\leq 20 \\ -20 \leq O_{\mathrm{HL}} &\leq 40 \\ 1 \leq \omega_{\mathrm{sw}} &\leq 12 \end{aligned} \quad (3)$$

Moreover, several inequality constraints were imposed (for $O_{\mathrm{FL}}, O_{\mathrm{HL}}, K_{\mathrm{FL}}$ and K_{HL}), given by

$$
\begin{aligned}
-40+\frac{A_{\mathrm{FL}}}{2} \leq O_{\mathrm{FL}} \leq 20-\frac{A_{\mathrm{FL}}}{2} \\
-20+\frac{A_{\mathrm{HL}}}{2} \leq O_{\mathrm{HL}} \leq 40-\frac{A_{\mathrm{HL}}}{2} \\
K_{\mathrm{FL}} \leq \max\{-O_{\mathrm{FL}}-\frac{A_{\mathrm{FL}}}{2}+50,-O_{\mathrm{FL}}+\frac{A_{\mathrm{FL}}}{2}+50\} \\
K_{\mathrm{FL}} \geq \max\{-O_{\mathrm{FL}}-\frac{A_{\mathrm{FL}}}{2}+20,-O_{\mathrm{FL}}+\frac{A_{\mathrm{FL}}}{2}+20\} \\
K_{\mathrm{HL}} \leq \max\{-O_{\mathrm{HL}}-\frac{A_{\mathrm{HL}}}{2}+40,-O_{\mathrm{HL}}+\frac{A_{\mathrm{HL}}}{2}+40\} \\
K_{\mathrm{HL}} \geq \max\{-O_{\mathrm{HL}}-\frac{A_{\mathrm{HL}}}{2}-5,-O_{\mathrm{HL}}+\frac{A_{\mathrm{HL}}}{2}-5\}
\end{aligned}
\tag{4}
$$

4 Optimization System

A scheme of the optimization system is depicted in fig 1. In order to tune the CPG parameters, we use a GA to search the optimal combination of the CPG parameters. GAs are population based algorithms that use techniques inspired by evolutionary biology such as inheritance, mutation, selection, and crossover [5]. GAs work with a population of points that represent potential optimal solutions to the problem being solved, usually referred to as chromosomes.

In this work, real representation of the variables was considered. So, each vector consists of a vector of 7 real values representing the decision variables of the problem. In our optimization system, we begin the GA search by randomly generating an initial population of chromosomes. The chromosomes were evaluated according to the fitness function defined in (2), in terms of robot body vibration, robot forward velocity and stability.

In order to handle the simple boundary constraints, each new generated point is projected component by component in order to satisfy boundary constraints (for all components of **x**, see (3)) as follows:

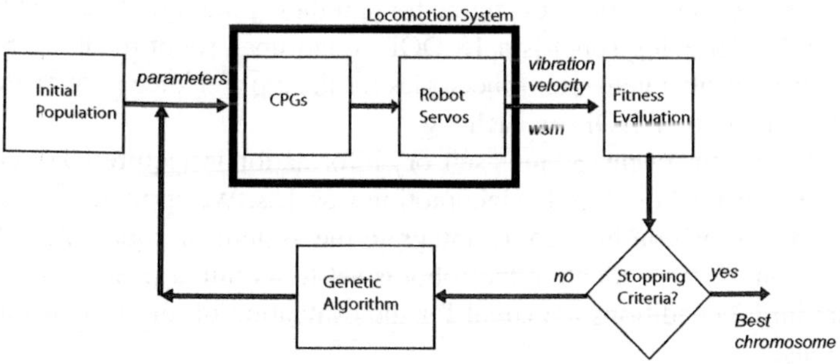

Fig. 1 Optimization Locomotion System

$$x_i = \begin{cases} l_i & \text{if } x_i < l_i \\ x_i & \text{if } l_i \le x_i \le u_i \\ u_i & \text{if } x_i > u_i \end{cases} \tag{5}$$

where l_i, u_i are the lower and upper limit of i component, respectively. A repairing mechanism is implemented to handle the inequality constraints. Therefore, any infeasible solution is repaired exploring the relations among variables expressed by the inequality constraints (see (4)), i.e., the values $O_{\text{FL}}, O_{\text{HL}}, K_{\text{FL}}$ and K_{HL} are repaired in order to satisfy the constraints. The application of this repairing mechanism to all infeasible solutions in the population, guarantees that all solutions become feasible.

We implement a tournament selection that guarantees that better chromosomes are more likely to be selected. Although selection assures that in the next generation the best chromosomes will be present with a higher probability, it does not search the space, because it just copies the previous chromosomes. The search results from the creation of new chromosomes from old ones by the application of genetic operators.

The crossover operator takes two randomly selected chromosomes; one point along their common length is randomly selected, and the characters of the two parent strings are swapped, thus generating two new chromosomes.

The mutation operator, randomly selects a position in the chromosome and, with a given probability, changes the corresponding value. This operator introduces diversity in the population since selection and crossover, exclusively, could not assure the exploration of new regions in the search space.

In order to recombine and mutate chromosomes, the Simulated Binary Crossover (SBX) and Polynomial Mutation were considered, respectively. These operators simulate the working of the traditional binary operators [4].

5 Simulation Results

In this section, we describe the experiment done in a simulated ers-7 AIBO robot using Webots [13]. Webots is a software for the physic simulation of robots based on ODE, an open source physics engine for simulating 3D rigid body dynamics.

The ers-7 AIBO dog robot is a 18 DOFs quadruped robot made by Sony. The locomotion controller generates trajectories for the hip and knee joint angles, that is 8 DOFs of the robot, 2 DOFs in each leg.

At each sensorial cycle (30 ms), sensory information is acquired. For each chromosome, the evaluation time for locomotion was 12s. We apply the Euler method with 1ms fixed integration step, to integrate the system of equations. At the end of each chromosome evaluation the robot is set to its initial position and rotation, such that initial conditions are equal for the evaluation of all chromosomes of all populations.

In all experiments, the optimization system ends when the number of generations exceeds 50 generations. We depict results when a population was established with

30 chromosomes. The SBX crossover and polynomial mutation probabilities were, respectively, 0.9 and $1/7$. An elite size of 3 chromosomes was implemented.

Table 1 presents the results obtained for several combinations of weights in the scalar function. As reference point considered has the following components: $F_a = 0$, $F_v = -150$ and $F_{\mathrm{WSM}} = -65$. Combinations of weights defining multiobjective problems with different number of objectives were considered:

- $T1$ to $T3$ correspond to the optimization of each one of the objectives;
- $T3$ to $T15$ are different combinations of weights of two objectives being simultaneously optimized;
- finally, $T16$ to $T19$ are combinations of weights in which the three objectives are simultaneously optimized.

In this table, the solutions obtained for each weights combinations are also presented, in terms of f_a, f_v and f_{WSM}. In the last column, the value of the scalar function is also presented ($f(\mathbf{x})$).

Table 1 Multiobjective results for different combinations of weights.

	W_a	W_v	W_{WSM}	f_a	f_v	f_{WSM}	$f(\mathbf{x})$
$T1$	1	0	0	0.0135	21.0471	14.6093	0.0136
$T2$	0	1	0	0.1213	138.5492	6.0067	11.4508
$T3$	0	0	1	0.0384	4.9457	64.6248	0.3752
$T4$	0.2	0.8	0	0.1058	125.1315	6.1785	19.8948
$T5$	0.4	0.6	0	0.1215	127.7335	7.2684	7.3599
$T6$	0.6	0.4	0	0.1181	134.967	6.3661	4.0242
$T7$	0.8	0.2	0	0.1118	129.3525	4.5230	4.1295
$T8$	0	0.8	0.2	0.1210	135.7304	7.6250	11.4750
$T9$	0	0.6	0.4	0.1125	114.7752	11.0080	21.5968
$T10$	0	0.4	0.6	0.1292	89.3642	19.1294	27.5223
$T11$	0	0.2	0.8	0.0307	17.9153	63.8483	26.4169
$T12$	0.8	0	0.2	0.0206	29.8379	54.8389	2.0322
$T13$	0.6	0	0.4	0.0144	3.3254	50.2399	5.9040
$T14$	0.4	0	0.6	0.0340	2.9505	63.7906	0.7256
$T15$	0.2	0	0.8	0.0249	0.5827	63.7955	0.9635
$T16$	1/3	1/3	1/3	0.1013	103.5880	11.3905	17.8680
$T17$	1/2	1/4	1/4	0.0698	95.7004	11.5376	27.1498
$T18$	1/4	1/2	1/4	0.1081	121.8231	9.1177	27.1498
$T19$	1/4	1/4	1/2	0.1259	66.8218	19.2712	22.8644

These solutions are also depicted in Figures 2a)-c). As expected, it can be observed that solutions $T1$, $T2$ and $T3$ are the extreme solutions of the Pareto front in Figures 2a), 2b) and 2c). The other solutions of the front represent different compromises of the objectives. Taking into account the information provided by these graphs, it is possible to choose a compromise solution that represent a different locomotion gait of the robot. Moreover, it possible to inspect the relationships between objectives.

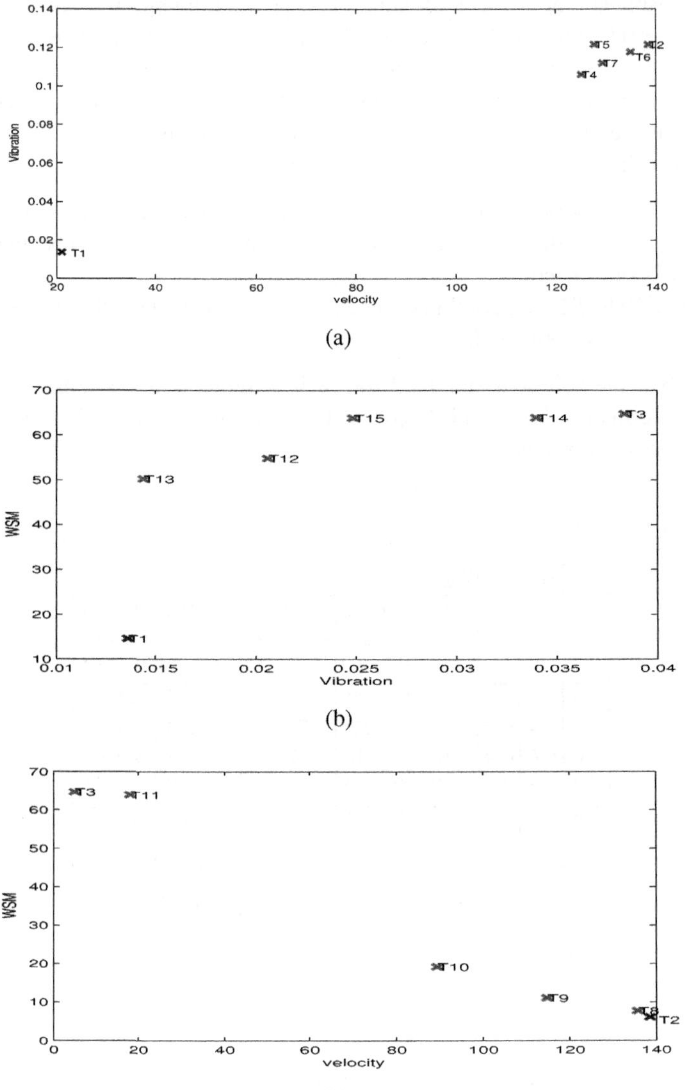

Fig. 2 Representation of the solutions in objective space. a) velocity versus vibration ($T1$, $T2$ and $T4$ to $T7$ combinations of weights). b) vibration versus WSM ($T1$, $T3$ and $T12$ to $T15$ combinations of weights). c) velocity versus WSM ($T2$, $T3$ and $T8$ to $T11$ combinations of weights).

In Figure 3, the 2D projections of 3D objective space is presented. Since the 3 objectives are conflicting, these solutions define a 3D Pareto surface. This multi-objective approach allows to select solutions that achieve the highest velocity for a slow, crawl gait and perceive the tradeoff in terms of vibration and WSM.

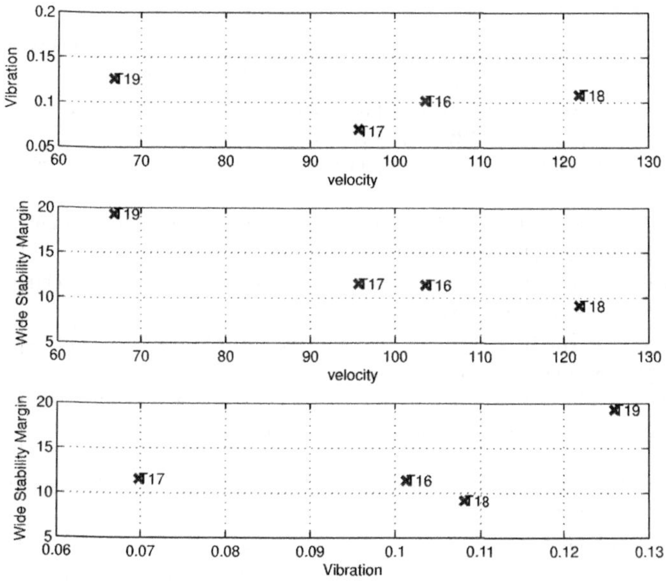

Fig. 3 Representation of the solutions in objective space: velocity, vibration and WSM ($T16$ to $T19$ combinations of weights).

6 Conclusions and Future Work

In this article, we have addressed the locomotion optimization of a quadruped robot. A locomotion controller based on dynamical systems to model CPGs, generates quadruped locomotion. These CPG parameters are tuned by an optimization system. This optimization system combines CPGs and a genetic algorithm which solves a multiobjective formulation of the problem. The goal is to optimize simultaneously, three conflicting objectives, namely the vibration, velocity and WSM.

Experiments were performed in the Webots robotics simulator. The multiobjective optimization was formulated considering a scalarization function based on the Weighted Tchebycheff method for different combinations of weights. The solutions obtained represent locomotion strategies that are different compromises of the objectives.

We also plan to use multi-objective optimization algorithms such as MEES [1] or NSGAII [3]. We will extend this optimization work to address other locomotion related problems, such as: the generation and switch among different gaits according to the sensorial information and the control of locomotion direction.

Acknowledgments

Work supported by the Portuguese Science Foundation (grant PTDC/EEA-CRO/100655/2008).

References

1. Costa, L., Oliveira, P.: An Adaptive Sharing Elitist Evolution Strategy for Multiobjective Optimization. Evolutionary Computation 11(4), 417–438 (2003)
2. Chernova, S., Veloso, M.: An evolutionary approach to gait learning for four-legged robots. In: Proceedings of IROS 2004 (2004)
3. Deb, K.: Multi-Objective Optimization using Evolutionary Algorithms. John Wiley and Sons, Ltd, Chichester (2001)
4. Deb, R.B., Agrawal, K.: Simulated binary crossover for continuous search space. Complex Systems 9(2), 115–149 (1995)
5. Goldberg, D.: Genetic Algorithms in Search, Optimization, and Machine Learning. Addison-Wesley, Reading (1989)
6. Golubovic, D., Hu, H.: Evolving locomotion gaits for quadruped walking robots. Industrial Robot: An International Journal 32, 259–267 (2005)
7. Hornby, G.S., Takamura, S., Yamamoto, T., Fujita, M.: Autonomous evolution of dynamic gaits with two quadruped robots. IEEE Transactions on Robotics 21, 402–410 (2005)
8. Grillner, S.: Neurobiological bases of rhythmic motor acts in vertebrates. Science 228(4696), 143–149 (1985)
9. Koo, I.M., Kang, T.H., Vo, G.L., Trong, T.D., Song, Y.K., Choi, H.R.: Biologically inspired control of quadruped walking robot. International Journal of Control, Automation and Systems 7(4), 577–584 (2009)
10. Kim, M.S., Uther, W.: Automatic gait optimisation for quadruped robots. In: Australasian Conference on Robotics and Automation (2003)
11. Kohl, N., Stone, P.: Machine learning for fast quadrupedal locomotion. In: The Nineteenth National Conference on Artificial Intelligence, pp. 611–616 (2004)
12. Matos, V., Santos, C.P., Pinto, C.M.A.: A brainstemlike modulation approach for gait transition in a quadruped robot. In:IROS, pp.2665–2670 (2009)
13. Michel, O.: Webots: Professional mobile robot simulation. Journal of Advanced Robotics Systems 1(1), 39–42 (2004)
14. Röfer, T.: Evolutionary gait-optimization using a fitness function based on proprioception. RobuCup, 310–322 (2004)
15. Saggar, M., Kohl, N., Stone, P.: Autonomous learning of stable quadruped locomotion. In: RoboCup2006: Robot Soccer World Cup X, Springer, Heidelberg (2007)
16. Santos, C., Oliveira, M., Rocha, A.M.A.C., Costa, L.: Head Motion Stabilization During Quadruped Robot Locomotion: Combining Dynamical Systems and a Genetic Algorithm. In: IEEE International Conference on Robotics and Automation (ICRA 2009), Kobe, Japan (2009)
17. Santos, C.P., Oliveira, M., Matos, V., Rocha, A.M.A.C., Costa, L.: Combining Central Pattern Generators with the Electromagnetism-like Algorithm for Head Motion Stabilization during Quadruped Robot Locomotion. In: 2nd International Workshop on Evolutionary and Reinforcement Learning for Autonomous Robot Systems, ERLARS 2009, St. Louis, Missouri, USA (2009)
18. Song, S., Waldron, K.: Machines that Walk: The Adaptive Suspension Vehicle. MIT Press, Cambridge (1989)

Author Index